WOODS, FORESTS, AND ESTATES

OF

PERTHSHIRE.

Frontispiece.

THE LAST OF "GREAT BIRNAM WOOD."

WOODS, FORESTS, AND ESTATE

OF

PERTHSHIRE

WITH

Sketches of the Principal Families in the County.

BY

THOMAS HUNTER,
EDITOR "PERTHSHIRE CONSTITUTIONAL AND JOURNAL."

ILLUSTRATED.

PERTH:
HENDERSON, ROBERTSON, & HUNTER.
EDINBURGH: WM. BROWN.　　GLASGOW: THOS. MURRAY & SC

MDCCCLXXXIII.

PRINTED AND BOUND IN GREAT BRITAIN BY
ANTONY ROWE LTD, EASTBOURNE

PREFATORY NOTE.

THE substance of this work originally appeared as a series of sketches in the *Perthshire Constitutional*, where the subject was found to awaken sufficient interest to induce me to issue the present volume. In its new form, the work has, to a large extent, been rewritten, and much important matter has been added. The main object I had in view was to trace the history of Forestry in Perthshire, which is largely the history of Forestry in Scotland, and to show the great improvement which has been effected upon the country by planting, chiefly during the last one hundred years. This naturally led to inquiries regarding estate improvements generally, and directed attention to the many old and notable trees in the county, as well as to the history of the various estates visited, and the families to whom they belong.

I take this opportunity to return my warmest thanks to the numerous friends—landed proprietors, factors, land-stewards, foresters, &c.—who have aided and encouraged me in my inquiries, only a few of whom I have had occasion to mention in the course of this work. I am specially indebted to the many proprietors and others who placed at my service valuable estate and family papers and rare works of reference, by which I have been enabled to ensure the greatest accuracy possible in the accounts of the various places and families.

The Illustrations have been engraved by J. M. Corner, Edinburgh, from photographs chiefly by Magnus Jackson, F.S.A. Scot., Perth, Photographer to the Scottish Arboricultural Society.

T. H.

PERTH, *October* 1883.

LIST OF ILLUSTRATIONS.

	PAGE
THE LAST OF "GREAT BIRNAM WOOD,"	*Frontispiece*
THE PERTH NURSERIES,	21
THE FALLS OF BRUAR,	38
DUNKELD CATHEDRAL,	46
HERMITAGE FALLS,	48
THE PARENT LARCHES,	52
GLEN TILT,	55
PASS OF KILLIECRANKIE,	59
DALGUISE HOUSE,	64
QUEEN MARY'S SYCAMORE,	102
ABIES DOUGLASII AT LYNEDOCH,	113
FREELAND HOUSE,	161
THE DUPPLIN ARAUCARIA,	217
BLAIR-DRUMMOND BEECHES,	253
TRIANGULAR GROUP OF SPANISH CHESTNUTS AT STRATHALLAN (*Diagram*),	319
THE GREAT BEECH HEDGE AT MEIKLEOUR,	380
TAYMOUTH CASTLE,	386
UPPER FALL OF MONESS,	392
SYCAMORE AT CASTLE MENZIES,	397
QUEEN'S VIEW AT LOCH TUMMEL,	421
FALLS OF TUMMEL,	422
KINLOCH-RANNOCH,	424
FORTINGALL YEW,	434
SCOTS FIR AT LAWERS,	464
WALLACE'S CAVE, INTERIOR AND EXTERIOR,	514
A "BIT" ON THE DEAN,	529
OLD ROMAN BRIDGE,	530
THE GREAT BEECH AT CARDEAN,	532
STENTON,	543
THE STEWART LARCH,	547

CONTENTS.

I.—INTRODUCTION.

Solemnising Influence of Woods—Their Effect on the Landscape—Climatic Influence—Commercial Importance,

II.—THE PAST AND THE PRESENT.

Extent and Character of the County—Perthshire originally a huge Forest—Destruction of Original Forests—Remains of Ancient Forests—Destruction of Woods during Feudal Times—Efforts at Improvement—Compulsory Planting—Acreage at present under Wood—Pecuniary Value of the Woods in Perthshire,

III.—THE PERTH NURSERIES.

Services rendered to Forestry by the Perth Nurseries—Their Formation and Situation—Cultivation of Forest and Ornamental Trees—The Latest Importations suitable for the Scottish Climate,

IV.—ATHOLE.

The Varied Attractions of Athole—Its Position in the History of Scottish Forestry—First Extensive Larch Plantations—Introduction of Larch into Scotland—Result of a "Social Crack" and a "Waught" from a "Pocket Pistol"—The Value of Larch—First Attempt at Mountain Planting in Scotland—Planting of Larch as a Forest Tree—"Humble Petition of Bruar Water"—Cost of Planting—Extent of Plantations—First Attempts to use Larch for Shipbuilding — Satisfactory Experiments — Beauty of Dunkeld — Walk through the Woods at Dunkeld and Blair-Athole—Blair Castle—The Athole Family—Destruction of Woods by Recent Storms,

V.—DALGUISE.

Situation—Ancient Ecclesiastical Residence—Steuarts of Dalguise—A Branch of the Royal Steuarts—Valuable Relics of Antiquity—Gigantic Araucarias — Other Remarkable Trees — "Glenalbert's Fam'd Abode"—Steuartfield—Remarkable Cairn and Cave—A Deserted Bridge—Plantations,

x   CONTENTS.

                                                                    PAGE
VI.—MURTLY.

"Great Birnam Wood"—Buried and Living Remains of the Ancient
Forest—Story concerning Neil Gow—The "Hanged-Men's Tree"
—Huge Beech and Chestnut Trees—The *Douglasii* Terrace—The
Stewarts of Grantully—Planting out of Flower-Pots—The *Deodara*
Sunk Terrace—Ancient Yews—The Finest Araucaria Avenue in
the World—Large Lime Trees—Huge *Wellingtonia gigantea*—
Summary of Remarkable Trees—The Plantations—The Trees of
the Future, ... 71

VII.—DUNSINANE.

The Historical connection between the Hills of Birnam and Dunsinane
—Macbeth's Castle—View from the Summit of the Hill—Original
State of the District—First Attempts at Improvement—"Trees
Planted for Behoof of the Poor"—Lord Dunsinane's Improvements
—Story regarding the Introduction of Macadamised Roads in the
District—Extensive Planting—Improvements by Subsequent Pro-
prietors, ... 86

VIII.—SCONE, LYNEDOCH, LOGIEALMOND, &c.

The Former Glory of Scone—The Stone of Destiny—The Ancient Abbey
—The Mansfield Family—The Modern Palace—Queen Mary's Sycamore—Trees Planted by King James VI.—Other Remarkable Trees
in the Policies—David Douglas—The "Boot Hill"—Appearance of
the Country a Century Ago—A Landscape Forester's Blunder—
The Plantations—Voracity of the *Hylobius abietis*—The Steading
at Balboughty—Preservation of Wooden Posts—*Douglasii* Plantations — An Experimental Plantation — Taymount — Lynedoch —
Logiealmond, ... 98

IX.—METHVEN.

Methven Wood a Lurking-Place of Wallace—The Battle of Methven
Wood—The Castle—Walk through the Ancient Wood—An Enor-
mous Beech—The Plantations—History of the Property—The First
Extensive Planter — Remarkable Trees — The Pepperwell Oak —
Ancient Church of Methven—The Bell Tree—The Black Italian
Poplar at Tippermallo, ... 117

X.—MONCREIFFE.

"The Glory of Scotland"—Remains of an Ancient Fort—Druidical
Circle—Woods Adorning Moncreiffe Hill—The Moncreiffes of Moncreiffe—Plantations on the Estate—Large and Remarkable Trees—
Largest Horse Chestnut in the Country—*Taxodium sempervirens*
as a Forest Tree—The Remains of the Old Chapel, . . . 128

CONTENTS. xi

PAGE

XI.—KILGRASTON.

Situation—The Mansion-House—An Apiarian Curiosity—Grants of Kilgraston — Policies — "Cromwell's Tree" — Other Remarkable Trees—Pitkeathly and its Mineral Wells—Drummonie House—Large Tulip Tree, &c., ... 139

XII.—INVERMAY.

The Beauties of Invermay—The Belshes of Invermay—Their Improvements—The Soldier and the Laird—The "Birks" of Invermay—The Mansion-House—The "Bell Tree"—The Old House—Notable Yew and Holly Hedges—A Remarkable Spruce Tree—The Walk by the May—Old Muckersie Church—The "Humble-Bumble"—The Plantations, ... 147

XIII.—FREELAND.

The Original Design of the Policies—The Village of Forgandenny—The Main Approach—The Old Home of the Ruthvens—The Den—Remarkable Trees — The Mansion-House — The Most Remarkable Elder Tree in Perthshire—The Woods—Natural-Sown Silver Firs —A Venerable Ash—Trees with an Unknown History, . . 154

XIV.—ROSSIE.

The Oliphants—The Founder of Rossie—The Mansion-House—The Policies—Large Silver Firs—The Dovecot Park—Cedar of Lebanon—Sycamore Trees—A Royal Souvenir—An Old Marriage Record, 168

XV.—CONDIE.

Situation of the Property—Destruction of the Mansion-House by Fire—Scenery of the District—Battle of Culteuchar—The Policies—Remarkable Trees—The Plantations, ... 173

XVI.—DUNCRUB.

History of the Rollo Family—The Old Thorn Tree at Dunning—The New Mansion-House—The Policies—The Pinetum—Memorials of a Royal Marriage—Remarkable Trees—The Old Castle of Kelty—Notable Trees at the Castle—The Plantations—The Village of Dunning, ... 177

XVII.—GARVOCK.

Situation—The Græmes of Garvock—How the Property was Acquired—The Mansion—Rose-Bush Planted by Prince Charlie—An Enormous Walnut Tree—Other Notable Trees—An Ancient Camp—The Plantations, ... 191

XVIII.—GASK.

Attractions of Gask—Roman Remains—Gask Woods a Hiding-Place of Wallace—Visit of Prince Charles—Jacobite Relicts—The "Auld House"—The "Auld Pear Tree"—The New Mansion—Chequered History of the Property—Destruction of the Woods during the Rebellion—An Exiled Laird and his Woods—The Policies—Notable Beech Trees—A Magnificent Walnut—Other Remarkable Trees—The Garden—Lawrence Macdonald the Sculptor—The Plantations, 196

XIX.—DUPPLIN.

The Interest Attaching to Dupplin—The Battle of Dupplin—Its Antiquities—The Kinnoull Family—Dupplin Castle—The Policies—The Queen's Visit to Dupplin—The Original Design of the Policies—The Pinetum—The Octagon—Remarkable Trees—A Mysterious *Wellingtonia*—Kinnoull Hill—The Plantations, . . . 207

XX.—ABERCAIRNY.

History connected with the Property—Notable Proprietors—The Mansion—Queen's Visit—The Old Mansion—A Venerable Ash—Design of the Policies—Remarkable Trees surrounding the Policies—The Nursery—Remarkable Trees within the Policies—The Den of the Muckle Burn—Lady Fanny's Grass Walk—The Shrubbery—Inchbrakie—Historical Yew—The Village of Fowlis-Wester—The "Tree of Fowlis"—The Araucaria in the Manse Garden—Druidical Remains—The Plantations, ... 224

XXI.—BLAIR-DRUMMOND.

History of the Property—The Moss of Kincardine—Discovery of Marine Remains—An Ancient Forest—Roman Relics—The Reclamation of the Moss—Lord Kames—The First Mansion-House—The New House—The Policies—Remarkable Oak and Beech Trees—A Memorial of Benjamin Franklin—Remarkable Spanish Chestnut, Sycamore, Lime, and Birch Trees—Remarkable Fir and Larch Trees—Doune Castle—The Silver Firs—Havoc caused by Great Storms—Progressive Growth of Common Trees—Burnbank—The Plantations, ... 244

XXII.—ARDOCH.

The Proprietors of Ardoch—The Policies—Ardoch House—Aged Sycamores—The Plantations and Remarkable Trees—The Roman Camp—The Walk along the side of the Knaick, . . . 261

CONTENTS.

XXIII.—KEIR.

Situation—Origin of Name—Stirlings of Keir—Peculiar Ways of Spelling the Family Name—Acquisition of the Property of Keir—Additions to the Family Estate—History of the Property—Burning of the Old Tower of Keir—Building of the present Mansion—The Woods in the 17th Century—Early Improvements—Improvements by Sir William Stirling-Maxwell—The Policies—Memorial Trees—Extensive Planting of the Newer Conifers—The Keir Araucaria—The Pinetum—Extent of the Property, . . . 272

XXIV.—KIPPENDAVIE.

Attractions and Extent of the Property—Stirlings of Kippendavie—The Great Sycamore of Kippenross—Other Remarkable Trees—"On the Banks of Allan Water"—Pinetum—Plantations—Sheriffmuir—Improvements, ... 284

XXV.—MONTEITH.

General Attractions of Monteith—The Earls of Monteith—The Forests of Monteith a favourite Resort of the Ancient Scottish Court—The "Law Tree"—Claimants for the Earldom—The "Beggar Earl"—Cardross—Cardross Mansion—Antiquities—Remarkable Trees—The Garden Policies—The Pinetum—The Glen—Cardross Noted for its Timber—Deforesting and Replanting—Big Wood of Cardross—*Abies Douglasii* the most Flourishing Variety—The Scenery—General Improvements—Efforts at Reclamation of Moorland—The Erskines of Cardross—Lochend—Lake of Monteith—Priory of Inchmahome—Queen Mary's Garden—Warning to Relic Hunters—The Old Spanish Chestnuts at Inchmahome—Castle of Talla—Rednoch—Blairhoyle — Glennie — Mondowie — Garden — Gartmore—Royal Visitors at Monteith, ... 290

XXVI.—STRATHALLAN.

An Arborous Sea—Strathallan Castle—The Drummonds of Strathallan—The Policies—Triangular Group of Spanish Chestnuts—Other Remarkable Trees—The Woods—Malloch's Oak—Tragic End of a Greedy Meal Merchant—Birks of Tullibardine—*Abies Douglasii*—Inexpensive and Effectual Method of Protecting Young Plants from Rabbits—Tullibardine Chair Tree—Old Castle of Tullibardine—Excavation Representing the Great "Michael" War-ship—The Nursery—Result of Protecting Plants from Late Spring and Early Autumn Frosts—An Ancient Target—Relict of Caledonian Forest—*Abies Douglasii v.* Larch—Ravages of Recent Storms, . . 316

XXVII.—DRUMMOND CASTLE.

The Attractions of Drummond Castle—The Founder of the House of Drummond—Origin of the Family Name—Glorious Death of the

xiv CONTENTS.

PAGE

First of the Drummonds—History of the Family—Early Connection with Forestry—Queen Annabella—How Glenartney Forest was Acquired—The Builder of Drummond Castle—The Castle itself—The View from its Towers—The Gardens—Drummond Park and its Monster Trees—Remarkable Trees at Pitkellony, Muthill, &c.—A Curious Phenomenon—The Plantations—The Trossachs—Extent of the Property and General Improvements, . . 328

XXVIII.—STOBHALL.

Antiquity of the Drummonds of Stobhall—Castle of Stobhall—Extent of the Property—Great Height of Trees in Stobhall Wood—Strelitz Wood—Peculiar Origin of the Name—Ancient Religious Houses—The Monks as Foresters—Linn of Campsie—Sir Walter Scott's Allusions to the District—Antiquities, ... 349

XXIX.—THE CAIRNIES AND GLENALMOND.

Early Home of the Newer Coniferous Trees—The History of the Properties—Lord Provost Hay Marshall—The Policies at the Cairnies—The Finest Specimens of *Abies Albertiana* in the Country—How the *Albertiana* got its Name—Curious Place for a Tree—Progress of the Different Varieties of the Newer Conifers—"The Kirk of the Wood"—An Interesting Old Scots Fir—The Plantations, . . 356

XXX.—GORTHY.

State of the Property Sixty Years Ago—Antiquity of the Family of Mercer—Ancient Connection with Perth—"Merchant Princes"—A Famous Provost—Purchase of Gorthy—Remains of the Old Caledonian Forest—A Gigantic Sycamore—Other Notable Trees—The Plantations—Struggles with the Beetle, ... 365

XXXI.—MEIKLEOUR.

Meikleour House—The Mercers of Meikleour and Aldie—The Policies—Remarkable Trees—Kinclaven Castle—Village of Meikleour—The Great Beech Hedge—The Plantations, ... 375

XXXII.—TAYMOUTH.

The Character and Extent of the Property—Poetic Allusions—The Campbells of Breadalbane—The Approaches to Taymouth—The Castle—The Policies—The Early Planters—Remarkable Trees—Trees Planted by Royalty—Drummond Hill—The Great Vine at Kinnell—Plantations in other parts of the Property, . . 382

XXXIII.—CASTLE MENZIES.

The Present and Former Castles—The Policies—Remarkable Chestnut, Sycamore, Beech, Oak, and Elm Trees—The Newer Conifers—The

WOODS, FORESTS, AND ESTATES

OF

PERTHSHIRE.

I.—INTRODUCTION.

Solemnising Influence of Woods—Their Effect on the Landscape—Climatic Influence—Commercial Importance.

THE beautiful appearance which woods give to the landscape of a country, and the part they play in the economy of Nature, render the study of them one of the most pleasing and interesting themes that can engage our attention. No one who has seen a forest can fail to be impressed with its majestic grandeur. As one looks upon the "green-robed senators of mighty woods," it is not difficult to realise, to some extent, the feeling which prompted the ancients to associate them with the worship of the Deity; and we can almost forgive the Greeks for believing that the woods were peopled with gods. There are few things better calculated to teach us how true it is that "man is as the grass of the field," than the sight of those vegetable giants that have survived the storms of centuries, and look as if they would still be "hale green trees," and lift their heads as proudly as they do now, "when a thousand years are gone." In a superstitious age, the awe-inspiring spectacle presented by a dense forest could hardly fail to awaken serious reflections in a thoughtful mind, but this solemnising influence of woods is an experience that extends beyond the superstitious Greek, who could only draw from the woods lessons that ministered to his terror. To borrow from Byron, we must all have felt, to some extent, that there is a peculiar "pleasure in the pathless woods," that makes us "love not man the less, but Nature more." We must all experience, more or less, a deep reverential feeling in the presence of a form of life which existed ages before we trod the earth, and which will continue to exist for ages after we have returned to our kindred dust. The reflections which such a thought suggest ought to make us truly humble; and, if we are wise, the lessons that are taught by the woods will not be unprofitable. The Scriptures themselves are full of lessons drawn from the woods in illustration of the truths sought to be impressed. The righteous man, we are told, is "like a tree planted by the rivers of water, that bringeth forth his fruit in his season, his leaf also shall not wither, and whatsoever he doeth shall prosper." As for the wicked, although we see him in "great power, and spreading himself like a green bay tree," we are assured that he will soon pass away. All through the sacred writings we find the woods drawn upon to point a moral, and not without effect. Poets and philosophers have each found in them an inexhaustible storehouse of imagery to embellish their thoughts, and the experience of one of the poets (Wordsworth) is that

"One impulse from a vernal wood
May teach you more of man,
Of moral evil, and of good,
Than all the sages can."

All our ideas of beautiful scenery are associated with woods. The landscape that is destitute of trees presents a barren and uninteresting appearance, while a country that is rich in arboreal features is as refreshing to the eye as is a sheet of water in an arid land. The majestic oak, with its grey rifted trunk, and its dark indented foliage, and the equally majestic beech, with its fine silvery bark and pale green leaves, add dignity and grace to the well-kept ancestral park. The light, graceful birch, which overhangs the mountain stream, imparts to the landscape that fairy-like charm which is so attractive to the lover of the picturesque; while the pine, with its straight tall stem and evergreen foliage, clothes the landscape with a pleasing uniformity.

But woods and forests are of far higher utility than merely contributing to the beauty of the scenery. Trees, as well as vegetation of all kinds, have an important geological influence, for, by keeping the rocks damp, these have a greater tendency to break up under the influence of frost. The climatic influence of trees, however, is more perceptible than their geological effect. By retaining the moisture in the ground, they keep the surface of the earth cool; and where there is a wet surface there is a greater condensation of vapour, the reverse being the case where there is a dry surface, or a scarcity of trees; so that woods play an important part in producing a moist climate. Some species of trees are known to have a greater effect upon climate than others. The *Eucalyptus globulus*, or Tasmanian Blue Gum Tree, is said to have marvellous properties as regards the drying of marshes and the prevention of malarious disease. There is a large amount of evidence to show that it actually possesses this power in a high degree; so that, not only is intermittent fever unknown where it naturally grows in abundance, although in situations and in a climate where the prevalence of this disease might be expected, but places at one time most subject to that malady, cease to be so when that tree is planted. Some unhealthy localities in the Cape of Good Hope were rendered perfectly salubrious, apparently through the influence of this tree, within a few years after plantations of it had been laid. Similar results have followed the introduction of this tree in Cuba, in Mexico, and in the south of France. In Cyprus, interesting experiments are also being made. We learn from the *British Medical Journal* for Feb. 7, 1879, that Prince Troubetkoy, who has taken up this matter energetically at Rome, says that careful observation has brought him to the conclusion that the *Eucalyptus amygdalina* is the most useful variety of the tree. It is not only picturesque, but of exceptionally rapid growth. Indeed, all the species of this wonderful tree are characterised by remarkable growth. Such is the rapidity of its growth, that seedlings raised in a hotbed, and planted out in the open air in the south of England, have been known to attain a height of 10 feet in the first year. In their native countries, some of them are said to attain a very great height, with trunks 16 feet in diameter. They are most useful timber trees, and are largely employed in shipbuilding. Although these trees, which belong to the great natural order *Myrtaceæ,* are almost exclusively natives of Australia and Tasmania, it is satisfactory to know that there is some probability of their being, to some extent at least, acclimatised in this country. We have it on the authority of Prince Troubetkoy, that the *Eucalyptus amygdalina* grows as well in a damp as in a dry and exposed soil, and also that it bears cold very well, seeing that at his villa in the neighbourhood of Rome it resisted a temperature of 6 degrees below zero (21° Fahr.).

However great may be the climatic influence of woods and forests, their commercial importance is far more apparent. Apart altogether from the timber, it is absolutely impossible to calculate the products of the forests. Pliny the elder (Caius Plinius Secundus), our chief authority so far as regards the trees and shrubs familiar to the ancients, considers trees bearing seed after their kind as the most precious gifts of Nature to mankind. "The forest," he says, "was that from which man first drew the aliment of his life; his early cave-dwellings were made habitable by the leaves of its trees, and by their bark or other fibres his clothing was supplied." The Arabs say that the palm tree has a use for every day there is in the year; and when one species of tree alone is of such utility, what might be said of all! The tree which, as has been shown, is of such importance in improving a malarious atmosphere, is also of great commercial value, the essential oil which it yields being employed in the manufacture of oil and spirit varnishes,—varnish containing this oil instead of spirits of turpentine being said neither to bloom nor crack. As a medicine, the leaves are said to be more efficacious than quinine in the cure of intermittent fever. *The Scientific American* recently gave drawings of a valuable Texan tree (Mesquit), and its podded fruit, which is said to be of great utility for fattening stock of all kinds. The natives pound the pod and keep it in stock, making cakes and a kind of porridge of the meal. The plant is leguminous, and belongs to the mimosa order, bearing a gum allied to that of acacia. The wood is said to be excellent fuel, and burns like anthracite. All over

Mexico, Arizona, and Texas, the tree flourishes, even in high and dry districts. Many familiar trees in our own country also yield valuable products. Ben Jonson speaks of the juice of the larch as possessing some mystic virtues. Pliny, in writing of the same tree, says :—" This tree is the best of the kind that bears resin." The substance known as Venice turpentine, and sold by druggists for a variety of medicinal and art purposes, is not brought from Venice at all, but is simply the hardened sap of the larch, obtained, notably in France, by tapping or perforating the trees with large augers—

> " The new-made trees in tears of amber run,
> Which harden into value by the sun."—OVID.

Even the Scots fir (*Pinus sylvestris*) is rich in a product the loss of which would entail enormous inconvenience and difficulty. We are all aware of the importance and usefulness of tar. The timber-merchant, the rope-maker, the shipwright, and the sailor are all indebted to it; and many articles, including submarine cables, owe to it much of their power to resist decay and deterioration. Lampblack is made in large quantities from the waste products resulting from the manufacture of common tar. It is from the Norway spruce (*Abies excelsa*) that the Burgundy pitch of commerce is made. The resin of the white spruce (*Abies alba*) makes excellent oil of turpentine,—the bark is made use of for tanning hides,—and anti-scorbutic spruce beer—a most valuable beverage for long sea voyages—is made from its branches and loppings. Food, as well as drink, is yielded by numerous members of the cone-bearing family, the Laplanders utilising the inner bark of the pine for breadmaking purposes, the result of their labours being known as *barkbroed*. The Siberian ermine-hunters also make use of the inner bark of the pine for the purpose of forming a substitute for yeast. The Canada balsam used in mounting objects for microscopical examination, is simply the juice of the balsam pine. The resin of the Lebanon cedar (*Cedrus Libani*), on which the ancients set so much value, is said both by Pliny and Vitruvius to have been used in the treatment of papyrus and the embalming of Egyptian mummies. The bark of the oak, as is well known, is largely used in tanning. An oil used in some parts of France instead of butter is extracted from the masts or nuts of the beech. An excellent wine is made out of the vernal juice of the birch ; and one variety, the birch of Jamaica, yields turpentine.

As showing the commercial—or, to speak more correctly, the pecuniary—value of tree planting in this country, a recent writer has calculated that the tree thinnings of an acre of land, worth only from 5s. to 10s. per acre, but planted with a mixture of larch, beech, pine, hazel, birch, and oak—the latter with a view to the growth of navy timber or trees of large size for building and other purposes—will at the end of from ten to fifteen years, according to local circumstances, repay the average expenses of planting, rent, and management during that period, together with compound interest at five per cent. ; and the profits of future falls may be estimated as follows :—In thirteen years, or at twenty-three years' growth, £24, 10s. per acre ; in thirteen years more, or at thirty-six years' growth, £39 per acre. After that period a triennial profit of about £12 per acre, until the oak is fit for navy or other purposes, for which timber of first-class quality is required, when the final clearance may be expected to fetch from £200 to £250 per acre.

We have said enough to show the great importance of looking after the woods and forests, and the interest which attaches to them. Their importance is well summed up in the old saying, that the man who plants even a birch tree little knows what benefit he is conferring upon posterity. If the man who makes two blades of grass grow where only one grew before deserves well of his country, in what esteem ought we to hold the man who plants a tree where none before existed, especially in a soil where nothing else would grow with profit ? The management of woods and forests is, therefore, a subject which, from its importance, ought to have attractions for all, whether pecuniarily interested or not ; and as Perthshire, besides owing very much of its beauty to arboriculture, has probably done more to promote forestry than any other county in Scotland, the history of its woods and forests possesses exceptional interest.

II.—THE PAST AND THE PRESENT.

Extent and Character of the County—Perthshire originally a Huge Forest — Destruction of Original Forests — Remains of Ancient Forests—Destruction of Woods during Feudal Times—Efforts at Improvement — Compulsory Planting — Acreage at present under Wood—Pecuniary Value of the Woods in Perthshire.

FROM whatever point of view we regard Perthshire, it is undoubtedly the finest county in Scotland. The scenery embraces the most choice spots, whether in the Highlands or the Lowlands, and for variety in mountain and valley, lake and river, rich woodlands and fertile fields, it cannot be equalled. In size it is exceeded only by the counties of Inverness, Ross and Cromarty, and Argyle, its area being about 2588 square miles, or 1,656,082 acres. Although, from its peculiar shape, a circle with a radius of 32 miles could be drawn which would nearly enclose it, its extreme length from east to west is 77 miles, and its extreme breadth from north to south, 68 miles. The county is divided into two distinct regions, the Lowland and the Highland. The former is noted for its fertility and the richness of its scenery, which is agreeably varied by judiciously-planted woods of larch, fir, oak, ash, beech, birch, &c. The greater part of the Highland region is open moorland, although large tracts have been recently planted with larch and Scots fir. A great part of the county is unfit for the raising of grain or green crops, only about one-fifth of the entire surface being under cultivation. The study of forestry is therefore of great importance, and receives much attention ; so much, indeed, that the county has justly been termed the great tree-growing county of Scotland. A recent writer, and competent authority (Mr Robert Hutchison of Carlowrie, Kirkliston), says :—" We may say of the forest trees of Perthshire generally, that, whether owing to the configuration of the county, and its inland position lending a more *continental* and less *insular* quality and value to its climate, there can be no doubt that they exhibit greater vitality and more rapid and better progress within a given period than the hardwood trees of any other county in Scotland as a whole."

Much as has been done in recent years to enhance the sylvan aspect of the county, there are still some parts now almost waste, which might be profitably planted with larch or other firs, and many of these places were not very long ago clothed with thriving woods and forests. In fact, almost the whole of Perthshire was at one time a huge and almost impenetrable forest. In the days of the aborigines a vast forest extended northward of the Forth and Clyde, covering all the territory of the Caledonii, and actually giving to them their name, the original meaning of which was "The People of the Coverts." That this forest was of vast extent, and was inhabited by a large population, is apparent from the term Caledonia having come to be applied to the whole of the country north of the Forth and Clyde; and latterly, in a general way, to the whole of Scotland. It is also apparent that this forest was as dense as it was extensive. Such an insuperable obstacle did it present to the progress of the Roman arms that, in 209, Severus was forced to take extraordinary measures to obtain access to the country, and deprive the hardy natives of their natural bulwarks. Instead of wasting his energies and sacrificing the lives of his soldiers in the almost impossible task of driving the natives from the depths of the forest, the Roman General is said to have felled the woods and constructed roads to meet the aborigines in their fastnesses. That this was an operation of no ordinary kind is testified by the fact, that he is believed to have lost no fewer than 50,000 men in destroying the forests and subduing the physical features of the country alone, and without feeling the stroke of the enemy. These native forests appear to have consisted principally of fir, birch, and oak; and buried remains of these are occasionally brought to the surface in Perthshire. In Balquhidder, large trunks of birch, as well as oak, trees are occasionally found, and evidently belong to a defunct forest. In some mosses in

Strathtay fossil wood has also been found. Remains of what was no doubt the classic forest of Birnam have been found at different times. While draining some hollows in Ladywell Wood, the Duke of Athole's forester, Mr M'Gregor, came across several noble oaks embedded in the earth. They were too ponderous to be easily removed, and were sawn across to allow the draining operations to be carried out, the wood being found very hard and sound. A short time ago several large limbs of oak, one which we measured being over two feet in diameter, were taken from the bottom of a lake in the American Gardens at Murtly Castle. The lake, which is 12 feet deep, was being cleaned out, when the tree was discovered protruding through the bottom. The wood was nearly as black as Irish bog oak, and was in a first-rate state of preservation. The bottom of the lake, which is mossy, abounds with the remains of birch, hazel, and alder, in tolerably good preservation. Remains of this ancient forest, we believe, might be found all over the Murtly property. Glenmore, a narrow vale chiefly in the parish of Fortingall, was, at a remote period, covered with the extinct forest of Schiehallion, and for a long time the roots of fir trees and the trunks of oaks furnished a profitable produce to the inhabitants of the district. The fir trees not only furnished excellent fuel, but it is said that when in a state of combustion they "emitted a light surpassing the brilliancy of gas." The oak trunks dug out from the soil were of a blackish colour, and though somewhat soft became very hard on exposure to the air. They were split up and manufactured into sharpening tools for scythes, and found a ready market in the locality. A large oak trunk, of a beautiful black colour, was, some years ago, found in the bed of the Tay, not far from Perth, where it had probably lain for ages. The wood was used for several ornamental purposes, amongst these being the making of three beautiful caskets, presented to the late Prince Albert, Mr Mackenzie, the late Premier of Canada, and Major-General (then Colonel) Sir John M'Leod, K.C.B., when they received the freedom of the city of Perth.

It is right to mention that one writer, at least, appears to have been under the impression that the country was originally of quite a different character from that which has just been described. In Wallace's *Nature and Digest of Ancient Peerages* (2nd edition, Edinburgh, 1785, p. 34) the following passage occurs:—" I must not, however, disguise that Scotland appears of old to have been wholly naked, or extremely open, and we have little reason to think that in any age, of which an accurate remembrance is preserved, this kingdom ever was much more wooded than it is now." Wallace, however, as we shall afterwards show, appears to have been misled by the nakedness which existed 300 or 400 years previously, and jumped to the conclusion that that was the original state of the country. Alluding, evidently, to the above passage, Tytler, in the "Historical Inquiry into the Ancient State of Scotland," contained in his *History of Scotland*, gives a different account of the appearance of the country. "No two pictures," he says, "could be more dissimilar than Scotland in the thirteenth and fourteenth, and Scotland in the nineteenth century. . . . . The face of the country was covered by immense forests, chiefly of oak, in the midst of which, upon the precipitous banks of rivers or on rocks which formed a natural fortification, and were deemed impregnable to the military art of that period, were placed the castles of the feudal barons. . . . . So erroneous is the opinion of a conjectural historian, who pronounces that there is little reason to think that in any age, of which an accurate remembrance is preserved, this kingdom was ever more wooded than it is now." The fact of the ancient remains of numerous gigantic forest trees being found in many parts of central Scotland speaks strongly in favour of the theory that this part of the country was in the time of the aborigines, a huge forest, and there can be little doubt but a large area of it would be cut down as a military necessity.

The work of destroying those ancient forests, however, was not confined to the conquering Romans. To such an extent was the work of destruction carried on during the feudal times, that it amounted to the grossest Vandalism. The common people, we are told, fancied that the growth of timber was an obstacle to the production of food, and the ancient barons do not appear to have been sufficiently enlightened to place any restraint upon the work of desolation. The low grounds were gradually divested of cover, and trees were only to be met with in the steep declivities of the glens, and other places that were deemed inaccessible to cultivation. After the destruction of those ancient woods and forests, the county must have had a most desolate and wretched appearance. Some idea of the desolation of the country may be gathered from the fact, that when the late Duke of Athole began those operations in forestry with which his name is so honourably associated, he had only 1000 acres of wood on his extensive property. The amenity of the county must also have been greatly lessened on account of the circumstance that, although the greater part of Perthshire was originally a forest, a large portion must have been nothing but a swamp. Along the valleys of the Tay and the Earn the ground lies very low, and before the existence of the barriers which now, to a large extent, protect the land from inundation, the floods which periodically descend must have submerged an extensive tract of country. The bed of the Tay, at no very distant date, embraced a much larger territory than it now does, and many places at present safe from inundations were quite recently not beyond the reach of destructive floods. Although the whole of the low-lying land is now well drained, the Carse of Gowrie, up till 1760, was, like many other tracts along the banks of the Tay, disfigured with large pools of water.

So naked and open was the country at this time that its condition engaged the serious attention of Parliament, and various Acts were passed with the view of covering the nakedness of the land,—a state of matters which may have led the "conjectural historian" into the error of supposing that there had at no time been extensive woods in the country. In the "Acts of the Parliament of Scotland" (14th Parliament, James II., 6th March, 1457), we find it enacted "Item, anent plantations of woods and hedges, and sawing of broom, the Lords thinks speedful that the King charge all his freeholds, baith spiritual and temporal, that in the making of this Whitsunday set, they statute and ordain that all their tenants plant woods and trees, and make hedges, and saw broom, after the faculties of their mailings in place condiment therefor, under sic pain as law and inlaw of the Baron or Lord shall modify." In the 6th Parliament of James IV., 11th March 1503, it is also enacted, " Item, it is statute and ordained anent policy to behalden in the country, that every Lord and Laird make them to have parks with deer, stanks, cunningares [rabbit warrens], dowcatts [dovecots], orchards, hedges, and plant at the least ane acre of wood, where there is na great woods nor forests." Again, we find in the fourth Parliament of James V., 7th June, 1535, " The above ratified ; and that every man having an hundred pounds land of new extent, where there is no wood, plant, and make hedges and haining, extending to three acres, and that the tenants of every merk land plant a tree." In 1616, at the close of one of those numerous insurrections fomented by the Macdonalds, Lords of the Isles, we find the leading island chiefs being bound over at Edinburgh, amongst other things, to build " civil and comlie " houses, or repair those that were decayed, and to have " policie and planting about them." Still further, we find it enacted, so late as the second Parliament of Charles II., 19th October, 1669, " All preceding Acts ratified, and that every heritor, liferenter, or wadsetter [bondholder], worth £1000 of valued rent, enclose yearly for ten years next ensuing, four acres of ground, and plant the same with oak and other trees, at three yards distance ; and other heritors of greater or less rents accordingly ; and that they uphold the same."

At the close of the Civil Wars, when the people became somewhat settled in their habits, more enlightened views regarding arboriculture began to prevail, and the landlords saw it to be to their advantage to do voluntarily what had hitherto been enforced by Acts of Parliament. From 1720 to 1735, the proprietors of various estates, accordingly, began to save what remained of the ancient woods, and to turn available land into new plantations. In this they were greatly encouraged in different ways. " The Society of Improvers," for instance, founded at Edinburgh, in 1723, and which included most of the Scottish nobles, made a very good start in this direction. One of its most prominent members, Thomas Hope of Rankeilor, in order to furnish an example of what might be done, leased from the city of Edinburgh a morass called Straton's Loch, and having drained it, " raised beautiful hedges and trees, where (in 1743) gentlemen and ladies resort." This land has now long been known as "the Meadows." "The Edinburgh Society for encouraging Arts, Sciences, Manufactures, and Agriculture," founded in 1755, went a step farther. In the first year of its existence, a prize of £10 was offered to "the farmer who planted the greatest number (not under 1000) of timber trees, oak, beech, ash, or elm, in hedgerows, before December 1756," and a prize of £5 to the second, "not under 500." A prize of £6 was also offered "to the farmer who should rear the greatest number (not under 2000) of young thorn plants before December 1758; and a sum of £4 for the second greatest number, not under 1000." The prize for planting the largest number of timber trees was awarded to Mr Alexander Baxter, farmer, Woodhead, Borrowstounness; and the second prize to Mr Alexander Walker, farmer, Auquhiry, Dunnottar, Kincardineshire. When the Highland and Agricultural Society was founded in 1784, another decided advance was made in the direction of improving and extending woodlands. The first attempt of this kind was in the year 1805, when premiums were offering for the raising of osiers and willows. In 1809, the Society, convinced that there was a good deal of ground, especially on the north-west coast of Scotland, which it would be advantageous both for proprietors and the country to have planted, offered honorary premiums to proprietors in this part of the country who should, betwixt February 1810, and 10th April 1812, plant the greatest extent of ground, after being properly enclosed; one half of the plants to be larch or hardwood. The premiums excited considerable attention, and the result was that a gold medal, bearing a suitable inscription, was awarded to each of the following gentlemen:—Alexander Maclean of Ardgower; Alexander Maclean of Coll; Ranald Macdonald of Staffa; Hugh Innes of Lochalsh, M.P.; and John Mackenzie of Applecross, all of whom had formed extensive plantations on their properties. In 1821 and 1822 honorary premiums were awarded for the greatest extent of ground planted and enclosed within the county of Dumbarton, the Isle of Skye, and small isles adjacent, as well as the Black Isle in Ross-shire. The first premium (a piece of plate valued fifteen guineas), for the islands was awarded to Lord Macdonald, who planted 149,600 trees; and a similar premium for the mainland was awarded to Colin Mackenzie of Kilcoy, who planted 501,000 trees on about 379 acres. A piece of plate, valued fifteen guineas, was also awarded to H. Macdonald Buchanan of Drumakill, Dumbartonshire, who planted 261,000 trees on about 50 acres; and a second prize, a piece of plate valued ten guineas, was awarded to Sir James Colquhoun of Luss, Bart., who planted 396,900 trees on above 61 acres. The first premium awarded to a tenant for planting appears to have been in 1823, when eight guineas were granted to Lachlan M'Lean, tacksman of Tallisker, Isle of Skye, "as a mark of the Society's approbation for his having planted a considerable extent of ground, after being properly enclosed, upon his farm." In the following year, we note that a piece of plate, valued fifteen guineas, was voted to Colonel M'Neill of Barra for extensive planting. Since this period numerous premiums have been awarded for planting, as well as a large number of prizes for papers on almost every subject connected with arboriculture, a variety of papers on this subject having been long a prominent feature in the "Transactions" of the Society. For upwards of a century, then, the woods and forests have been receiving considerable attention, and nowhere have they received more attention than in Perthshire. In every corner of the county plantations adapted to the soils and climate now adorn the landscape, the most common being the various varieties of the pine, although oak, ash, sycamore, and beech also grow abundantly. The pine predominates in the higher lands; the oak in lower lands and the valleys of the Grampians, where the climate is congenial and the soil light and dry; the ash grows spontaneously by the side of almost every stream or lake; the alder abounds in the swamps; and the birch climbs boldly up the sides of the Highland glens. Less than a hundred years ago, many of the hills, particularly in the Sidlaw range, were only covered with heath, where now they carry loads of valuable timber.

In this work of improvement a number of the larger proprietors have taken a prominent part. Amongst the more distinguished of these was the Duke of Athole, already referred to, whose accession occurred in 1774. On succeeding to the property, that nobleman planted the enormous number of 15,473 English acres of woods, or somewhere about twenty-seven millions of trees. A fragment more than the half of these were larches, about 1000 acres were oak, the remainder being principally Scots fir, spruce, and birch. On the Breadalbane, Scone, Murtly, Abercairny, Blair-Drummond, Dupplin, Moncreiffe, and other extensive properties, a great deal of planting has been going on during the same period, of all of which we shall have more to say afterwards.

It is only within the past few years that any notice was taken in the Agricultural Returns of the country of the extent of woods and forests. A return of the acreage of woods and forests in Scotland appears to have been obtained in 1812, when Perthshire was returned as having 203,880 acres under wood, being by far the largest acreage of any county in Scotland, the next highest being Aberdeen, with 148,800. Of these 203,880 acres, 61,164 are represented as having been planted, the remaining 142,716 acres being natural woods. What, perhaps, may be termed the first regular notice of woods and forests in the Government returns occurs in the report for 1869, but it is rather remarkable, that, although we have in this report returns from five foreign countries, there are no returns for Great Britain or Ireland. A return of the woods and plantations in this country was first obtained in 1871. The acreage of woodland was ascertained by the officers of Inland Revenue employed to collect the agricultural returns from the best sources of information available in their respective districts, as " a column for this information could not have been usefully added to the form issued to occupiers of land, as woods are not often held by farmers." The result of the special return for woods, coppices, and plantations showed a total acreage so occupied in Great Britain in 1871 of 2,175,471, of which 1,314,316 acres were in England, 126,625 acres were in Wales, 734,530 in Scotland, and 324,285 in Ireland. In 1872 the collecting officers were instructed to examine the first returns, with the view of testing their general accuracy, and in consequence of this revision the acreage for England was slightly increased, and that for Scotland slightly reduced, the amended return standing at 1,325,675 acres for England, 126,823 for Wales, and 734,490 for Scotland. The total acreage in Scotland under wood in 1812 is returned as 913,695, there being thus a falling-off of no less than 179,205 acres during the intervening sixty years. The falling off in Perthshire alone was 120,355 acres, being the difference between 203,880 acres and 83,525, as ascertained in 1872. As compared with 1812, 28 out of the 33 counties in Scotland show a decrease of the area under wood in 1872, and 16 show an increase. The land under woods and plantations is not considered to vary sufficiently to require a new return to be taken annually, and the acreage ascertained in 1872 was reported up till 1880. In that year, however, a special return of woods and plantations was obtained. This return showed that since 1872 the acreage had increased from 2,187,000 acres to 2,409,000, or nearly 10 per cent. In 1881 an amended return was obtained, and several corrections made. The following table, compiled from the agricultural returns, although it shows no variation in the acreage of woods until 1880, will be interesting, as showing the great alterations now in progress in Perthshire in the acreage of orchards and land used by nurserymen:—

Year.	Orchards, &c. Acreage of Arable or Grass Lands, but also used for Fruit Trees of any kind.	Nursery Grounds. Land used by Nurserymen for growing Trees, Shrubs, &c.	Woods. Coppices or Plantations, excepting Gorse Land and Garden Shrubbery.
1871	2110	—	83,525
1872	1098	239	83,525
1873	743	86	83,525
1874	789	94	83,525
1875	302	100	83,525
1876	265	114	83,525
1877	378	74	83,525
1878	349	89	83,525
1879	350	100	83,525
1880	327	92	91,333
1881	398	105	94,563
1882	394	99	94,563

The table given below, also compiled from the Agricultural Returns, shows the acreage of wood grown in Perthshire, as compared with the other counties in Scotland, Orkney and Shetland being excluded, as no return from that part of the country have been published. It will be noticed from the table that Perthshire shows the third largest acreage under wood, Inverness topping the list with 162,201 acres, Aberdeen being second with 103,156, while the acreage of Perthshire is 94,563.

Counties.	Orchards, &c. Acreage of Arable or Grass Lands, but also used for Fruit Trees of any kind.	Nursery Grounds. Lands used by Nurserymen for growing Trees, Shrubs, &c.	Woods. Coppices or Plantations, excepting Gorse Land and Garden Shrubbery.
Aberdeen	34	214	103,156
Argyll	3	2	42,741
Ayr	162	97	22,177
Banff	12	4	28,188
Berwick	31	21	13,376
Bute	1	18	3,454
Caithness	—	—	210
Clackmannan	10	3	2,028
Dumbarton	5	12	7,926
Dumfries	52	189	31,162
Edinburgh	125	477	11,354
Elgin or Moray	26	93	50,130
Fife	36	40	19,471
Forfar	52	104	30,287
Haddington	148	6	10,474
Inverness	26	77	162,201
Kincardine	—	12	27,880
Kinross	—	—	2,576
Kirkcudbright	36	25	19,741
Lanark	531	39	18,780
Linlithgow	5	11	4,899
Nairn	12	3	13,241
Peebles	—	6	10,177
PERTH	398	105	94,563
Renfrew	33	64	5,424
Ross and Cromarty	18	10	43,201
Roxburgh	26	59	14,679
Selkirk	2	2	3,228
Stirling	42	56	12,483
Sutherland	—	—	12,260
Wigtown	6	76	8,009
Total for Scotland	1,832	1,825	829,476

When we come to look at the pecuniary value of the woods and forests of Perthshire, we find we have a very extensive interest to deal with. From the statistics before us it is impossible to arrive at anything like an accurate estimate of the money value of the wood growing in the county, but an approximate calculation may be made that will be sufficient for our purpose. According to the latest returns, there are 94,563 acres of plantations, these including both young and old wood, as it is impossible to ascertain from the returns the relative proportion and value of each. We believe, however, we are taking a very moderate estimate when we say that the woods all over the county should average £35 an acre. This would make the total value nearly three millions and a half. It will thus be seen that the interest which we are considering is a most important one from a monetary, as well as a national, point of view—a fact which will become the more apparent as we proceed to deal with the details.

III.—THE PERTH NURSERIES.

Services rendered to Forestry by the Perth Nurseries—Their Formation and Situation—Cultivation of Forest and Ornamental Trees—The Latest Importations suitable for the Scottish Climate.

It is undeniable that Perthshire is, to a very large extent, indebted to the Perth Nurseries for the splendid timber now existing in its woods and forests; and it is only proper that we should commence our descriptive sketches with an account of those grounds where so many of the trees which now adorn the county have been trained. An old writer, referring to the magnificent wood on Kinnoull Hill, incidentally remarks that to the Perth Nurseries "the whole vicinity of Perth owes, in a prime degree, the wealth and exuberance of plantation and shrubbery which so extensively beautify it." This is not only true as regards the neighbourhood of the city of Perth, but it might be applied with equal force to the county generally, and, indeed, to a large part of Scotland, both north and south of Perthshire. Supposing the Perth Nurseries had done no more than brought into existence the arboreal wealth which embellishes the neighbourhood of the "Fair City," they would have accomplished something to be proud of, and would have deserved an honourable place in any history of the woods and forests of Perthshire. Without the waving forests that surround the town on every side,—without the magnificent ornamental trees that grow even within its borders,—Perth would not have been entitled to the designation of the "Fair City." The everlasting hills, which look down upon it from "a' the airts the wind can blaw," would no doubt have enveloped it with a rugged grandeur, even although unadorned with growing timber; but how naked and barren would have been the prospect! The great charm of the magnificent view to the north from the centre of Perth Bridge is due to the luxuriant foliage that skirts the eastern bank of the river, and extends as far into the policies of Scone as the eye can reach ; while to the south the eye is gratified by the refreshing sight of the dark green pines which clothe the sides and summit of Kinnoull, and by the variegated clumps and isolated trees which everywhere enrich the landscape. What, need we ask, would even Kinnoull Hill be without its plantations of pine ? These are some of the arboreal features which surround Perth ; and for these we are indebted, in a prime degree, to those Nurseries we propose to describe. It must be borne in mind, however, that what they have done for the immediate surroundings of the city is absolutely insignificant compared with what they have done for the county and the country at large.

We have said that the country at large has benefited by the establishment of these Nurseries, but we might even go further, and say that they have obtained a world-wide celebrity for their splendid collections of trees, including coniferæ, ornamental shrubs, deciduous trees, rhododendrons, herbaceous and alpines, not to speak of florists' flowers, roses, stove, greenhouse, orchidaceous plants, &c. The grounds were originally laid out in 1766, by Mr James Dickson of Hassendean-burn, in Roxburghshire, and at a very early period they obtained distinction for the introduction of new and useful plants. The first collection of double Scotch roses was raised from seed at the Perth Nurseries, whence they were distributed over the country. For some time back the newer kinds have taken their place, but the double Scotch roses are again coming into favour, and are at present largely inquired after. The superb Scarlet Hawthorn (*Cratægus oxyacantha punicea*) was likewise raised here. These Nurseries are also entitled to the distinguished honour of being one of the first, if not the very first, to distribute one of the most valuable roots the Scottish farmer has received from foreign countries,—viz., the Swedish turnip,— a number of seeds having been sent to the firm in an envelope by the great botanist, Linnæus, about the year 1772.

The Nurseries occupy a beautiful situation, and command a view which cannot be excelled anywhere in the neighbourhood of Perth. Being immediately adjacent to the picturesque Hill

THE PERTH NURSERIES.

of Kinnoull, they have the advantage of a most delightful background; and, with a wide stretch of the valley of the Tay in front, the eye is carried over an extensive tract of rich country. Before our feet lies the ancient city of Perth, the whole of which is seen at a glance, with the "ample Tay" flowing past, bright and transparent. A bend in the river directs attention to the Palace of Scone, with its heavily-timbered park. Away in the distance we look over the battlefield of Luncarty, to the slaty hills of Logiealmond, and "great Birnam wood," backed up by the snow-capped Grampians. Turning round a little, we catch a glance of Dunsinane Hill, which, like the "great wood," has been rendered classic by the genius of Shakespeare. To the left, the view is arrested by Moncreiffe Hill, whose fertile slopes and pine-clad summit greatly enhance the beauty of the scene.

Entering the Nurseries by the Bellwood approach, we gradually ascend the side of the hill, and enjoy the invigorating atmosphere, redolent of the ozone from the flourishing pines of Kinnoull. Here, we note, that the Nurseries are not only remarkable for the beauty of their situation, but also for the utility of the site for planting purposes. Lying on the northern and western sides of the Hill of Kinnoull, they are fully exposed to all the cutting blasts, which impart to the young trees that hardiness and vigour so necessary for our Scottish climate. The soil is variable. In some places a nice, friable loam prevails, and in other places the ground is sandy, and occasionally stiff-clay. In many places there are but a few inches of soil, with whinstone cropping up, here and there, above the surface. This variety is most serviceable for practical experiments, and is of great advantage to intending planters. We noticed that in the loam and sandy ground the trees had a most thriving appearance, and those that were lifted carried large masses of fibrous roots.

The impetus which has lately been given to planting in the United Kingdom in consequence of the rapid disappearance of the forests in North America, Sweden, and Norway, has led to special arrangements being made at the Perth Nurseries to supply the large and increasing demand for timber trees. Great exertions are made during the season to get the stock into the necessary state for safe removal; and when it is mentioned that, at a recent stocktaking, it was found that there were over ten millions of trees of all kinds in these Nurseries, some idea may be formed of the magnitude of the operations entailed to bring them to a marketable condition. The immense number of seedling beds of ordinary kinds of fir is something astonishing, and no one can visit the grounds without being interested in the enormous varied stock of all kinds, from the tiny seedling to the largest trees grown for single specimens for planting in parks and pleasure grounds. Although the proprietors have always aimed at enriching their collection by adding all the new introductions in every department,—especially if they are sufficiently hardy to stand the climate,—still the arboricultural experience of upwards of a century has not been thrown away. They have found that the larch fir seeds, of which large quantities are annually imported from the Tyrolese mountains do not thrive in these Nurseries, and they consequently rely upon home-saved seed. This has also been their experience with native Scots fir. The imported seed is found to be useless, in consequence of its inability to stand the climate. They are, therefore, careful to sow nothing but home-grown seed.

In addition to the cultivation of enormous quantities of forest trees, such as larch, spruce, silver, and Scots firs, *Pinus Laricio, P. Austriaca, A. Douglasii*, sycamores, beech, ash, oak, &c., thorns or quicks, for fences, are largely raised. The recent introduction of so many fine and hardy conifers from the Crimea, California, British Columbia, the Rocky Mountains, the Himalayas, China, and Japan, has created a great demand for those best adapted for ornament and timber. With an enterprise worthy of the reputation of these grounds, the present proprietor, Mr John A. Anderson,—who has been associated with the Nurseries for nearly half-a-century, and has been enabled to do more to find out the truly hardy varieties of trees of all sorts than probably any other person living,—has endeavoured to secure selections of all the latest importations, so as to be able, from actual cultivation, to speak with confidence of their respective merits. It is satisfactory to learn that most of the importations from the countries named have stood the test of the recent severe winters unscathed.

As the timber of many of those trees to which we have referred as having successfully stood the test of the extraordinary winters of 1878 and 1879 especially is of great value when grown in their native habitats, we will endeavour to give some idea of the estimation in which the most important of them are held, which may, in some measure, serve as a guide to planters, and at the same time supply much useful preliminary information regarding many of the trees of which we shall afterwards have occasion to speak. Of course, many years must elapse before their utility in this country can be finally demonstrated; but the experience already obtained at the Perth Nurseries and elsewhere is of the very greatest value, and will help to determine the future profitableness of the different varieties if extensively planted.

The first of the trees to which attention may be directed is *Abies Albertiana*, which is found growing 150 feet high in California, with a trunk from 4 to 5 feet in diameter, the wood being soft and tough. It thrives in a deep rich soil, and in mossland if not too wet. The tree is named in memory of the late lamented Prince Consort, whose skill and zeal in the collection of coniferous plants at Osborne are beyond all praise.

Abies Douglasii is found growing in the forests of California to the enormous height of 300 feet. The wood is heavy, of the same colour as the yew, and is superior to the best red deal. It may be interesting to note that the principal flagstaff at the International Exhibition of 1862 was taken from a specimen of this tree 309 feet high! Several magnificent specimens are to be seen at Murtly Castle, Scone, and other places in Perthshire, where they have shown a wonderful adaptability to the climate of this locality. It has been found that they will not succeed in stiff soils,—a light, rich soil being most suitable.

Abies Engelmannii and *Abies Parryana* are both from the Rocky Mountains and the valleys of California. They are exceedingly hardy, being found at the highest altitudes in the Rocky Mountains as dense bushes. They grow quickest in rich soils, but at home they grow in any soils, except stiff clay.

Abies Menziesii is a beautiful tree from California and the western slopes of the Rocky Mountains. It is a noble tree when in good condition, but requires a deep rich soil and some shelter from cutting winds.

Abies Morinda (syn. *Smithiana* and *Khutrow*) is an importation from the Himalayas, where it is found associated with *Pinus excelsa* and *Cedrus Deodara*, at an altitude as high as from 7000 to 10,000 feet. It is therefore eminently adapted for mountain planting; and like its two associates, it should, if possible, be protected from south-west winds. We shall have some additional remarks applicable to this tree when we come to speak of the *Cedrus Deodara*.

Abies orientalis is one of the handsomest of the spruces, and very hardy. It is a native of the Crimea, and is not a large tree, only attaining a height of from 70 to 80 feet. Its value as a timber tree is not yet known.

Abies Tsuga (syn. *Sieboldii*) is a quite recent introduction. It bears a close resemblance to *Abies Pattoniana*, but is yet quite distinct. If grown in deep, rich, friable soil, it has every appearance of proving a handsome tree.

Araucaria imbricata, a really noble tree, is found growing on the sides of the Andes of Chili. It must be planted in an elevated situation, as it is frequently injured by severe winters when planted in low and moist situations. In its native regions it attains a height of from 80 to 150 feet, and produces the heaviest timber of all the conifers. The wood is also very durable.

Cedrus Atlantica (the silver or Mount Atlas cedar) is also a handsome and noble tree. It is found growing on the Atlas range in Northern Africa, to the height of 100 feet. It is believed to be best adapted for elevated situations. The timber is durable, and of considerable value.

Cedrus Deodara (the Indian cedar), as has been already mentioned, grows on the Himalaya Mountains, at a very high altitude, where it attains an immense size. It reaches a height of from 150 to 200 feet, and in many instances the side branches are as large as ordinary trees. Along with *Pinus excelsa* and *Abies Morinda*, it requires a moist soil and situation. In their native state the finest trees are found growing on lofty plateaus, sheltered from the south-west winds by immense masses of projecting rocks. In spring and summer the perpetual snow of the Himalayas is melted to a considerable extent by the sun, and flows down the sides of the mountains, converting the plateaus into bogs. This has evidently been overlooked in the planting of the tree in this country, and, consequently, it has not answered the expectations that were formed of it by planters. In many instances it has died at the top, gradually decaying downwards, after attaining the height of from 25 to 30 feet. The timber of this tree, when matured, is of great value. It is found in Hindoo temples above one thousand years old, and still perfectly sound, no insect being found about it. It is susceptible of a very beautiful polish.

Cupressus Lawsoniana—one of the most elegant, beautiful, and graceful of trees—can be planted anywhere, and thrive in almost any situation. It is a native of California, where it grows 100 feet high, and 6 feet in circumference. The timber is of excellent quality, easily worked, and emits a strong aromatic fragrance. This is a tree which it is desirable should be largely planted. *C. Lawsoniana stricta viridis* is an upright variety, and equally valuable.

Cupressus Nutkaensis (*Thujopsis borealis*) is a most valuable and hardy tree, of quick growth, symmetrical habit, and of large size, being found 100 feet high in British Columbia and Vancouver's Island. It is also suitable for any soil.

Cryptomeria elegans is a very beautiful Japanese tree, of rapid growth. In the winter months the foliage is of a rich bronze hue, and presents a pleasing contrast to the sombre colour of many of the *Pinus*. This is a tree that might also be largely planted with advantage.

Juniperus sinensis, a native of China, is one of our best trees for ornamental purposes. It is very hardy, and adapted for any soil. This tree, as well as its variety, Young's golden Chinese juniper, is so highly valued that no collection should be without some specimens.

Libocedrus decurrens, often erroneously called *Thuja gigantea*, is a native of California. It is a very handsome tree, growing from 50 to 60 feet high, and is best adapted for elevated positions, being found in its native country at an altitude of from 4000 to 5000 feet.

Picea concolor and *P. concolor violacea* are two recent introductions from California, evidently allied to *P. lasiocarpa*, yet distinct. They have at present every appearance of being hardy and valuable.

Picea firma (the Japanese silver fir) is a noble tree, of considerable size, and is found at an elevation of from 2000 to 4000 feet. The timber is soft, and is much used by Japanese upholsterers.

Picea grandis, a magnificent tree found on the banks of Fraser's River, California. It is of quick and robust growth, and attains a height of from 200 to 250 feet. The quality of the timber has not yet been ascertained.

Picea lasiocarpa (syn. *Lowii* and *Parsonsii*) is one of the most beautiful trees known. Of quick growth and very hardy, it forms a noble tree in California, attaining the height of 250 feet. The timber of all the silver firs will, no doubt, be valuable when matured.

Picea nobilis (the noble silver fir) is also one of the most magnificent of trees. It is of rapid growth, and forms immense forests on the mountains of California. This tree attains a height of 200 feet, and the timber, as is pretty well known, is very valuable

Picea Nordmanniana is also a grand fir, which seems to thrive everywhere. When young, it is particularly handsome, and attracts universal admiration. It is a native of the Crimea, where it attains the height of about 100 feet. The wood is hard and good.

Picea polita is an ornamental Japanese variety.

Pinus aristata is a small tree from the snowy mountains of North America. It grows from 30 to 40 feet high, and is a very useful ornamental variety.

Pinus Benthamiana, a very large and noble pine, found in California, growing to the height of upwards of 200 feet. The timber is very good, and the tree is considered to be specially adapted for the pinetum in this country.

Pinus excelsa is such a magnificent tree that the Hindoos call it the "King of the Pines." It is found growing with *C. Deodara* at an altitude of from 5000 to 12,000 feet above the sea, and attains a height of 150 feet. The timber is white, soft, and strong; and the tree produces a large quantity of clear, fragrant, and pure turpentine.

Pinus Jeffreyi forms a very handsome tree. It grows to a height of 150 feet in California, and the timber is very good.

Pinus monticola is one of the most beautiful and hardy of the Californian pines, with fine glaucous foliage. It is found growing on poor soil on the sides of the Rocky Mountains, and attains a height of about 100 feet. The timber is white, fine in grain, and tough.

Pinus parviflora, a small growing tree from Japan, and having beautiful glaucous foliage, is very suitable for ornamental purposes.

Pinus ponderosa is a large and splendid tree found in many parts of California. It grows to a height of 100 feet. The wood is hard, very durable, and so heavy that it sinks in water.

Retinospora obtusa is a magnificent evergreen tree from Japan. It is found in the forests on the mountains of the island of Niphon, growing to a height of from 70 to 100 feet. The timber is white and finely grained, and is very beautiful when polished. It is likely to prove exceedingly valuable in this country.

Retinospora pisifera is another very fine Japanese conifer that should be largely planted.

Retinospora plumosa is one of the finest introductions from Japan, being hardy and of very graceful habit.

Retinospora plumosa aurea is another very beautiful variety of *R. plumosa*. All the *Retinosporas* thrive and grow freely in any soil, except wet bog.

Sciadopitys verticillata (the Umbrella Pine) is also a beautiful plant from Japan, where it is said to grow to a height of 140 feet. From its unique appearance, it forms a very valuable addition to our collection of conifers.

Thuja Lobbii (syn. *Menziesii*) is evidently the true *Thuja gigantea*, the plant usually known by that name being *Libocedrus decurrens*. It is one of the fastest growing trees known, is of large size, and the foliage is of a fine shiny green. It is a native of California, and, like many other trees from that country, deserves to be largely planted here.

Thujopsis dolobrata variegata ranks amongst the handsomest trees known. It is of rapid growth, and in its native country (Japan) its timber is very highly valued. It is said to attain a height of more than 200 feet in its native habitats. Professor Thunberg calls it the most beautiful of evergreen trees, and worthy of extensive cultivation, not only as a single specimen, but as a permanent timber tree. It adapts itself to almost any soil and locality, but high altitudes and moist places seem to suit it best.

Thujopsis Standishii is also a very fine tree, and, from its rapid growth, is likely to prove valuable in this country.

Wellingtonia gigantea is a magnificent tree from California. It was found growing in a valley on the slopes of the Sierra Nevada by Douglas in 1831. It is one of the most distinct and beautiful of trees, perfectly hardy, and well adapted for hill-planting in this country, its rapid growth being also a great recommendation.

Our object in so fully describing the above selection from the unique collection of coniferous plants grown in the Nursery has been to present a list of those most likely to prove useful and hardy trees in Perthshire and the surrounding counties. Many beautiful trees are omitted on account of their not being sufficiently hardy to be useful for our climate. There are, for instance, the glorious Cedar of Lebanon and the Redwood of California (*Taxodium sempervirens*), two splendid trees, to be found in many places in the south and west of England, although suffering from frost with us.

As the object of this work is mainly to describe the condition of arboriculture in Perthshire, it would be somewhat foreign to our purpose to give a detailed account of those portions of the Nurseries occupied with fruit trees, roses, &c., not forgetting the extensive glass structures and their valuable contents. We may state, however, that we found throughout the whole of these departments the same care bestowed in the selection and cultivation as in the arboricultural section. Any one paying a visit to these extensive grounds will enjoy a treat that will long be remembered, and will meet with the utmost courtesy from the principals and officials connected with the establishment.

IV.—ATHOLE.

The Varied Attractions of Athole—Its Position in the History of Scottish Forestry—First Extensive Larch Plantations—Introduction of Larch into Scotland—Result of a "Social Crack" and "a Waught" from a "Pocket Pistol"—The Value of Larch—First Attempt at Mountain Planting in Scotland—Planting of Larch as a Forest Tree—"Humble Petition of Bruar Water"—Cost of Planting—Extent of Plantations—First Attempts to use Larch for Shipbuilding—Satisfactory Experiments—Beauty of Dunkeld—Walk through the Woods at Dunkeld and Blair-Athole—Blair Castle—The Athole Family—Destruction of Woods by Recent Storms.

THERE is no district in Perthshire, indeed no district in the country, that presents more varied attractions than Athole. Those who delight in historic scenes will find in Athole all their hearts can desire, as here some of the most stirring events in early Scottish history have been enacted. The ancient town of Dunkeld, with its magnificent cathedral, carries the mind back to the remote period of the Picts, when, as is supposed, it was a seat of royalty. Coming to the romantic Pass of Killiecrankie, and the ancient Castle of Blair, we tread upon ground that has been traversed by many Highland armies; where some of the most desperate battles that have ever taken place upon Scottish soil have been fought; and where rest the remains of not a few of Scotland's heroes. To the lover of Nature, too, Athole presents unrivalled charms. Although situated in the most northern portion of Perthshire, it possesses, in many parts, the characteristics of Lowland as well as of Highland scenery; and in some places, notably at Dunkeld, the Highland and the Lowland landscapes are so beautifully blended that the scenery becomes positively enchanting. Everywhere the face of the country is of the most picturesque description. Lofty mountains tower in all directions, many of them rising to close upon 4000 feet above the level of the sea. Numerous glens of surpassing beauty ; extensive lakes and classic streams and waterfalls ; deep, solemn forests, lonely moors, and fertile plains, combine to make the district without a rival in natural beauty. It is impossible to deny that much of this beauty is due to those " solemn forests " to which we have referred, and to the judicious way in which the planting of the property has been carried out. Athole, like the rest of the country, suffered materially during the troublous times, by the cutting down of the woods ; but although at no very remote period there was scarcely such a thing as a tree to be seen, the property now affords such an excellent illustration of what can be done in a very short time in the way of forest-planting, that arborists point to it with no ordinary interest and pleasure.

As there is no district which occupies a more interesting or a more important place in the history of forestry in this country, than Athole, so there are no names more honourably associated with successful planting than the Dukes of Athole. John, the fourth Duke, like most proprietors of his day, found that his extensive patrimony was almost denuded of that sylvan beauty which so greatly enhances the value of an estate, and contributes to the internal strength of the nation. Besides adding to the richness of the scenery of a land which Nature has favoured in no ordinary degree, his Grace, with that business-like capacity which has always characterised the Dukes of Athole, saw that by judicious planting he would, at no very distant day, develop a source of wealth so prolific as almost to baffle calculation. He had the great honour of being the first to plant larch (*Larix Europea*) to any extent in this country. The larch is mentioned by Parkinson in his *Paradissus* as early as the year 1629, but at that time it was " nursed up but with a few, and those lovers of rarities." Till towards the middle of the 18th century, however, no one appears to have thought of planting it with the view of profit, or to have regarded it as calculated to afford timber for naval and other purposes. It is supposed that Goodwood, the property of the Duke of Richmond, was the first place at which larch was planted as a forest tree in this country, but there it was only in small numbers. Duke John, however, rightly concluded that if the larch grew naturally in elevated situations, among the central alpine ranges of the continent of Europe, it would also grow with vigour in any other alpine country, the mean temperature of which was not materially below that of its native site, and the mountains of which could afford due shelter from the influence of sea air. The experiment, at all events, was well worthy of a trial. We learn from a rare account of the larch plantations, prepared in 1832 from papers and documents attested by the fourth Duke's own hands—and to which we are much indebted—that his Grace entered into the project with the greatest enthusiasm, and with the liveliest expectations of success. In 1815, before he commenced to plant the great larch forests with which his name is associated, he remarked that "if one-fourth part of the product of 2,600,000 larches arrive to maturity in 72 years, by the time the present century expires, it will supply all the demands required by Great Britain for war or commerce. The success which has attended my efforts will probably induce, and indeed has induced, many already to plant to an extent which will not only meet the wants of Great Britain, but enable her, possibly within a century, to export wood to an immense amount. Under these circumstances, the prices of wood for shipbuilding may, and probably will, be much decreased at the same time. The grounds I have planted, and intend to plant, I consider admirably calculated to produce the *best wood ;* and I think, too, that my plantations will be the first in the market for a number of years, to any considerable extent ; and, lastly, the greater quantity, though of less price, will make up and probably be productive of an income to a much greater amount than that of any subject in the kingdom. The price of larch-wood will, no doubt, always be regulated according to the demand ; but I have no hesitation in saying that the price, when the wood is thoroughly known, will long continue superior to the best foreign timber, and little inferior to the oak."

The tree in which Duke John took such a lively interest was introduced into Scotland in 1738, when Mr Menzies of Culdares, in Glenlyon, brought a few small plants, which had been secured in the Tyrol, in his portmanteau from London. Five of these he left at Dunkeld, and eleven at Blair-Athole, for

Duke James, the grandfather of the "Planting Duke," as he was familiarly called. The five were planted in the lawn at Dunkeld House, the soil being of an alluvial gravelly nature, abounding with rounded stones, and in a sheltered situation, at an elevation of 40 feet above the Tay, and 130 feet above the level of the sea. Two of the five were felled by Duke John in 1809, and one had been cut by mistake about twenty years before, and made into mill axles. Of the two felled in 1809, one, containing 147 cubic feet of timber, was sent to Woolwich Dockyard, and formed into beams for the repair of the *Serapis* store-ship. The other, containing 168 cubic feet, was bought on the spot by Messrs Symes & Co., shipbuilders, Leith, at 3s. per foot, or £20, 4s. the tree. The two parent larches which remain are still growing in great vigour, and will be referred to afterwards. The eleven which were planted at Blair-Athole, at an elevation of 500 feet above the level of the sea, measured, in 1817, from 8 to 12 feet in girth; now, those that remain are not far behind their contemporaries at Dunkeld in size. Although these were actually the first larch trees introduced into Scotland, a few were also planted about the same time by other proprietors, and some rather curious stories are told as to how these proprietors obtained possession of the plants which are now no mean rivals of the Dunkeld trees. A very good story is told as to how these interesting trees were introduced into Belladrum, in Ross-shire. It is said that the then proprietor of Belladrum, who possessed keen arboricultural tastes, visited the Duke of Athole at Dunkeld House about 1738. In the course of a walk through the grounds one morning he came across the gardener on his way to plant a number of young larches, and, evidently fully alive not only to the scarcity of the plants, but also to their importance, the laird of Belladrum determined, by hook or by crook, to obtain possession of some of them. He politely asked the gardener to oblige him with a few plants, but the gardener was equal to the occasion, and as politely refused. The gardener was then tempted by a sum of money to part with a portion of his treasure, but the sight of the "filthy lucre" had the effect of making the gardener the more determined in his refusal. The laird, however, was not to be thwarted in his object, and, changing his tactics, he commenced a " social crack " with the gardener, and produced his " pocket pistol." It is almost needless to say that the influence of the " crack," and a " waught or twa" out of the " pistol," was so potent that the laird carried off his plants in triumph. They are still standing, and are almost equal to the larches at Dunkeld. In 1740, six larch plants were brought up in the greenhouse at Dunkeld, but not appearing to thrive in it they were planted out. They had evidently suffered from the warmth of the house, and, although they have become fine trees, they are much smaller than many larches that were planted after them. It is not mentioned in the documents of his Grace whether they had been raised from the seed, although the presumption is that they had been so reared. Between 1740 and 1750, Duke James planted 350 larches at Dunkeld, at an elevation of 180 feet above the level of the sea, and 873 at Blair, among limestone gravel, in a sheltered situation, at an elevation of 560 feet above the sea, the ground being worth from 20s. to 30s. per acre. Situated as they were in the neighbourhood of Dunkeld House and Blair Castle, they were evidently intended more as a trial of a new species of tree than for forest timber. But in 1759, Duke James planted 700 larches, over a space of 29 Scotch acres, intermixed with other kinds of forest trees, with the view of trying the value of the larch. This plantation extended up the face of a hill from 200 to 400 feet above the level of the sea. The rocky ground of which it was composed was not worth above £3 a year altogether, and covered with loose and crumbling masses of mica-slate. This may be considered the first attempt at mountain-planting in Scotland. According to the fashion of the time, the trees were arranged in rows, and they converged towards a small piece of water in the centre like radii. This concluded the whole attempt at planting by Duke James. Before he died, however, in January 1764, he had tried the quality of the larch as timber, and was quite satisfied of its superiority over other firs, even in trees of only eighteen or nineteen years' standing. This plantation throve in the most satisfactory way. In 1816, John, the fourth Duke, cut out a larch, aged 57 years, for naval purposes, and it was found to contain 74 cubic feet of timber, and was sold, exclusive of all expense, at 2s. 6d. per foot, or £9, 5s. the tree. "I don't believe," said Duke John, speaking of this plantation, "there is another species of tree in the 29 acres, oak included, except a few spruces, that would bring a guinea. Some of the spruces might contain from 30 to 40 feet, and be worth 2s. a foot."

John, the third Duke, who was the first to conceive the idea of planting larch by itself as a forest tree, succeeded to the property in 1764. In 1768 he commenced by the planting of three acres with larches alone on Craigvinean, above the wood which Duke James planted on the same hill in 1759, at an altitude of from 100 to 200 feet above it, or 500 or 600 feet above the level of the sea, on soil that was not worth 1s. per acre. It is to him we are also indebted for the planting of the sides of the hills about Dunkeld, and which contributes so largely to the beauty of that charming locality. He enclosed a considerable extent of ground for the planting of mixed wood, including 190 acres on Cragiebarns, 30 acres at Callie, 30 acres at Haughend, 90 acres on Craigvinean, 25 acres of the Hermitage plantation, and several small clumps, amounting in all to 5 or 6 acres. Besides this, he planted about 300 acres at Blair-Athole, forming a total of 665 acres, of which 410 acres were finished before his death in 1774. The greatest obstacle to the progress of planting was the scarcity, and consequent dearness, of the larch plants. A few plants were raised by his Grace himself from the cones gathered from some trees at Blair-Athole, but this supply did not exceed 1000 plants in a season. All that could, therefore, be obtained for planting did not exceed fifty plants per acre in the large plantations; and the rest of the quantity, amounting to 4000 plants per Scotch acre,—that being the allowance of plants to the acre at that time,—was made up of Scots fir, and the different kinds of hard wood. The larch was planted at a height not exceeding 600 feet, and the Scots fir at 90 feet above the level of the sea. Another difficulty which the Duke had to encounter was from the broom, furze, juniper, and heath, which flourished abundantly in the region allotted to the larch, and which had not been entirely eradicated before the planting began. The broom, though indicative of a good soil for larch, is a troublesome plant to young trees, its long switch-like elastic twigs whipping their tops violently in windy weather; while the furze, with its thick-set spiny branches, smothers, or draws up prematurely, the young trees. These, and many other obstacles to successful planting, would no doubt have been speedily removed by the Duke had he not frequently to be absent from home during the short time he occupied the property.

Such were the state and extent of the larch plantations in Athole when John, the fourth Duke, succeeded to the title in 1774. The first object of the new Duke was to plant the 225 acres which formed part of the plantations that were left unfinished by his father. Owing to the difficulty of obtaining larch plants, this, with the planting of some larches about the Loch o' the Lows, occupied him till the year 1783. Hitherto, the larch had been chiefly planted along with other trees; but, observing its rapid growth and hardy nature, the Duke determined on extending it to the steep acclivities of mountains of greater altitude than any that had yet been tried. He commenced, between 1785 and 1786, by enclosing 29 acres on the rugged summits of Craig-y-barns, and planting a stripe, almost entirely of larch, among the crevices and hollows of the rocks where the least soil could be found. As at this elevation some of the larger kinds of natural plants grew, the ground required no previous preparation or clearing. The expense of enclosing and planting at this time was £1, 19s. 1½d. per acre. From 1786 to 1791 the Duke planted 480 acres at Dunkeld, the greater part of which was, owing to the difficulty of procuring a sufficient number of plants, only sprinkled with larch, from 6 feet to 30 feet asunder; and 200 acres at Blair-Athole, which were planted wholly of larch at 6 feet apart. The number of larch plants consumed in these plantations in the five years was 500,000. Wages rose at this period, and, there being a greater substitution of larch for Scots fir, the expense of planting was considerably increased, the planting, with the enclosing, amounting to £2, 10s. 6d. per acre. The pitting alone cost 10s. 6d. per acre. In the eight years from 1791 to 1799 the Duke still continued to diminish the number of Scots fir, and to increase that of the larch. It was between these years that so much was done to beautify the banks of the Bruar Water, 70 acres being planted around the beautiful waterfall rendered classic by our national poet. The Bruar is, no doubt, indebted, in the first

THE FALLS OF BRUAR.

instance, to Burns for much of the sylvan beauty of its banks; his keen eye for the beautiful detecting the one thing needful to give grandeur to the scene, and his ready pen representing the wants of Bruar Water in a way that would command attention from any one, far less so keen an arboriculturist as Duke John—

> "Would then my noble master please
>   To grant my highest wishes,
> He'll shade my banks wi' tow'ring trees,
>   And bonny spreading bushes.
> Delighted doubly then, my Lord,
>   You'll wander on my banks,
> And listen mony a grateful bird
>   Return you tuneful thanks."

Duke John lost no time in answering the "Humble Petition of Bruar Water," by adorning its banks with Scots fir, larch, spruce, &c., a policy which has been continued by the present Duke. The most of the trees originally planted at Bruar have been blown down by gales between 1879 and 1883, and the ground is now being replanted. Up till recently, however, the whole of the wood along the Bruar side was in a most thriving state, and, could the poet have gazed upon it his heart would have been delighted to see—

> "Lofty firs, and ashes cool,
>   The lowly banks o'erspread,
> And view, deep bending in the pool,
>   Their shadow's wat'ry bed !
> Here fragrant birks, in woodbines drest,
>   The craggy cliffs adorn ;
> And for the little songster's nest,
>   The close embow'ring thorn."

Even up to this period, his Grace had to contend against a great scarcity of plants, as the value of the larch as a timber tree was beginning to be appreciated. At Logierait, Inver, and Dunkeld, the space altogether planted extended to 800 acres, 600 of which were entirely of larch. The expense of planting this piece of larch was the same as the last, and the plants consumed amounted to 800,000,—a number which was obtained with much difficulty. As the larch still continued to adapt itself to the most exposed regions, his Grace resolved to push entire larch plantations to the summits of the highest hills. The Scots firs planted 900 feet above the sea, were "beet up" ten years afterwards with larch as an experiment. In 1800, when the Duke was anxious still farther to extend his larch plantations, the effect of this experiment confirmed him in the opinion he had previously entertained of the very hardy nature of the larch plant. In a period of nearly forty years these Scots firs had only attained a height of 5 or 6 feet; while the larch, planted amongst them, ten years afterwards, rose to a height of from 40 to 50 feet. Before this it was supposed that larch was incapable of vegetating at an elevation of 900 feet. In the same year (1800) a favourable circumstance concurred with the result of the above experiment to give an impulse to the commencement of a great undertaking in planting. In that year several of the farms in Dowally parish fell out of lease, and, as they were all in a miserable condition, his Grace took them into his own hand to improve them, and to build suitable farm-houses and offices. This circumstance gave the Duke the command of a range of mountains forming the background to the farms which his Grace had taken into his own hands, extending from the edge of Craig-y-barns over a space of 1600 Scotch acres, including a commonty, the rights of which the Duke bought up. This range was situated from 900 to 1200 feet above the level of the sea. The soil presented the most barren aspect, and was so thickly strewed with fragments of rock that vegetation of any kind scarcely existed upon it. "To endeavour to grow ship timber," remarks his Grace, " among rocks and shivered fragments of schist, such as I have described, would have appeared to a stranger extreme folly, and money thrown away. But in the year 1800, I had for more than 25 years so watched and admired the hardihood and the strong vegetative powers of the larch, in many situations as barren and as rugged as any part of this range, though not so elevated, as quite satisfied me that I ought, having so fair an opportunity, to seize it." During the same period in which this range was planted, his Grace also planted 400 acres in other situations,—150 acres at Haughend, and 259 acres about the Loch o' the Lows. These make a total of 2409 Scotch acres, 1800 of which consisted solely of larch, and 300 of the larch occupied a region far above the growth of the Scots fir. The enclosing and planting of these woods occupied the long period of fifteen years,—the delay, as formerly, arising from the difficulty of obtaining larch plants, and which only permitted them to be planted to a thickness of from 1500 to 1800 per acre.

As the Duke entertained no doubt whatever of the successful growth of the larch in very high situations, he still farther pursued his object of covering all his mountainous regions with that valuable wood. Accordingly, a space to the northward of the one last described,—known as Loch Ordie, from the beautiful lake of that name, which is within the forest,—containing 2959 Scotch acres, was immediately enclosed, and planted entirely with larch. This tract, lying generally above the region of broom, furze, juniper, and long heath, required no artificial clearing. An improved mode of planting was employed here,—that of using young plants only, two or three years' seedlings, and put into the ground by means of an instrument invented by the Duke instead of the common spade. This change of arrangement facilitated the operation, and, at the same time, greatly increased the supply of the plants, so as to enable the whole ground to be planted in the three years from 4th December 1815 to 2nd December 1818. The increased number of plants per acre, and the high price of the plants, enhanced the cost of planting and fencing this forest from 10s. 6d. per acre to 16s. 8d. In 1824, the growth of the larch in Loch Ordie forest having greatly exceeded the most sanguine hopes and expectations of the Duke, he determined on adding to it an extensive adjoining tract, consisting of 2231 Scotch acres, denominated Loch Hoishnie. The preparations of fencing, clearing where that was necessary, making roads, and procuring plants from different nurserymen, occupied the time till October 1825, when the planting commenced, and was carried on in good earnest, the whole being finished by December 1826. In this case the fencing and planting cost 15s. per acre. It is recorded that there was no plantation which his Grace had executed that gave him so much satisfaction in the work as that of the forest of Loch Hoishnie. The planting of this forest appears to have terminated the labours of the Duke in planting, and he could now look back with pardonable pride upon the splendid results of his labours. That the contemplation of what he had accomplished gave him no ordinary pleasure is well

brought out by the following note in his memorandum-book :—
"Drove up to Loch Ordie, and home by the back of Craig-y-barns, every way much gratified with the growth of the larch and spruce,—a very fine, grand picturesque drive, not to be equalled in Britain ! The extent of the drive through woods of my own planting, from one to forty years old, is fifteen miles." This was certainly something to be proud of. Although his labours may here be said to have been completed, it is highly probable, both from his known desire to extend his larch woods, and from the following entry in his private diary, that he intended still further to prosecute them :—

Wednesday, 28th July 1824.—Mr Urquhart, the nurseryman at Dundee, went up to Loch Ordie with me. He is to furnish 50,000 one-year-old transplanted larch and 1,000,000 of seedlings, 500,000 of which he engages to transplant, and to be ready to put into the ground in the autumn of 1825, and to deliver the remainder as seedlings. These, with 100,000 transplanted seedlings, to be furnished by Donaldson at Dunkeld, and 500,000 more in autumn 1825, with what I shall try and collect otherwise, will make a good beginning in my new forest of 6000 acres.

The following table will show at a glance the extent of the larch plantations executed by the different noble Dukes :—

	No. of Larches exclusive of the other Plants Mixed with them.	No. of Larches Planted without Mixture.	Acres of Entire Larches.
Duke James planted at Dunkeld and Blair-Athole in 1738	16	—	—
Do. do. to 1750	350	—	—
Do. do. to 1759	1,575	—	—
Duke John planted at Dunkeld and Blair-Athole from 1766 to 1774	11,400	—	—
Duke John the Fourth, 1774 to 1783	279,000	—	—
Do. 1783 to 1786	—	43,500	29
Do. 1786 to 1791	20,000	480,000	450
Do. 1791 to 1799	560,000	240,000	600
Do. 1800 to 1815	250,000	2,250,000	1800
Do. 1816 to 1818	—	5,922,000	2961
Do. 1824 to 1826	—	4,038,880	2231
Totals,	1,122,341	12,974,380	8071

The total amount of larch plants, mixed or unmixed with

other kinds, is thus brought up to the enormous number of 14,096,719 plants ; and if 2000 plants are allowed per acre for the amount mixed with other kinds of trees, these would occupy a space, if planted alone of larch, of 533 acres ; so that the whole extent of ground occupied by larch amounts to 8604 Scotch acres, or 10,324 imperial acres.

Although Duke James has the honour of first introducing the larch into the woods of Athole, the properties of the new tree were not so well known by him as to stimulate him to an extention of its culture. He followed the fashion of the day, of placing trees in parallel rows, in diverging rays from a centre and in quincunces. Gusts of wind, having free access through these alleys, blew down many thousands of the Scots firs, and broke as many more, their heavy heads and superficial hold of the soil rendering them unfit to resist the effects of a strong gale. Although Duke James had little opportunity of introducing improvements in planting, the more extensive experience of John, the fourth Duke, resulted in many improvements in the mode of planting trees in general, and the larch in particular. The result of that experience has been the introduction of a simple, cheap, and efficacious mode of inserting larch plants into the ground. That experience has also determined the proper age at which the larch should be planted, so that it may reach maturity at the earliest possible period ; and the number of plants that ought to be allotted to an acre. To Duke John, indeed, is due the credit of discovering most of the important facts now known about the treatment of the larch.

We have already said that the main object of Duke John in planting such extensive forests of larch was his conviction that it was highly suitable for shipbuilding purposes. Besides testing its suitability for this purpose, it was extensively employed for ordinary building purposes with satisfactory results. About the year 1800 the tanning properties of larch bark were tried, at the Duke's request, by a tanner in Perth, and succeeded tolerably well, although the tanner complained that it had not half the strength of oak bark. The first attempt to use larch for the purposes of navigation was in 1777, when a number of fishing cobles on the Tay were constructed. Previous to that they were made of Scots fir, and at the end of three years had to undergo a thorough repair. Fifteen years afterwards, ferry-boats were constructed of larch instead of oak, and proved to be more buoyant. The satisfactory nature of these experiments induced the Duke to bring the subject under the notice of the Admiralty, and in 1800 a large quantity of larch was employed in the repair of the store-ship *Serapis*. Like many other new introductions, the new wood had to contend with much prejudice amongst the workmen in the dockyard. The prejudice, however, was said to arise in consequence of the workmen having repaired a Russian vessel that was built of Russian larch a very short time before the Duke's larch arrived, but on the latter being cut up it was found to be of such a very different texture from the former, that they were ultimately obliged to praise it. After the lapse of a year or two, a survey of the *Serapis* was made, when a most favourable report of the state of the larch was given. The most important trial of larch as a tree for naval timber was made from 1816 to 1820, in the building of H.M. frigate *Athole*. The *Niemen* frigate, built of Baltic fir, was also built at the same time, the *Niemen* being launched on 20th November 1820, and the *Athole* on the following day. Minute inspections were made at different times as to the state of the larch in the *Athole*, and all were very laudatory of its qualities as ship timber, it being reported that the larch became harder and more durable by age in a ship; it held iron as firmly as oak, but, unlike oak, did not corrode iron; iron bolts might be driven out afterwards perfectly clean; it did not shrink; the *Athole* had been caulked but once in four years; it possessed the valuable property of resisting damp, inasmuch as the pump-well was as dry as the cabin. A communication from the Admiralty of date 13th December 1827, founded upon a report on the condition of the *Athole* and *Niemen* frigates, states that the *Athole* at that time would "only require very small repairs, whereas the *Niemen*, built of Baltic fir, was found so very defective as to be proposed either to be broken up or taken to pieces." At the same time as the *Athole* was being built, the Duke caused Mr Brown, of Perth, to build a brig entirely of larch. The smaller vessel was, of course, first ready, and on being launched on 6th August 1819, was very appropriately named the *Larch*. The history of this brig, of which there are ample details, afforded very satisfactory evidence of the utility of larch for the construction of ships in the merchant service. She was registered at 171 tons, though she could carry a cargo of 300 tons dead weight. She sailed almost over the entire globe, and never so much as carried away a spar for a period of eight years, till at length she was wrecked on the island of Tendra, in the Black Sea, through the ignorance of the pilot, on 27th November 1827. She is described as having been a fine sea boat, a fast sailer, and so tight that she always brought her cargoes of dried fruit from the Mediterranean in excellent condition, as was testified by her consignees. Her wrecked hull was sold to some people in Odessa, and though it lay high on the beach of the island of Tendra for two years, exposed to the vicissitudes of summer and winter, before she was launched off again, such were the strength and toughness of her timbers that she never went to pieces. A number of other ships were built of larch during the Duke's lifetime, some of them being built of wood planted by the Duke himself. Amongst the other properties noted as showing the larch to be a valuable tree for shipbuilding, is that of its being slow of kindling by fire. Though hot embers be thrown on a floor of larch, it will not suddenly get up in a blaze, like other kinds of fir. It is also said that a shot-hole through larch closes, and does not splinter.

Duke John was never tired of trying experiments as to the nature of the larch, both practical, physical, and chemical, and the results of his experiments afforded him satisfactory proof of its durability, strength, elasticity, and resilience. So confident was the Duke in the satisfactory results of his experiments, that in the last years of his life he planted 6500 Scotch acres of mountain ground solely with larch. Altogether, his Grace planted 12,478 Scotch or 15,573 imperial acres, which consumed the enormous number of 27,431,600 plants. His Grace was altogether an extraordinary man, and, from his patriotic experiments, he deservedly occupies a conspicuous place in the list of those who have rendered eminent service to the country ; while the magnificent forests with which he clothed the previously desert and dreary ranges of the valleys of the Tay and the Tummel entitle his name to be revered by all lovers of the beautiful in Nature, and by natives of Perthshire in particular, on whose county he has shed so great a lustre.

No one crossing the bridge which spans the Tay at Dunkeld can fail to be impressed with the beauty of the scene. Indeed, the oftener one visits this ancient city the more convinced will he be of the impossibility of finding, in all Scotland, a greater variety of scenic grandeur within the same space. Nature has

DUNKELD CATHEDRAL.

been most lavish in her favours, but Art has largely contributed to make Dunkeld what it is. How dull and uninteresting would be the river at our feet, clear and ample though it be, were it not for the noble trees of almost endless variety in colour, form, and disposition, which skirt its margin, here and there feathering down into the dark water, and concealing its boundaries! How unattractive would those towering mountains be were it not for the forests that clothe them to the top! How incomplete would the picture be without that grand Cathedral, hoary with antiquity, and embowered amid umbrageous grandeur! What would Dunkeld be even without its famous gardens, the produce of which has figured with no ordinary credit in all the important horticultural shows in the country! Without in the meantime looking into the gardens, we direct our steps to the woods.

In passing Little Dunkeld Bank, our attention was directed to several fine specimens of the *Pinus laricio* growing alongside the Scots fir, from which the foreigners stretch away with a markedly superior growth. In no more satisfactory way, indeed, could there be afforded an illustration of the advantages that attend the cultivation of this recently-introduced pine—a tree which has an interesting history, having, it is said, been propagated from seed accidentally obtained from Corsica at the time of the first Napoleon, and which recommends itself to foresters by its quick growth, and its remarkably clean timber, though as yet it has shown itself rather troublesome in transplanting. We next come to Ladywell Nursery. The nursery extends to 4½ imperial acres, and previous to 1868 was an arable field belonging to Ladywell Farm. This nursery is intended for the supply of the woods in the Dunkeld, Strathord, and Middle Districts of Athole, and contains a large stock of the best-known forest and ornamental trees. In one portion of the nursery we noted an interesting experiment : Mr M'Gregor, his Grace's head forester, endeavouring to rear *Picea Nordmanniani* under the shade of larger trees, that they may be put out at a size sufficiently large for immediate effect.

Leaving the nursery, we proceed to Ladywell Wood, a fine plantation of 350 acres, planted principally with Scots fir and larch about the year 1850, and after a crop of larch and fir. The appearance of the larch would seem to confirm the prevailing belief amongst foresters, that it does not succeed very well after conifers of any kind. Where the larch has retained its vigour, however, its superior growth has carried it from 15 to 18 feet above the heads of the surrounding Scots firs ; but, unfortunately, it is only in isolated places that it is thriving so satisfactorily, the greater number of the trees having had to be cut down in consequence of disease. It was apparent from the "bleeding" condition of a number of the trunks, that the disease was to be attributed to the ravages of a species of aphis.

48   WOODS, FORESTS, AND ESTATES OF PERTHSHIRE.

We noticed that Mr M'Gregor cuts off the dead branches of the Scots firs and conifers of all sorts, thus, besides preventing black knots in the wood, improving the general appearance of the tree. Before leaving Ladywell Wood, Mr M'Gregor pointed

HERMITAGE FALLS.

out a rather curious formation of heights and hollows, and stated that in the course of the process of draining the hollows the remains of several old black oak trees were found. Where these trees interfered with the drainage operations they were cut across, and allowed to lie. The wood was still hard and sound.

After passing through the Hermitage Coppice, we paused to admire the picturesque falls and the romantic scenery at the Hermitage Bridge. Here there are several notable trees. At the north end of the bridge there is a very fine cedar of Lebanon (*Cedrus Libani*), and within thirty yards of it is a particularly good specimen of the hemlock spruce (*Abies Canadensis*), girthing 10 feet at 4 feet from the ground, and 80 feet high. The tanning properties of this tree are very great, and in America it takes the place of oak bark. The relative price of the bark of the hemlock spruce and the oak bark is as 15s. per ton to £6 per ton. It is almost impossible to conceive the enormous resources of America in this respect. Lower Canada, which furnishes bark for the Boston market, will be, it is admitted, the greatest field for hemlock in America for half a century to come. We recently noticed in an American paper that one firm alone has fourteen tanneries in Canada, using 50,000 "cords" of bark per year, each "cord" being from 17 to 19 cwt., "to tan the sole leather required for the boots and shoes of four millions of people in the British provinces." The same firm own in land and hemlock growth and bark in Maine, New Brunswick, and Canada over one million acres. The extent of territory for this bark in America is said to be equal to two States the size of Massachusetts, or over 10,000,000 acres of land. With such resources as these, and with a glutted market already, there does not appear to be much hope of an improvement in the trade for oak bark, or, indeed, much prospect of tanners making a fortune in this country. A Scots fir, also in the neighbourhood of the Hermitage, girths 11 feet 6 inches at 5 feet from the ground.

On leaving the Hermitage we crossed the old Strathbraan Road, one of General Wade's, and, like the rest of his roads, constructed more from a strategical than an engineering point of view. We were next conducted to Craigvinean Wood ("The Craig of the Goats"), the upper portions of which afford magnificent views of the scenery around Dunkeld, of the valley of the Tay to the north and west, and of the more distant hills in all directions, presenting altogether of the these views which leave Dunkeld without a rival in Highland scenery. This is an old wood of mixed Scots fir and larch, but was replanted in 1868, almost entirely with Scots fir, the experience gained in Ladywell Wood preventing Mr M'Gregor planting any larch. Although neither the beetle nor caterpillar have been met with here, the trees are not without their enemies. Mr M'Gregor pointed out the remarkable way in which a number of Scots firs suddenly died away. After examination it was found that the trees owed their death to the borings of an insect. Although a few of the Scots firs have met with an untimely end, the plantation as a whole is in a fine healthy state. In a few places where the ground was boggy, drains were dug from 15 to 18 inches deep, and some spruce planted, these also thriving very well.

Coming down by the banks of the Tay, a splendid row of Spanish chestnuts at once arrests attention. The row is planted in regular order at the foot of a knoll called "Torval," thus signifying that the grove had, at a remote period, been associated with worship. One of those chestnuts has a beautiful spiral habit of growth. It is very rare indeed that a tree is seen with such a complete twist, there being no fewer than six distinct spirals, each of them clearly traceable to the very top of the tree. The other trees in the group are perfectly straight in the fibre. They are all of large size,—one, taken at random, girthing 11 feet 5 inches at 3 feet from the ground, and 90 feet in height, with a fine clean bole of 25 feet. Crossing the Tay, we enter the policy grounds of Her Grace the Dowager Duchess of Athole. The first most noted objects in arboriculture are to be met with at the King's Seat, believed to have been a Caledonian stronghold, and on which are still to be traced the outworks of an ancient fort. Here there is a clump of splendid old Scots firs of about 150 years' growth. The bark of all these trees is remarkably clean, and has those lizard-like scales peculiar to the indigenous Scots fir well developed. The cleanliness of the bark is readily accounted for by the fine gravelly soil on which the trees rest, affording good drainage, and keeping the trees in first-rate health. The clean appearance of the trunks is also supposed to be due to the effects of the storms from the opposite side of the valley. The slopes of the King's Seat being very steep, the storms, coming down the gully in the hills on the opposite side of the Tay, strike against this abruptly-rising ground at times with great fury; and it is supposed that, being thus weather-beaten, the cleanness of the bark, as compared with trees in more sheltered situations, is sufficiently accounted for.

On the banks of the river, in front of what is known as the North Terrace, and opposite the "King's Ford," is one of the largest and most beautiful larches on the property. It lies in a hollow, and its great height is, therefore, not at first sight discernible. On going down the bank to measure the girth, we noticed a peculiar mark about 5 feet up, and, on explanations being asked, Mr M'Gregor stated that the mark indicated the flood in the river upon 7th February 1868, when the Tay rose to about 20 feet above its ordinary level, inundating all the low-lying grounds, and reaching to the mark upon the larch. The girth was found to be 10 feet 1 in. at 5 feet from the ground, with a straight height of 98 ft. of splendid bole, there being altogether about 200 cubic feet of saleable timber. Some very fine silver firs are also to be seen in this neighbourhood. At the "King's Park" there are some fine clumps of Scots fir and beech on the tops of numerous knolls. An oak which was met with along the river-side at this place girths at the narrowest point, being 4 feet from the ground, 15 ft. 2 in. At 3 feet from the ground it girths 15 ft. 9 in. It has a fine bole of 12 ft., then branches into five huge limbs, each of them being the size of ordinary trees. The spread of branches was found to be exactly 99 ft.

On the opposite bank of the Tay is the oak under whose kindly shade the celebrated Neil Gow was in the habit of retiring with the violin, and where it is believed he composed some of his finest pieces. The tree is pointed out to visitors as "Neil Gow's Oak"—

> "Famous Neil,
> The man that played the fiddle weel."

We next come to the American Gardens, which are beautifully laid out with luxuriant specimens of rhododendron, kalmia, and other analogous shrubs, finely sheltered by the surrounding trees. In the course of the walk through these gardens, we noticed a fine specimen of *Pyrus aria*. There is also the fine old rhododendron, *Ponticum*, rising to a height of 8 or 10 feet, and a *Magnolia acuminata*, said to be one of the largest in

THE PARENT LARCHES.

Britain. It girths 3 ft. 4 in. at 4 feet from the ground, and is 40 feet high. A beech, with a peculiar graft, is also worthy of some attention. After growing straight till about 5 feet from the ground it splits, and rejoins the main trunk about 4 feet

ATHOLE.

higher, the two stems becoming incorporated by a process of natural grafting.

We next reach the two parent larches at the west end of the Cathedral. Although they were originally treated as greenhouse plants, they have proved so hardy that they are universally recognised as the best and largest specimens that exist, although they are closely approached by the ones at Monzie. Planted in 1738, their progress has been watched with ever-increasing interest. In the summer of 1831, the one was 11 feet and the other 12 feet in girth at 4 feet from the ground. In 1878, the larger one girthed 14 feet 9½ inches at 5 feet. The present girth of the largest one is as follows :—At the base, 27 feet; at 1 foot from the ground, 22 feet 7 in.; at 3 feet, 18 feet 9 inches; and at 5 feet from the ground the girth is 14 feet 11 inches. There are only four moderately strong branches throughout the entire tree, which has a height of fully 100 feet. The other tree girths 20 feet 2 inches at the base; 17 feet 11 inches at 1 foot up; 14 feet 7 inches at 3 feet; and 13 feet 9 inches at 5 feet from the ground. Both of these trees are in splendid health, and are still making wood. Near to the parent larches is a beautiful variegated plane tree (*Acer pseudo platanus albo variegatum*), with a very fine upright habit of growth, and measuring 9 feet 3 inches at 4 feet from the ground, with a height of 65 feet. To the west of the Cathedral are four fine old yews; one, right opposite the centre of the west door, with a nice upright habit, girths 7 feet 3½ inches at the narrowest part, and rises to a height of 35 feet. Further west, and probably also within the limits of the old graveyard connected with the Cathedral, is another yew, 10 feet 6 inches in circumference at 1 foot from the ground, being the narrowest part, and rising to a height of 40 feet. A magnificent oak also stands near the parent larches. It measures 12 feet 6 inches in circumference at 5 feet from the ground, with a splendid bole of 30 feet, the height of the tree being altogether about 100 feet.

Coming next to the New House Park, we get a peep at that part of the grounds where the larch was got for the building of the *Athole*. On the road to this place there is as fine a specimen of the *Pavia flava* as can be seen anywhere. At the Bishop's Drive, a beautiful grass walk, there are some fine beech, larch, spruce, silver fir, and other trees. One of the silver firs girths 17 feet in circumference, and has a height of 100 feet. Mr M'Gregor pointed out a larch, which, although not so large as the parent ones, is, from its remarkable symmetry of growth, considered to be the best specimen of *Larix Europœa* in this country. At 5 feet from the ground it girths 11 feet 2 inches, and at 1 foot, 13 feet 10 inches. The tree rises to a height of 120 feet, all of it being clean bole, and good wood from the ground to the top, there being altogether about 300 cubic feet of saleable timber. There is also a fine Spanish chestnut in the New House Park. At 5 feet from the ground it girths 14 feet 10 inches, and at 1 foot, 17 feet 6 inches, the bole being 15 feet, and the entire height 70 feet. An old walnut tree, now showing signs of decay, girths 12 feet 2 inches at 5 feet from the ground, the height being 80 feet. Coming back to our starting-point, we find ourselves at Dunkeld Gardens, which, under the management of Mr P. W. Fairgrieve, have achieved a celebrity second to none in this country. Everything here is in the best of order, and at each turn there is evidence of the highest cultural intelligence.

Proceeding to Blair-Athole, we find that the more one sees of Athole the more will he admire it. The policy grounds here were originally laid out by Duke James, about 1742, in the Dutch style, but have been improved from time to time by his various successors to the dukedom. Like the grounds at Dunkeld, they are adorned with many grand trees, the larches being the most conspicuous. Immediately on entering at the grounds along the side of the Tilt, we noticed a very interesting belt of old native Scots fir and a fine line of larch trees. These trees, although not so old as some of those at Dunkeld, are very large. One, taken at random, girths 10 feet at 5 feet from the ground, and 9 feet at 7 feet, with saleable timber to 90 feet. By a short detour we get a view of the lower falls of the Fender, where the stream descends from among the rocks and woods above in a series of cascades, boiling and eddying, and falling in a shower of foam into the dark Tilt below. On regaining the carriage-way, we cannot help commenting upon the beautiful pasture on the south side of the Tilt, a mile further up the river, due, no doubt, to the limestone bottom. We next reach an eminence, from which a splendid view is got of the far-famed Glen Tilt. This

GLEN TILT.

glen is not only considered one of the finest in Scotland, but it has a character peculiarly its own. It is distinguished from most other Highland glens no less by its extreme depth, narrowness, and prolongation, than by the wildness of its upper extremity, and the highly-ornamented beauty of that part which approaches Blair-Athole. Proceeding along the Blairuachder Drive to Blairuachder Wood, we get a really good view of Blair Castle, the residence of the Duke and Duchess of Athole. As far back as history takes us, Blair Castle has been the seat of the families holding the Athole property. It was the original patrimony of a family which gave kings to Scotland from Duncan to Alexander III., and is one of the places earliest mentioned in Scottish history. The Athole family traces its lineage back to Andrew Murray of Tullibardine, great-grandson of Sir John de Moravia, Sheriff of Perth in the beginning of the reign of Alexander II. He aided Edward Baliol in his invasion of Scotland in 1332, and contributed to the victory gained at Dupplin, on the 12th August of that year. He was, however, taken prisoner at Perth, on 7th October 1332, and beheaded. From him descended Sir David Murray of Gask and Tullibardine, who was knighted at the coronation of King James in 1424. His grandson, Sir William, was steward, forester, and coroner, within the Earldom of Strathearn and Lordship of Balquhidder for life, and was a commissioner for a treaty with the English in 1495. Sir John of Tullibardine was a great favourite of James VI. He was made Master of the King's Household in 1592; was knighted and raised to the peerage of Scotland, by the title of Lord Murray of Tullibardine, by patent dated 25th April 1604; and further advanced as Earl of Tullibardine on 10th July 1606. William, the second Earl of Tullibardine, married Dorothy, eldest daughter of John Stewart, the fifth and last Earl of Athole of that name, and resigned the earldom of Tullibardine to his younger brother, Patrick. John, son of William, the second Earl, was, on 6th August 1628, served heir to his mother's grandfather's grandfather, John Stewart, first Earl of Athole. The earldom was confirmed to him and his heirs by patent dated 17th February 1629, "in right of his mother, nearest and lawful eldest heir of the late John, Earl of Athole, brother uterine of James II.," and was also granted anew to him and his heirs without prejudice to his hereditary right. He took an active part in the civil war against the Parliament. He died in June 1642, and was succeeded by his son John, second Earl of Athole of this house, K.T., and, like his father, was a royalist. To such an extent did he carry his opposition to the Parliament, that he was excepted by Cromwell out of his Act of Grace. After the Restoration, in 1661, he was made hereditary Sheriff of Fifeshire, Justice-General of Scotland in 1663, Captain of the King's Guards in 1670, Keeper of the Privy Seal in 1672, and an Extraordinary Lord of Session in 1673. He succeeded to the Earldom of Tullibardine in 1670, upon the death of James, fourth Earl, and was created Marquis of Athole by patent dated 17th February 1676. He died 6th May 1703, and was buried in the Cathedral Church of Dunkeld. Of his three sons, John succeeded as second Marquis of Athole, Lord Charles was created a peer of Scotland by the title of Earl of Dunmore, and Lord William succeeded as second Lord Nairne, having married Margaret, the only daughter and heiress of Robert, Lord Nairne, who was so created by patent in 1681, to himself for life, and after his decease to Lord William Murray. John, the second Marquis, was also a man of considerable note. In 1704 he was advanced to the dignity of Duke of Athole, K.T. He was chosen one of the sixteen representatives of the Scottish Peerage, 1710–1713; an Extraordinary Lord of Session in 1712; and he proclaimed George I. as king at Perth in 1714. His son having joined in the insurrection of 1715, he obtained an Act of Parliament "for vesting the honours and estate of John, Duke of Athole, in James Murray, Esq., commonly called Lord James Murray, at the death of the said Duke." His eldest son, John, fell at Malplaquet, in 1709, and the second son, William, who took part in the insurrection, was attainted in 1716, and died a prisoner in the Tower of London in 1746. The fifth son, Lord George Murray, was also attainted for his share in the rebellions of 1715 and 1745, but his eldest son, John, afterwards became the third Duke. James, the second Duke, as already indicated, succeeded his father by Act of Parliament. On the death of the tenth Earl of Derby, in 1736, he succeeded, as heir through his grandmother, to the Barony of Strange and the Lordship of Man. His only surviving daughter and heiress, Charlotte, married her cousin, whom we have referred to as the third Duke of Athole. On account of his father's attainder, a petition was presented to the King, claiming the title, and on this petition being referred to the House of Lords, it was sustained. All the titles, dignities, and honours of the house of Athole were thus vested in him, including the sovereignty of the Isle of Man, a dignity which was purchased by the Lords of the Treasury in 1765. He was member of Parliament for Perthshire from 1761 to 1764. His son, John, was "the planting Duke," already referred to. He was created Baron Murray of Stanley and Earl Strange in the peerage of Great Britain, and inherited the Barony of Strange on the death of his mother. John, the fifth Duke, died unmarried, and was succeeded by his nephew, George Augustus Frederick John, as sixth Duke, whose father, Lord James, a major-general in the army, was created Baron Glenlyon of Glenlyon, in the peerage of the United Kingdom. The sixth Duke married the only daughter of the late Henry Home Drummond of Blair-Drummond, the mother of the seventh and present Duke, Sir John James Hugh Henry Stewart Murray, K.T. The present Duke is hereditary Sheriff of Perthshire, and Lord-Lieutenant of the county. In October 1863 he married Louisa, eldest daughter of the late Sir Thomas Moncreiffe, Bart. of Moncreiffe, by whom he has had four sons and three daughters.

Blair Castle, the home of the Athole family, is situated on an eminence rising from the plain watered by the Garry, and on all sides it is surrounded by the grandest Highland scenery. To the south-west Schiehallion raises its prodigious shoulders, and towards Glenerrochie we have Beinnchualoch terminating the view. The Ben-y-Gloe range rises to the north-east, and northwards are the mountains of Athole Forest. The latter, however, is in no way associated with forestry, being a wild mountainous range, and a forest only in the sense of the chase. Looking along the valley of the Garry, we have a magnificent view of the whole country from Struan Point to the Pass of Killiecrankie, the southern spur of Ben-y-Vrackie terminating the view in that direction. Bounding Blair Parks, to the north-west of the Castle, is Craig Urrard, with its splendid forest of pine. In Blairuachder Wood there is a remarkably fine deer fence on the Corrymony principle. The fence is 5 feet high, with standards (where straight) at 14 yards apart, the standards being of T-iron, 1 inch by 1½ inches. The droppers are the invention of Mr M'Gregor, and consist of using No. 6 wire instead of the wooden droppers in the original Corrymony fence. The fence is a most efficient

PASS OF KILLIECRANKIE.

and economical one. At a place with the (at least to southerns) unpronounceable name of Glaiceachlaidhe, signifying "The hollow of a sword," there is a very thriving plantation. The previous crop was larch, and was perhaps the finest crop of this valuable timber ever grown upon the Athole property. The ground has been replanted with a mixed crop of pine and hard wood. The pines are somewhat eaten by vermin, but the hard wood is making rapid progress, and promises to be a crop of no ordinary value. At "The Den," a little farther on, we came across a precipitous bank, about half a mile in length, sloping to the Banvie, deservedly worthy of notice, carrying, as it does, a splendid crop of full-grown timber, consisting of larch, Scots fir, and spruce. The trees average from 8 to 10 feet in circumference, and tower majestically to the height of from 100 to 120 feet, with straight, clean stems. This was altogether the finest lot of timber we saw at Blair-Athole.

Amongst the more notable trees in the policies and gardens is a splendid spruce growing by the side of the Banvie, and rising to the enormous height of 140 feet, as measured by a line from the top, and containing about 420 cubic feet of timber. A larch girths 12 feet 10 inches, at 5 feet from the ground, and another girths 11 feet at 5 feet from the ground, both trees being 100 feet in height. No one visiting this part of the policies would willingly miss the opportunity of seeing the ruins of the old church of Blair-Athole,

"Where the bones of heroes rest."

In a vault under the aisle lie the remains of Viscount Dundee, who was killed in the battle of Killiecrankie, in what is now Urrard Garden. Here are also interred the remains of the late Duke of Athole, George Augustus Frederick John Murray, who died 16th January 1864, and over whose tomb his Duchess has placed a beautiful allegorical monument to his memory, by Sir John Steell, R.S.A., an appropriate feature in the design being a tree, broken across the middle. Blair-Athole Nursery, which consists of $2\frac{1}{2}$ acres of all kinds of plants common to the country, is situated in this neighbourhood. Coming down by Blair Castle, we noticed in the lawn fronting the Castle a *Picea nobilis*, planted by the Prince of Wales on 23rd September 1872, and a *Picea Nordmanniana* planted by the Princess of Wales at the same time. Both of them are beautiful, healthy, and free-growing trees. Passing along "Hercules Walk," so called from a fine statue of the Grecian hero with which it is adorned, we cannot but express our admiration of the shrubbery border. Near "Hercules Walk" there is a magnificent larch, measuring 13 feet 6 inches at 5 feet from the ground. It is a fine healthy tree, with a wide-spreading branchy head, although it has suffered considerably from storms. The gardens at Blair-Athole, it may be remarked, are amongst the finest of the kind in the country, their artistic design, and the charming effect produced by the artificial lakes, presenting a scene of great beauty.

The whole of the Athole woods have suffered very severely from the disastrous gales of recent years. Throughout large tracts of country trees have been blown down in enormous numbers, and an extent of ground left blank which will keep the foresters busy replanting for a considerable time. The following table, prepared by Mr M'Gregor, will give a good idea of the number of trees blown down in the Athole woods from 3d October 1860 to the gale of 6th March 1883:—

Districts.	Oct. 3, 1860.	Dec. 28, 1879.	Oct. 14, 1881.	Nov. 21 and 22, 1881.	Jan. 6, 1882.	Mar. 6, 1883.	Total.
Blair*	3,755	9,931	3,001	4,723	3,666	8,837	33,913
Middle	8,031	4,236	6,401	3,101	2,051	...	23,820
Dowally†	...	50,142	530	3,454	4,350	...	58,476
Dunkeld	59,531	17,252	3,229	4,583	10,413	...	95,008
Strathord	3,884	1,442	54	466	155	...	6,001
	75,201	83,003	13,215	16,327	20,635	8,837	217,218

* Blair district includes Bruar and Struan Woods.
† Dowally is included in Dunkeld for 3d October 1860.

V.—DALGUISE.

Situation—Ancient Ecclesiastical Residence—Steuarts of Dalguise—A Branch of the Royal Steuarts — Valuable Relics of Antiquity—Gigantic Araucarias — Other Remarkable Trees — "Glenalbert's Fam'd Abode" — Steuartfield — Remarkable Cairn and Cave — A Deserted Bridge—Plantations.

DALGUISE is one of the prettiest little properties in the district of Athole, and occupies the western bank of the Tay, from its confluence with the Tummel at Logierait to the ancient city of Dunkeld. The advantages of its situation may be pretty shrewdly guessed when we remark that it was the property of the See of Dunkeld from the year 1060, the old ecclesiastics being notorious for selecting the richest lands in the district for their private possession. The present proprietor of the estate is Charles Horace Durrant Steuart, and he is the representative of a family with royal blood in its veins. The progenitor of the family was Sir John Steuart of Cardney, second son of King Robert II., by Mariotta, daughter of Sir John de Cardney of that Ilk, and a sister of Robert Cardney, who was made Bishop of Dunkeld in 1396. He married Jane, daughter of Sir John Drummond of Stobhall, and sister of Annabella, Queen of Robert III. His first charters appear to be dated in the year 1383, when he inherited, by his mother, the barony of Cardney and other lands. His descendants enjoyed the lands of Cardney till 1792, when they became the property, by purchase, of John, Duke of Athole. John was knighted by his nephew, King James I., at his coronation, in 1424, and is said to have been imprisoned along with Murdoch, Duke of Albany, in 1425. He was succeeded by his son Walter, who acquired the lands of Cluny, Stormont, Petty, &c., in Perthshire. He, again, was succeeded by his son John, designed of Cluny, and who became Heritable Forester of the Bishopric of Dunkeld. He had several sons, from the third of whom, John, sprang the family of Dalguise. John, the first of Dalguise, obtained possession of this property in 1543, chiefly, as is supposed, through the exertions of his uncle, who was Bishop of Dunkeld about that period, and to whom the venerable Cathedral owed much of its ancient splendour and decoration. He married Elizabeth, daughter of Alexander Steuart of Grantully, by whom he had two sons, and on his death, in 1570, he was succeeded by his eldest son, Alexander, who married Beatrix Forbes, daughter of James Forbes, vicar of Little Dunkeld, by whom he acquired the property of Ladywell, near Little Dunkeld, now in the possession of the Athole family. He afterwards purchased the lands of Kingcraigie, which were also sold to the Duke of Athole. On his death, in 1616, he was succeeded by his eldest son, John, commonly called John More Macalester on account of his great stature. He was much respected in the district, and was specially distinguished for his success in settling differences amongst his neighbours, which gave rise to the saying that when his tenants were at some of Montrose's battles, "they should have said that if their laird had been there he would have made up matters between parties." He purchased part of the lands of Middle Dalguise, which he gave as a patrimony to his second son. John More Macalester Steuart died in 1653, and was succeeded by his eldest son, Alexander, whose eldest son, also Alexander, died in 1669, before his father. In 1675, the son of John More Macalester was succeeded by his grandson, John, who married Isabella, only daughter of John Steuart, portioner of Middle Dalguise, by whom he got part of these lands. This, the sixth laird, was chamberlain to several successive Bishops of Dunkeld, and became a zealous supporter of Montrose. He died in 1706, and was succeeded by his son, John, whose fifth son, Thomas, became treasurer of the Bank of Scotland, and is distinguished as the inventor of the calculation of interest by decimal arithmetic, and the author of many useful tracts. We have had the pleasure of examining a handsome little volume containing the original calculations as prepared by Mr Thomas Steuart's own hand, and which came into possession of the family some time ago quite accidentally, it having been sold amongst a bundle of loose pamphlets. The caligraphy is exceedingly beautiful, very close, but as plain as print. The arrangement of the tables displays much ingenuity, and the entire volume is a splendid monument to the patience of this representative of the house of Dalguise. The work consists in all of 96 pages, and embraces his whole system of calculation, with explanations, while there are also four tables showing "what day of the week any day of any month falls upon for 4299 years from the birth of Christ, both according to the old and new stiles." One of the fly-

DALGUISE HOUSE.

leaves bears the following inscription:—"To John Robert Steuart, Esquire, from his uncle, Thomas Steuart, Edinburgh, 26th April 1790." John, the seventh laird, engaged in the rebellion of 1715, fighting as an officer of cavalry at Sheriffmuir. For this he nearly lost his estate, which was only saved by the good offices of James, Duke of Athole. He was, however, subjected to fine and imprisonment for his share in the attempt to restore the Stuarts. The fact of his being "out" in 1715, was very prejudicial to his property, so much so, that at this period it was in a very low state. However, by good management, he succeeded in overcoming his difficulties, and lived to make ample provision for his children. He carried out several important improvements upon the property, and built the House of Dalguise, the oldest portion of which bears the date 1716. Considerable additions have since been made to the house, but the oldest portion still retains its original architectural features, which are thoroughly characteristic of the times in which it was built. He also added to the family possessions, by purchasing the lands of Easter Dalguise. He died in 1776, in the 88th year of his age, and was succeeded by his second son, John, who added further to the property by the purchase of the estates of Ballo and Glenalbert. His fifth son, David, was, in 1781, elected Lord Provost of Edinburgh. The Lord Provost was an industrious book-collector, and in his possession originally was the first edition of the Latin Bible, in two large folio volumes, one of the earliest books executed from moveable types, and supposed to have been printed by Guttenberg & Faust in 1450. He also possessed the Roman Breviary, beautifully printed on the finest vellum at Venice in 1478, by Nicholas Jenson, and finely illuminated. Both of these valuable volumes are now in the Advocates' Library. John died in 1785, when he was succeeded by his son, Charles, the ninth laird of Dalguise, who was twice married, first to a daughter of Robert Steuart of Ballechin, who died without issue, and second, to a daughter of Laurence Oliphant of Gask, by whom he had two sons and one daughter. He also did much to improve the estate. He had the honour of being appointed a Deputy Lieutenant of Perthshire, on the institution of that office in 1794. He was succeeded by his eldest son, John Steuart, who was also a Deputy-Lieutenant and Magistrate for the county. He was sent to the Cape of Good Hope in 1829, as High Sheriff of the colony, and was afterwards raised to the office of Master of the Supreme Court. The present proprietor, the eleventh laird, is his grandson, and succeeded, on his grandfather's death, at the beginning of 1882. He was also resident at the Cape of Good Hope for many years, and honourably distinguished himself in the late Basuto war, serving on the staff of Colonel Carrington. He only returned to this country in the month of June 1882, when he and his lady, Leila Mary, daughter of John Wright, Esq., M.D., Wynberg, Cape of Good Hope, to whom he was married on 9th April 1882, received a truly Highland welcome to their new home. They have a son, John Nairne, born 15th January 1883. It will be seen, from the outline we have given of the pedigree, that the Steuarts of Dalguise are connected matrimonially with several families related to the royal blood of Scotland, while, it may be added, they are similarly related to many other families of distinction.

From its great antiquity and long connection with the most important families in the country, the Steuarts of Dalguise have been the means of preserving some of the most valuable relics of the past. The collection of family portraits at present in Dalguise House is very extensive, and the new laird has shown the importance he attaches to these heirlooms by having them thoroughly renovated. These portraits include one of Prince Charles, King Robert the Bruce, and many other historical personages. The collection of old swords and other weapons is also most extensive, and these are likewise being put in the best of order by Mr Steuart. One of the most interesting of these weapons is a sword which belonged to Prince Charles, and which bears his initials and the insignia of royalty on the blade. The sword is evidently not one which had been intended for ornament, as its whole construction shows that it was meant to be used upon the necks of his foes, if they came within reach. The targets, like the sword of Prince Charles, seem to have been made not merely for full dress, or as ornaments for the hall, but for use, and appearances indicate that they have been used. The most valuable of all the relics, however, were lent by the late proprietor to the Museum of the Society of Antiquaries of Scotland, on account of their national interest. These consist of two genuine old Highland targets, now very rare, and two harps, one of which belonged to Queen Mary, and the other is known as the "Lamont Harp." The first we hear of the harps is their being sent to Edinburgh in 1805 by General Robertson of Lude, at the request of the Highland Society, and examined by a sub-committee appointed for the purpose. The family tradition is that for several centuries past the first of these two harps has been known as the "Clarshach Lumanach," or the "Lamont Harp," and that it was brought from Argyleshire by a daughter of the Lamont family on her marriage with Robertson of Lude, in the year 1464. It is supposed that even at this remote period the harp was an old, knocked-about, battered, broken, and mended instrument, with a pre-traditional story that cannot now be traced. It is a plain, substantial instrument, made more for use than ornament, rather fitted for the wandering minstrel than for noble or royal hands. The other harp, which has long been known as that of Queen Mary, is of much more elaborate construction. The tradition regarding it is that when Queen Mary was on a hunting excursion in the Highlands of Perthshire in 1563, she had this harp, and presented it to Miss Beatrix Gardyn, daughter of Mr Gardyn of Banchory, whose family is now represented by Mr Garden of Troup. She appears to have been married to one of the ancestors of the present family of Invercauld,—distinguished, according to the custom of the Highlands, from his size by the appellation of "Findla More,"—and from her both the families of Farquharson and Lude are descended. In this manner Queen Mary's harp came with one of her female descendants into the family of Lude. These harps were secured as a loan for the Society of Antiquaries, through the instrumentality of the late Mr Charles D. Bell, F.S.A., who, in a paper read before that society in 1881, fully describes their appearance and their history, so far as can be ascertained.

On approaching the mansion-house, one of the first objects to attract the attention of the visitor is a beautiful cross, surmounted by a unicorn, standing in the centre of the lawn, and shown in the illustration on page 64. Nothing is known regarding its history, but, from its form and design, it may be inferred that it was originally intended as a village cross. From this cross to the garden only a few yards intervene, and here we first behold some of the arboreal treasures of the property, which, for its size, is one of the most thickly-wooded estates in the county, very little of the timber having been cut for a long series of years. The trees in the garden are purely ornamental, and those most worthy of notice are three gigantic araucarias, planted about 1845. They are each about 37 feet high, and the girth of the largest of them is 7 feet 6 inches at the ground, 5 feet 10 inches at 1 foot up, 5 feet at 3 feet, and 4 feet 2 inches at 5 feet. Two of them, including the one of which the girth has been given, grow upon a terrace, with a flower-bed on either side ; and the other one grows in a hollow immediately below, where there are no more than 6 inches of soil, the subsoil being large stones and gravel, evidently a river deposit. All of these trees, however, are now past their best. After the year 1861, when the rime was exceptionally severe, the trees began to show signs of going back, and about three years ago they commenced to cast their lower branches. The one in the hollow, however, although several of the branches have been dropped, still retains the lower ones, and is tolerably well furnished. At the foot of the garden there is one of the original larches,—a gift from the then Duke of Athole,—and nearly the same age as the ones at Dunkeld. It has attained to a great size, but has been considerably shattered by the storms of winter. It has a grand round bole of about 35 feet, with a girth at 1 foot of 15 feet, and 11 feet 7 inches at 5 feet. An old oak near the house girths 12 feet at the narrowest part of the bole. This tree originally had an immense spread of branches,—one of the limbs, since broken off, being 76 feet long, and weighing 5 tons. There is another very good oak in the Horse Park known as the " Garter Oak." This park was at a former period covered with coppice, and while the coppice was being cut down, the laird of Dalguise, observing the fine proportions of this tree, offered the party to whom the coppice had been sold, a pair of the garters then worn with the knee-breeches, if he would allow that tree to remain. The offer was accepted, hence the name of the tree. There is a fine bole of 24 feet, with a girth of 13 feet 10 inches at 1 foot, and 11 feet at 5 feet from the ground. The finest ash on the property grows at a little gate leading to the main road, the girth at 1 foot from the ground being 17 feet 6 inches ; and at 5 feet, the narrowest part of the bole, the girth is 15 feet. At " Glenalbert's famed abode," the scene of Mrs Brunton's well-known novel *Self-Control*,—which opens at Balnaguard Inn,—there are a considerable number of trees above the average size, most of them being quite close to the road-side. Amongst the first of these to attract attention are two very good elms, directly opposite each other, and which were made to serve the purpose of pillars in the erection of a triumphal arch, through which the Queen and Prince Albert passed when they first visited Scotland in 1842. A grand beech tree has a girth of 16 feet 4 inches at 1 foot, and 13 feet 1 inch at 5 feet, with a bole of 20 feet,—the cubic contents being altogether about 180 feet. Two Spanish chestnuts of large size grow within 6 feet of each other. The larger of the two girths 14 feet 10 inches at 1 foot, and 13 feet 3 inches at 5 feet up; while the other girths 13 feet 4 inches at 1 foot, and 12 feet 3 inches at 5 feet,—the bole of each of the trees being about 12 feet. Two old walnut trees at Glenalbert approach almost the same size,—the larger girthing 13 feet 3 inches at 1 foot up, and 11 feet 1 inch at 5 feet; and the other girthing 12 feet 4 inches at 1 foot, and 11 feet 5 inches at 5 feet. A line of gigantic silver firs extends along a bank from Glenalbert Burn to Easter Dalguise Burn, the largest of which girths 13 feet 4 inches at 1 foot, and 11 feet 11 inches at 5 feet from the the ground. Almost all of those trees are about the same size. There are also a number of large beech trees at Milton, but none of them approach the dimensions of the one at Glenalbert, —the best of the Milton lot girthing 14 feet 8 inches at 1 foot, and 12 feet at 5 feet from the ground, with a bole of 18 feet.

At Steuartfield, where there is an excellent shooting lodge, 57 acres of Scots fir and spruce were planted in 1881. The young plants are thriving amazingly, and, as the country is rather naked at this part, a very great improvement will be observable in the course of a few years. Within a short distance from Steuartfield, there is a somewhat remarkable cairn, which has long excited the curiosity of those residing in the district. It is situated on the summit of Elrigg Mohr, the highest hill in the parish. Impelled by a traditionary story about treasure being deposited in the cairn, several attempts have been made to get to the bottom of it, but hitherto without success. The stones comprising it are comparatively small, and unlike any of those in the immediate neighbourhood. On the top of another hill, there is also a remarkable cave, traditionally said to extend to Kinnaird. The story is told in the district that a piper was once sent in to discover the extent of the cave. He struck up a tune upon his pipes on entering, and the sound gradually became fainter, until it died away altogether, and the piper never returned. The Wolf's Bridge has an interesting story attached to it. At the period when wolves had been almost exterminated in Scotland, the wife of a reaper was crossing this bridge on her way to a neighbouring field with her husband's dinner, when she was confronted by a wolf. She was too far from assistance to make her danger known, but she heroically produced a large knife from her basket, and defended herself with such good effect that the ferocious animal fell dead at her feet. This is said to have been the last wolf seen in the district. At Easter Dalguise there is an important relic of the former bed of the Tay, in the shape of a substantial stone bridge in first-rate repair. The river presently flows in a totally different channel, although there is always a certain quantity of water underneath the old bridge.

The plantations upon the property extend in all to about 700 acres. The greater part of the woods consist of mature timber, chiefly larch, ranging from 60 to 70 years old. The wood is of splendid quality, and many of the trees are of remarkable size for their age. The plantations are being carefully managed, and there is an evident desire to improve them still further. The new laird has entered upon the management of the estate with considerable enthusiasm, and with a determination to effect whatever alterations may be necessary to improve the property. He has already set to work with the view of increasing its amenities, and in a short time, we believe, his energy in this direction will make itself felt to good purpose.

VI.—MURTLY.

'Great Birnam Wood"—Buried and Living Remains of the Ancient Forest—Story concerning Neil Gow—The "Hanged-Men's Tree"—Huge Beech and Chestnut Trees—The *Douglasii* Terrace—The Stewarts of Grantully—Planting out of Flower-Pots—The *Deodara* Sunk Terrace—Ancient Yews—The Finest Araucaria Avenue in the World—Large Lime Trees—Huge *Wellingtonia gigantea*—Summary of Remarkable Trees—The Plantations—The Trees of the Future.

WHEN we mention that what Shakespeare has immortalised as "Great Birnam Wood" forms part of the Murtly property, it will be acknowledged that the woods and remarkable trees upon this estate have a special attraction both for the general reader and for the lover of trees. Both are alike familiar with the story which forms the argument of Shakespeare's greatest tragedy, and both are alike interested in the locality in which the leading incidents are depicted. Not a few of those who have climbed the slopes of Birnam Hill, a favour which Sir Douglas Stewart allows to tourists, and gazed with admiration upon the magnificent country which lies to the south and east, must have been a little puzzled to know how the movements of the Earl of Northumberland's army could be observed by the messengers of the usurper all the way from Dunsinane Hill, which is fully twelve miles distant. Standing upon the pinnacle of Birnam Hill, and stretching the eyes towards Dunsinane, one is forced to confess that the eyes of Macbeth's soldiers must have been very much better than the eyes of people now-a-days, or that "living grove" which they saw advancing towards their encampment must have been of Brobdignagian proportions, if they saw it starting from the sides of Birnam Hill. Two explanations, however, can be given which seem simple enough. Birnam Hill is not confined to that eminence which has become such a favourite resort of tourists, but extends for several miles round the highest peak. In ancient times, too, it is highly probable that Birnam was the generic term for the whole district south-east of the hill, to the banks of the river opposite Dunsinane, while Dunsinane would be the general name given to the district extending from Macbeth's stronghold to the river. The outposts of the opposing armies—provided the whole of the tyrant's army were not encamped on Dunsinane Hill—would thus be stationed on the opposite banks of the Tay, so that Macbeth's spies could easily detect the unusual movements in the neighbourhood of Duncan's Hill, where there were some rude fortifications, the remains of which have recently been restored. The other, and more probable, explanation is that at the period referred to the country between the river bank and Dunsinane Hill was a huge jungle, and, as the whole of Macbeth's army are believed to have been within their fortified camp, they would only be able to detect the unusual movement within "this three mile," as Shakespeare reports the messenger to have informed the King. Supposing this theory to be correct, Macbeth's messengers would not be able to see effect given to the command—

> "Let every soldier hew him down a bough,
> And bear't before him; thereby shall we shadow
> The numbers of our host, and make discovery
> Err in report of us."

They would, however, have early intimation of the jungle in front of Dunsinane Hill showing unwonted stir, and would immediately communicate the startling intelligence to their sovereign, who, relying upon the prediction of the witches, remained in fancied security within his fortified camp, until "Great Birnam Wood" should march against him—

> "Be lion mettled, proud: and take no care
> Who chafes, who frets, or where conspirers are,
> Macbeth shall never vanquished be until
> Great Birnam Wood to high Dunsinane Hill
> Shall come against him.
>   *Macbeth.* That can never be.
> Who can impress the forest; bid the tree
> Unfix the earth-bound roots?"

Firm in his assurance that the forest could never come

against him, Macbeth rested in peace, placing implicit confidence in the prophecy of the witches—

> "I will not be afraid of death or bane,
> Till Birnam Forest come to Dunsinane."

Whether the witches were honest in their predictions, or whether they were among the "conspirers" regarding whom the King was advised "to take no care," is a question for the critics to discuss. It is at all events certain that Macbeth was not prepared for the peculiar strategy adopted by his enemies. No sooner does the faithful messenger report—

> "As I did stand my watch upon the hill,
> I looked towards Birnam, and anon methought
> The wood began to move,"

than the tyrant strikes him down, and threatens to hang him alive upon the next tree till he starves, if his speech proves false. The information, however, proved too true for the murderer of the "gracious Duncan" and the wife and children of Macduff, and, notwithstanding his refusal to yield to young Malcolm,

> "Though Birnam Wood be come to Dunsinane,"

he is made to bite the dust.

The great forest from whose trees the soldiers of Duncan cut the boughs to cover their numbers has now entirely disappeared, but the eye can still bear testimony to its reality. Not only are buried remnants of it frequently found, as mentioned in the chapter on "The Past and the Present," but there are still a few gigantic living specimens that excite the wonder of all who behold them. The most famous of these are the oak and the sycamore (see frontispiece) growing on the south bank of the Tay immediately behind Birnam Hotel. These two trees are believed to be close upon one thousand years old, and are worthy of the admiration which is bestowed upon them by tourists and students of Nature from all parts of the world. The girth of the oak at 5 feet from the ground is 18 feet; and the sycamore, which stands about 80 feet west of the oak, is even larger, being 19 feet 8 inches round the stem at 5 feet from the ground. Seventy years ago, a measurement of the oak was taken by the celebrated traveller, Dr E. D. Clarke, who reports its girth as being 17 feet; so that it has only grown one foot during that period. These remnants of "Great Birnam Wood" lie in the direct line of the march of the English army which came to place Malcolm on the Scottish throne in place of Macbeth, and it is not taxing the imagination too much to suppose that branches may have been cut from these identical trees to conceal the army on its advance from Dunkeld Ferry to the stronghold of the tyrant. In connection with the oak, a very good story is told concerning Neil Gow. One day Neil was away on Birnam Hill on a drinking carousal in some of the smugglers' dens which then abounded in the neighbourhood, and, as the night turned out very stormy, he remained until a late hour next morning. When the storm had somewhat subsided, Neil made for the ferry, and shouted for the boatman, who lived on the opposite bank, to come and take him across. On account of the storm, however, Neil was not successful in making his voice heard, and, after having exhausted his strength by shouting, he sat down at the foot of this oak, and, under the soothing influence of the "mountain dew" he had so freely imbibed on Birnam Hill, slept with unnatural soundness. The river by this time was in flood, and rising rapidly. Presently the water tipped the bottom of the tree under whose friendly shade the famous musician was sleeping. Then it gradually reached the sleeper, who awakened with a most uncomfortably chilly feeling about his legs, just as he was about to be washed into the powerful current of the river. On coming to his senses, he found that he had missed his way, and was at the wrong ferry altogether. This incident might well have led to the composition of the favourite tune, "Neil Gow's Farewell to Whisky."

There is another ancient oak, also in the line of the march of the English army from Birnam to Dunsinane, about two miles south-east of those trees already referred to, and very near Birnam Hall, now rented from Sir Douglas Stewart by Mr J. E. Millais, the celebrated artist. Locally, it is known as "The Hanged-Men's Tree," and is adjacent to the hill upon which Duncan held his courts of justice, and half a mile distant from Duncan's camp. Whether the King actually used this tree for the execution of criminals or enemies is not known, but for many generations it was used by the lords of Murtly for the hanging of thieves and other offenders, and it is highly probable that they only continued a practice first commenced when Duncan was encamped in the immediate neighbourhood. The tree measures over 10 feet in girth, and is very well proportioned.

Murtly Castle, a fine glimpse of which can be had from the railway, is beautifully situated on the south bank of the Tay, in the heart of a landscape that is most magnificent, and has elicited the warm admiration of Her Majesty, as recorded in *Leaves from the Journal of our Life in the Highlands*. Close to the river are several huge beech trees of from 700 to 800 years of age. The eight largest of these girth as follows at 5 feet from the ground :—10 ft. 6 in., 15 ft. 8 in., 15 ft. 3 in., 14 ft., 13 ft. 6 in., 13 ft. 5 in., 10 ft. 10 in., and 10 ft. 3 in. It may be interesting to note here that this species of tree had a good deal to do with the building of the iron frames of the ancient Britons, who grew to be veritable giants upon beech mast and the fat of swine, which, again, were reared upon acorns. There are also two Spanish chestnuts of great antiquity. These are supposed to have been amongst the original trees introduced into this country by the monks, and girth respectively 17 feet 7 inches and 15 feet 10 inches. Sir Douglas Stewart, the owner of this property, is the representative of the Stewarts of Grantully, who trace their descent from the same Stewarts from whom the royal Stewarts are descended. Grantully has been in possession of this family for over 450 years. The first charter of Grantully and Aberfeldy is dated March 30th, 1414, and granted by the Earl of Douglas to his cousin and shield-bearer, Alexander Stewart, son of the Lord of Lorne. A brother of his was ancestor of the Earls of Traquair. This title, now dormant, is claimed by Sir Douglas Stewart. Sir John Stewart de Bonkill, killed at the battle of Falkirk in 1298, was another distinguished ancestor. Sir William Stewart, born 1566 or 1567, was remarkable in the history of his family. He was the page of King James VI., and brought up with him, and high in his favour. The King gave him the

Barony of Strathbraan in 1606. He purchased the baronies of Murtly and Airntully from the Abercrombies in 1615, which were then far more extensive than they are now, having included Ballathie, Inncruytie, Tullybeagles, &c. Sir Thomas Stewart, son of Sir William, was another distinguished member of the family. He took a leading part in the public affairs of his day, and was treated with great respect and deference by people of rank and position. John Stewart, son of the above, commonly called Old Grantully, born 1643, added largely to the family possessions in Forfarshire, but had to sell much of this acquired property to pay a heavy fine for having invited the old Pretender to dine with him. He entailed the estates. He was succeeded by his cousin, Admiral Sir George Stewart of Balcaskie, son of Sir Thomas of Balcaskie, Senator of the College of Justice, by Lady Jean Mackenzie, daughter of George, first Earl of Cromartie. Then his younger brother, Sir John, followed for five years. He married—1st, Elizabeth, daughter of Sir James Mackenzie of Royston, and 2nd, Lady Jane Douglas, sister and heiress of the Duke of Douglas. About the succession to the latter there was a famous lawsuit. Then his son, Sir John, succeeded for twenty-five years. Then his son, Sir George Stewart, had possession for thirty-eight years and died in 1827. He married Catherine, daughter of John Drummond of Logiealmond, and eventually heiress of that estate. Sir George was succeeded by his son, Sir John Drummond Stewart, married to a daughter of Francis, Earl of Moray. He commenced the grass drives, and built the new castle which his successor, Sir William, helped to injure by cutting out the beams. Sir John was succeeded by Sir William Drummond Stewart, who in his earlier days went to America, and lived for some years with the North-American Indians. He added greatly to the grass drives, which are a distinguishing feature and a great beauty of the grounds of Murtly, extending for miles in different directions, but he charged them as improvements to the entailed estate. He also embellished the grounds by planting fine pines, &c. Sir William Drummond Stewart sold Logiealmond, his mother's property. In his latter days he became entangled by a Texan adventurer, to whom he tried to leave the estates. The attempt, however, was unsuccessful, and Sir Douglas Stewart has possessed them now for many years.

Besides the more ancient monarchs of the forest which we have been describing, there are a large number of remarkable trees that are comparatively young. These are not isolated specimens here and there, but are to be met with in clusters and in avenues as well as in solitary specimens. One of the most interesting sights in the policies is the magnificent *Douglasii* terrace or avenue, consisting entirely of *Douglasii* pines of from 27 to 33 years old. They rise to a height of from 50 to over 70 feet,—have a circumference, at 2 feet from the ground, of from 8 to 10 feet,—and a spread of branches of from 30 to 80 feet. The soil in some places is a light sandy loam of average quality, resting on deep gravel, while in other places the ground is rather marshy. In many places, however, the soil is remarkably dry, yet the trees have a vigorous growth, and the foliage is of such a beautiful dark green that at a little distance it is apt to be mistaken for a common yew. The situation is moderately exposed, and in some instances almost devoid of protection, and the height above the sea-level is from 200 to 300 feet. In the immediate neighbourhood of the *Douglasii* Terrace is an excavation known as the *Deodara* or Sunk Terrace, which has been made at great expense, and planted out with rhododendrons of almost every hue, presenting in their season a blaze of colour that is really dazzling. This terrace is approached by massive flights of stairs, the work of the present baronet. Near this place is a beautiful specimen of what is believed to be *Abies Douglasii elegans*, or *glauca*, although it does not fully answer to the description. It is about 65 or 70 feet high, and is splendidly clothed with a purple green and closely set leaf. At the north end of the Sunk Terrace is a line of the most handsome specimens of *Cupressus Lawsoniana* to be seen anywhere. They are from twenty to twenty-five years old, 30 feet in height, 9 feet spread of branches, and are beautifully clothed to the ground, with scarcely a branch awanting. Parallel with the Sunk Terrace is a fine old yew walk. The walk opens with about twenty very old trees, probably from 300 to 400 years of age. All of them are of large size, and have magnificent boles. At 5 feet from the ground they average about 2 feet in diameter, and the boles range from 9 to 15 feet high,—the average height of the entire trees being about 40 feet. The lower part of the avenue is composed of younger trees of about 15 feet in height, and trimmed in an arch-shape. This avenue leads directly to the door of the picturesque old chapel, to which it forms a most appropriate walk. In the flower garden there are also a number of yews fully as old as those in this avenue. The wood bears traces of having been drawn upon in olden times to furnish bows for the archers. One of the largest of these trees measures 210 feet in circumference of branches, the lower ones lying flat on the ground. This great tree is quite circular, and presents a beautiful natural bell shape, no attempt having been made at trimming in any way.

There are also several other avenues of interest. Lying between the *Douglasii* Terrace and the Sunk Terrace is a line of very fine specimens of *Abies Pattoniana* or *Hookeriana* of compact growth and fine foliage. Leading out from the front of the old castle is a beautiful avenue of *Cedrus Deodara*, remarkable alike for beauty of foliage and uniformity of growth. There is, however, no part of the grounds that will strike the visitor as being more noteworthy than the Araucaria Avenue, leading from the old castle to the new chapel. This avenue is said to be the finest of the kind in the world, and one can readily believe it. As single specimens of this peculiar ornamental conifer, each tree seems perfect, and collectively they present a sight as delightful as it is unique. All the trees are of large size, and several of them have produced a crop of cones. There are some very fine old silver firs near the new chapel. This is a remarkably beautiful structure and much admired. It is joined to the old Mortuary Chapel by a very handsome Byzantine Arch. It was built for Roman Catholic worship, but is now used only for Protestant services. In the Lime Avenue leading to the old castle, there are several trees worthy of special mention. Ten of them are of the following dimensions :—10 ft. ; 9 ft. 10 in. ; 9 ft. 7 in. ; 9 ft. 3 in. ; 9 ft. 2 in. ; 9 ft. 1 in. ; 9 ft.; 8 ft. 11 in. ; 8 ft. 6 in. ; and 8 ft., and about 100 to 120 ft.

high. The Lime Avenue was planted in 1711. While speaking of lime trees, we might mention that at Grantully, also a part of the Murtly estate, there are some very large limes, one of them girthing 17 feet.

Those looking for remarkable trees are certain to have their attention directed to a square figure of ornamental pines, a little to the north-east of the old castle. Amongst the most conspicuous of these are some splendid specimens of *Cupressus Lawsoniana* and other Californian pines. They are all extremely healthy, and are distinguished for their uniformity and symmetry of growth. This is rendered the more remarkable on account of the soil being of the very poorest quality; but their magnificent growth may be attributed to the pureness of the atmosphere and the openness of the subsoil, which consists of gravel to a very great depth. Amongst other interesting Californian pines, we observed, a little to the east of the Castle, a fine specimen of *Pinus monticola*, 50 feet high, and beautifully clothed to the ground.

The list of remarkable trees in this arboreal paradise is so extensive that many excellent specimens must be omitted in our enumeration. Before closing this part of the subject, however, mention must be made of a magnificent *Wellingtonia gigantea*, 35 years old, and 55 ft. high, which, if sawn up for the common uses of pine, would produce 9 railway sleepers, and a quantity of ¾-inch boards. Summarising the list of more remarkable trees, we find there are eight beech trees having a circumference, at 5 feet from the ground, of from 10 ft. 3 in. to 16 ft. 6 in.; seven Scotch elms, from 12 ft. 7 in. to 13 ft. 6 in.; two oaks, 18 ft. and 10 ft. 4 in.; two Spanish chestnuts, 17 ft. 7 in. and 15 ft. 10 in.; one horse chestnut, 12 ft. 5 in.; two sycamores, 19 ft. 6 in., and 10 ft.; ten limes, from 8 ft. to 10 ft.; twelve silver firs, from 8 ft. 7 in. to 12 ft. 4 in.; one larch of 9 ft. 8 in.; three Scots firs, from 8 ft. 6 in. to 10 ft. 7 in.;—in all, forty-eight trees, containing in the aggregate, at a random estimate, about 14,000 cubic feet of measureable timber, besides about 300 loads of firewood.

We have hitherto confined ourselves to the remarkable trees on the Murtly property, but when we come to examine the plantations, we will also find much that is worthy of notice. Pennant has jocularly remarked that Birnam Wood has never recovered from the march of its ancestors to Dunsinane, or, as a poet expresses it—

> "Huge Birnam towers above the tide,
> All bright with morning glow;
> But scarce a tree adorns his side,
> Where forests waved long, long ago."

A great change, however, has come over Birnam since the days of Pennant, and the joke is at present devoid of force. Instead of the nakedness which existed at no very distant date, the slopes of the hill are now thickly covered with thriving plantations, which bid fair to prove worthy of the reputation of "Great Birnam Wood"—

> "What's been before may be again,
> Perhaps to crush some tyrant's reign
> With Britain's thunder on the main."

The planting has not been confined to the classic slopes of Birnam, but has been extended all over the property, there being altogether about 4000 acres of plantation. These plantations are divided into twenty-five sections, besides many belts and clumps, which, in the aggregate, are of considerable extent. All the woods are traversed by regular roads, rides, and drives, and in most cases are quite suitable for the transport of timber, some of the roads being first-rate, some second-rate, while a good many are of an ordinary kind.

The plantations are generally well fenced with stone dykes, wood four-barred fences, wire on wood posts, wire on iron posts, and some upright wooden fences. The latter are very easily upset by winds, and continually get out of order, those giving the least trouble being the stone dykes and the wire on iron posts. Lying as these plantations do, about 140 miles from the Atlantic Ocean and about 70 miles from the German Ocean, the effects of sea breezes are little felt, even at high altitudes, so that good timber is grown over 1000 feet above sea level. Humid winds from the sea lose the greater part of their moisture before reaching this distance (except when at a great velocity), and in consequence the climate is found to be very favourable for most varieties of the indigenous timber trees of northern Europe.

The main geological formation on which seventeen of the sections stand is the Old Red Sandstone (Devonian), and the soil is in most cases a light loam. In some instances the soil is sharp and sandy, readily drying up, especially when under cultivation. That under timber being for the most part covered with mosses, grasses, and heath, is protected from dry winds, and, in consequence, timber stands the effects of dry weather better than other crops. The remaining eight sections are principally on the Lower Silurian formation, and in many places the clays are stiff, but not barren. The soils, however, are very damp, and in low and level places require draining. Although these are the general formations, other strata, such as greenstone, calciferous sandstone, &c., crop up, but bear a small proportion to the gross area under wood. They, however, influence the soil considerably, and the difference is observable in the growth and health of the timber. The kinds particularly affected are the finer varieties of coniferæ, as, though they grow the same length of shoots in a season, their shoots are more slender and their foliage of a lighter colour than trees grown on a more congenial soil. In one plantation of 40 acres, 160 to 200 feet above sea level, and with a northern exposure, are some very fine specimens of *Abies Douglasii, A. Albertiana, A. Morinda, A. Pinsapo, A. magnifica, A. Menziesii, A. nigra, Pinus cembra, Picea nobilis, P. grandis, P. Lasiocarpa, P. Nordmanniana, Cupressus Lawsoniana, C. Funebris, C. Lambertiana, C. Macrocarpa, Cedrus Libani, Libocedrus decurrens,* &c. All of these have been judiciously planted as regards soil, situation, and exposure. Most of the specimens are handsome and large for their age,—*A. Douglasii, A. Menziesii,* and *P. nobilis* being particularly so. The oldest of the trees in this plantation are about twenty-five years of age, and they are now 40 feet high, with a girth, at 3 feet from the ground, of from 3 to 7 feet. They have been rather thickly planted, many of the branches already interlacing; so that in a short time every second tree will have to be removed. The planting here has been on the grouping system, and excepting some *Wellingtonias* planted in a sterile, sandy soil, it has been admirably managed. The main crop on the plantation of which we are at present speaking is of various ages, from one to one hundred years old, and consists of oak, ash, elm, beech, birch, and alder, and, in smaller quantity, of hazel, Scots fir, larch, spruce, and silver fir,—the two last being about one hundred years old, and of good dimensions. The main crop, however, is not in a very thriving condition, having been neglected in the earlier stages of growth. Another section of 260 acres, chiefly Scots fir, twenty-five years old, affords an illustration of the unprofitableness of spontaneous growth, the state of about 60 acres proving that when planting can be cheaply done, it is a loss of money to allow the ground the necessary time to replant itself.

There is nothing calling for special remark, except of a very technical kind, in any of the other ordinary plantations. A few remarks, however, may be made regarding the ornamental plantations, which cover many acres, exclusive of the rides, terraces, &c., throughout the estate. The soil varies considerably, but, on the whole, it may be said to be of a loamy sand, with, occasionally, from a few inches to a few feet of peaty soil. It is found that in this soil all the pines, without exception, thrive admirably, and are making large annual growths,—the trees being of a better form, larger, and having their foliage of a darker green where peat exists. This is specially the case with *A. Douglasii*, *A. Menziesii*, and other varieties of *Abies*, as well as all the *Piceas*. Even when merely planted with peat soil about the roots, they are doing well. The planting, for the most part, has been very judiciously done; but, as in many of the ordinary plantations, the planters have been guilty of cramming the trees too closely together. In dry, hot soil, *A. Menziesii* is affected with red spider, and the *Wellingtonia* has a stunted appearance; but cases of this kind are rare. The most numerous specimens in the ornamental plantations are *Abies Douglasii* and *Araucaria imbricata*, and reference has already been made to them amongst the notes on the more remarkable trees.

If there be one lesson more than another which can be drawn from a visit to Murtly Woods, it is the evidence they furnish as to what are to be the timber trees of the future. Mr M'Kenzie, the late head forester, is of opinion that, sooner or later, *Abies Douglasii* will take the place of our larch, and *Abies Menziesii* that of our other pines. The timber of *A. Douglasii* is said to be as durable as larch, when in contact with the soil as a fence or other post or stob, it being in most cases equally full of resin. It is also considered much prettier for decorative work and house carpentry. " Roofs," says Mr M'Kenzie, " on the Gothic principle, such as the famous roof over Westminster Hall, London, if done with the wood of the Douglas fir, could not for beauty be excelled even by the best oak, as it has, naturally, a rich mellow colour when of large dimensions. It also lacks the bad qualities of the larch, as it does not twist or warp, and is not, so far as has yet been ascertained, liable to any disease in this country, while it produces nearly double the bulk of timber in a given time." The tree (*Abies Menziesii*) which Mr M'Kenzie believes will yet replace our other pines is also a rapid grower, and, although yet a young tree in this country, promises to reach from 80 to 100 feet, or about the same height as in its native habitats. " On account of its knotty quality, it would, like *A. Douglasii*, be useful for house carpentry, and any wood-work to be varnished; but the same quality prevents it from being used for beams, or where any great strength is desirable, as knotty wood is not strong; but if grown close in masses it would be more suitable, as, generally speaking, its fibre is more elastic and tenacious than fir or Norway spruce. Neither does it rot so readily as Norway spruce when in contact with the soil, but I have no experience of its qualities when in continual contact with water." What is really wanted now-a-days, to supply the rapidly-increasing demand, is a tree that will grow quickly into good useful timber, and a comparison of the growth of *A. Douglasii* and *A. Menziesii*, with the growth of larch, Scots fir, and Norway spruce, as to be seen at Murtly, is very much in favour of the former. To plant an acre of *Abies Douglasii*, 10 feet apart, with plants 18 to 24 inches high, and with larch of the same height to 5 feet over all, would cost about £7, 10s., including everything, *Abies Menziesii* costing about the same figure. This, no doubt, is a large sum, but the returns would much exceed the ordinary returns from an average acre of timber. To thin out the crop of larch to one-fourth of the whole at the age of 15 or 20 years, allowing for ordinary casualties, would give, after paying for felling, &c., £30 to £40. The remainder cut out at 35 years old, would yield about the same sum of clear revenue, and at 75 years old, the principal crop would be worth £500 or £600, under ordinary conditions. The returns of an acre of *Abies Menziesii* would be about £100 less, as the timber, although of the same bulk, would not be so valuable. The data for the opinion that the newer varieties are more profitable, are taken from a careful measurement of the different trees, compared with common trees grown within a few yards of the newer ones. We find *Abies Douglasii*, at from 27 to 33 years of age, measuring 45 to 60 feet in height, having a circumference, at 2 feet from the ground, of from $6\frac{1}{2}$ to $8\frac{1}{2}$ feet, with a spread of branches from 20 to 70 feet,—the soil being remarkably dry, and the situation moderately exposed. *Abies Menziesii*, at from 25 to 30 years old, grown under similar conditions, reaches a height of from 45 to 55 feet, have a circumference, at 2 feet from the ground, of from 5 to $8\frac{1}{2}$ feet, and a spread of branches of 24 feet. Comparing this with the larch, Scots fir, and Norway spruce growing alongside of those just mentioned, we find the results as follows :—Larch, 25 years old, 30 to 40 feet high ; circumference at 2 feet from the ground, 2 feet 4 inches ; Scots fir, of the same age, 20 to 30 feet high, 2 feet to 2 feet 6 inches in girth at 2 feet from the ground ; Norway spruce, 30 to 35 years old, 30 to 40 feet in height, and $3\frac{1}{2}$ to 4 feet in circumference at 2 feet from the ground.

There is here undoubtedly a great difference in the bulk of the timber produced, much more than would repay the difference in the price of planting an acre. To the eye the newer and the older varieties present a very striking contrast. In some places, where the older conifers were planted as nurses for the younger, the difference between the two is so great that it requires not a little faith on the part of the uninitiated to believe that the larch, Scots fir, or Norway spruce are not much younger than the *Douglasii* or the *Menziesii*, the former being mere handsticks compared with the gigantic trees they were originally intended to protect from the winter's storms. The experienced eye, however, will at once see that the nurses though infinitely smaller are much older. These facts open up a very interesting inquiry, and if experience at other places proves that the newer varieties referred to are so much more profitable, larch and Scots fir, as well as Norway spruce, will become somewhat rare,—a circumstance which would give a different complexion to several very large tracts of wooded country.

We have by no means exhausted all that might be said regarding the woods of Murtly, but as anything we have to add would be of a more technical nature, it would have little interest for the general reader. All over the estate the defects of former years are being remedied as rapidly as circumstances will permit. Plantations that are too crowded are being thinned, and the coppice is receiving the most approved treatment for facilitating its growth. Drainage is being carried out wherever that is necessary, and nothing is allowed, as far as that is possible, to stand in the way of the successful cultivation of the timber. There are still, however, large tracts of hill land that might yet be planted. These have all been duly reported upon, and in the course of time action may be taken, Sir Douglas Stewart not having been sufficiently long in possession of the property to have all his wishes in this respect carried out ; but the commencement which he has already made is an earnest of what we may expect in the future.

VII.—DUNSINANE.

The Historical Connection between the Hills of Birnam and Dunsinane—Macbeth's Castle—View from the Summit of the Hill—Original State of the District—First Attempts at Improvement—"Trees Planted for Behoof of the Poor"—Lord Dunsinane's Improvements—Story regarding the Introduction of Macadamised Roads in the District—Extensive Planting—Improvements by Subsequent Proprietors.

DUNSINANE HILL and neighbourhood are so closely associated, historically, with "Great Birnam Wood," that a description of what is to be seen in this locality naturally follows an account of the woods on the Murtly property. It is scarcely within our province to dwell upon the interesting historical associations of the district; but, still, these are so inseparably mixed up with our narrative, that a brief reference is indispensable. In writing upon Birnam Wood, we ventured upon an explanation of the difficulty which has frequently been experienced as to how Macbeth's messengers upon Dunsinane Hill could see "Great Birnam Wood" start on its journey towards the Castle of Macbeth. We then pointed out that at that time the whole of the country between the River Tay and Dunsinane Hill was one vast jungle,—the predominating growth, it is believed, being hazel, birch, broom, and briar. Supposing, as is also believed to be the case, that Macbeth's army were confined within the walls of Dunsinane Castle, no movement in Birnam Forest could possibly be observed; but no sooner would the advancing foe reach the higher grounds of Stobhall, than an unusual bustle would be detected in the brushwood. Not a man would be seen, but to the astonished eyes of the "messenger," as he "looked toward Birnam," the whole of the jungle would appear as if moving towards the encampment of the King, who, observing the excited appearance of the soldier, would naturally command him to "tell his story quickly."

DUNSINANE.

> "*Messenger.* Gracious my lord,
> I shall report that which I say I saw,
> But know not how to tell it.
>   *Macbeth.*   Well, say, sir.
>   *Mess.* As I did stand my watch upon the hill,
> I look'd toward Birnam, and anon, methought,
> The wood began to move."

This instantly brought to Macbeth's mind the recollection of the prophecy of the witches, but the story of the soldier seemed so improbable that the tyrant exclaimed, "Liar and slave," and struck him. Following the popular version of Shakespeare, we learn that the messenger adhered to his original story, despite the cruelty of his Sovereign.

> "*Mess.* Let me endure thy wrath, if't be not so;
> Within this three mile may you see it coming;
> I say, a moving grove.
>   *Macb.*   If thou speak'st false,
> Upon the next tree shalt thou hang alive,
> Till famine cling thee: if thy speech be sooth,
> I care not if thou dost for me as much.—
> I pull in resolution; and begin
> To doubt the equivocation of the fiend,
> That lies like truth: "Fear not, till Birnam wood
> Do come to Dunsinane;"—and now a wood
> Comes toward Dunsinane.—Arm, arm, and out! -
> If this, which he avouches, does appear,
> There is nor flying hence, nor tarrying here.
> I 'gin to be a-weary of the sun,
> And wish the estate o' the world were now undone.—
> Ring the alarum bell:—Blow, wind! come, wrack!
> At least we'll die with harness on our back."

Macbeth, having fully realised the truth of the messenger's statement, made every preparation for defence. It is evident that very little time was left for deliberate action, as within a short time after being discovered, the foe threw down the "leafy screens" which had disguised their approach, and stormed the castle with comparatively little difficulty.

Unless there had been some treachery in the camp, it is hardly possible that Macbeth's stronghold could have been so "gently surrendered." The Castle stood at a height of upwards of 1100 feet above the level of the sea, and at an elevation of 800 feet above the base of the hill. On three of its sides the hill is very difficult of access, and the defences of the Castle were high outer rocks, and where these were not available, there was a wall with a deep fosse. Excavations have been made at different times, with the view of discovering the extent of the fortifications, and in the course of the last of these, in the spring and summer of 1854, it was ascertained from several openings made in the ruins that the outer wall of the main building was an immense mass, 21 feet broad, surrounding the summit of the hill, "strongly built with large blocks of stone, regularly put together, no lime being visible, but much red mortar pulverised by the action of fire; and the inner wall, distant from the outer about 24 feet, seemingly only encircled the hill. The intermediate space between the walls was full of what must have been building material, probably indicating that the two walls had fallen against each other, and formed this mass of ruins." The workmen, digging within the inside wall, down to the rock, cleared out an underground apartment, 13 feet long by 6 wide, and found it paved with a rude dressed freestone, bedded upon the rock with red mortar; and from that small apartment a passage led out, towards the centre of the area, 18 feet long by 6 broad, and paved in the same way, with the walls on each side standing entire, about 5 feet high, rudely built of freestone, having four small entrances diverging from the ground floor,—two on the right, and two on the left,—about $2\frac{1}{2}$ feet high by 2 feet wide, leading in through the walls, and forming communications with the other underground apartments on each side of the passage. In one of these small entrances some human remains were found in a wonderfully good state of preservation. Beyond the passage where the human relics were discovered, a small apartment was reached, measuring $5\frac{1}{2}$ feet long by 5 broad, "and squared in the building, filled with burnt materials and remains of bones, indicating a place of concealment or a burial vault, with no apparent ingress or egress but the small 2-feet entrance above referred to." The workmen removed a piece of the side wall of this small room to the right, and came upon a floor of pavement jointed in a rather more tradesmanlike manner, but still undressed plinths from the quarry, no tool-mark having been seen on any of them. Several other remains were also discovered, and the narrator of the exploration concludes with a statement

that gives a fair idea of the remains as they are now to be seen:—
" Let us imagine a wall, 21 feet thick, surrounding an oblong-circular area of 40 poles of ground, intersected with vaulted passages and small underground apartments of every shape and size (with the exception of an open court of considerable extent occupying the centre), and we have probably an idea of the ground floor of Macbeth's Castle, as it now stands after the lapse of 800 years." Amongst the other objects of antiquarian interest in the district is a spot known as "The Lang Man's Grave." The tradition of the district is that Macbeth, finding it impossible to escape from Macduff, threw himself from the top of the hill, was killed upon the rocks, and buried at the "Lang Man's Grave." The tumulus, on being examined, however, proved to be a druidical stone that had toppled over, and no relic was discovered that could throw any light upon the tradition.

Although very little can now be seen of the remains of the fortifications, a visit to the summit of Dunsinane Hill is amply repaid by the glorious prospect it affords of a wide expanse of surrounding country. Beneath our feet lies the beautiful Valley of Strathmore,—no longer a jungle hiding the advance of a relentless foe, but laid out with fertile fields, and ornamented with flourishing plantations, thriving clumps, and isolated trees of considerable size and beauty. Lifting our eyes from the rich plain below, we instinctively seek out and mark the eminence historically associated with Dunsinane, and which is hardly distinguishable from the mass of towering mountains. To the north and west, the blue Grampians look frowningly upon the peaceful valley, and many well-known peaks can be recognised—Ben-y-gloe, Ben Vrackie, Faragon, Ben Lawers, Schiehallion, &c.

> " Yonder Ben Lawers' mighty crest
> O'erlooks Breadalbane to the west,
> Schiehallion—Nature's throne : where rest
>   The thunder cluds :
> An' Ben-y-gloe, that claims behest
> Owre Atholl's woods."

The view in this direction, however, is not limited to mountain scenery, but takes in the environs of Forfar and Brechin, and the country up to the borders of Aberdeenshire. The entrance to Glenlyon, the woods of Airlie, the slopes surrounding Dunkeld, and the towns of Coupar-Angus and Blairgowrie, are also distinctly visible. Turning westwards, the eye takes in the whole range of the rich green Ochils to the blood-stained field of Sheriffmuir. Southwards, we look upon the boundaries of Fifeshire, and note the more prominent Lomonds, one of which shelters the venerable palace of the Scottish kings, and the other reflects itself on one of the most famous trouting lochs in Scotland, rendered classic by the romantic escape of the unfortunate Queen Mary from one of the islands which adorn its bosom. Stretching the eyes still further, we can see the Firth of Forth gleaming in the sunshine; and on a clear day we can even distinguish the well-known Arthur's Seat, looking down upon "Edina, Scotia's darling seat." To the south-east, we have a glimpse of the widening Tay, from the gossamery bridge, whose partial fall struck so much terror into the heart of the country, till it approaches the German Ocean.

Such is a very imperfect portrayal of the magnificent view which can be obtained from the summit of Dunsinane Hill. The prospect is, if anything, superior to that which can be got from Birnam; but as Dunsinane is away from railway communication, visitors are, consequently, few and far between. The interest which attaches to the remains of so ancient an historical relic as the ruins of the Castle of the great Macbeth, probably the most stupendous undertaking of that age, ought to be sufficient to attract larger numbers; while the trouble of ascending the hill, which has a very gentle slope, is repaid ten times over by the splendour of the view, north, south, east, and west.

Dunsinane cannot claim to have ever been anything of a forest, but sufficient evidence can be produced to show that it is not devoid of interest to the arborist, and to all who delight in the improvement of landed property. The estate has been exceptionally favoured in having some distinguished improving landlords, and several of them were characterised for keen arboricultural tastes. Through their exertions, what was marsh or jungle in the days of Macbeth has become productive fields or valuable woods, and several magnificent crops of the latter have been reaped.

We have already alluded to the original appearance of the

country around Dunsinane Hill. Unlike the larger tract of the county with which we have hitherto been dealing, this district seems to have been originally destitute of timber. The soil was of a damp, marshy nature, and in its wild state, where the land was not altogether a bog, it was thickly clothed with brushwood, in which broom, briar, hazel, and birch predominated. The work of reclaiming such soil as this must have been no easy task; and, considering that a commencement was only made within comparatively recent years, the results which have been achieved are simply marvellous. So far as can be ascertained, the first attempts at systematic improvement were made about the beginning of last century, the proprietor being Sir William Nairne, who had the baronetcy conferred upon him by Queen Anne in 1704. Sir William inaugurated the improvements by planting hardwood trees very extensively upon the drier portions of the marshy land. Beech was the favourite wood, but this variety was well mixed with oak, sycamore, and elm. A great many trees were also planted about the mansion-house and the villages round about Dunsinane, beech being again the principal tree planted. At that time the villages were much more numerous than they are now, several of them having since become quite extinct. The chief of these villages, Kinrossie and Collace, however, are still of respectable dimensions, and can boast of great antiquity,—the former being mentioned as early as the days of Alexander III., who succeeded to the Scottish throne in 1249. From this we infer that the village must have been a place of considerable importance in the infancy of Scottish history,—an importance which may be attributed to its lying almost midway on the main road between Scone and Dunsinane. The trees which were planted round these villages by Sir William Nairne were simply for ornament, and many of them were allowed to grow to a considerable size. The arable land of the property was also enclosed to a large extent with timber, and many of the trees which were then planted are still to be seen. Amongst the more interesting of these trees are those which adorn the churchyard of Collace, and which were planted by Sir William in the year 1736, "for behoof of the poor." According to modern ideas, this seems rather a unique provision for the poor; but in those days, when the poor were supported entirely by the landlords and the kirks, and when parochial boards were undreamt of, there was nothing very peculiar in the laird making such a donation. It was doubtless desirable to have the last resting-place of the parishioners embellished with a few robust, green trees, as these would at once tend to enliven a place full of sorrowful associations, and, by way of contrast, suggest the short-lived character of man. The laird was the man to whom the people would look to adorn their burying-place in this way, and the question would naturally arise as to whom the trees would belong when they were matured. There would be some difficulty in the successor of the laird seeking to claim them, and it was most unlikely that he would ever do so. Supposing the kirk-session, again, were to cut down the trees, there might be some doubts as to what purpose they could apply the proceeds. It was, therefore, a wise act to stipulate that the trees should be grown for the benefit of the poor, as it will ensure that the timber will not be allowed to rot in the ground, and no difficulty will be experienced in disposing of the price of the trees when they have reached maturity, while at the same time a respectable sum may be realised for a most laudable purpose whenever it is most required.

Although Sir William Nairne appears to have done a great deal towards the improvement of the property, Lord Dunsinane was a much greater enthusiast in the same direction. He was also a Sir William Nairne, being the fifth Baronet, and a younger son of the second Baronet. He studied for the bar, was admitted an advocate in 1755, and was appointed Sheriff of Perthshire in 1783. Three years afterwards he was promoted to the Bench by the title of Lord Dunsinane. He purchased the property of Dunsinane from his nephew, Sir William Nairne, on whose death his Lordship succeeded to the baronetcy. As soon as he entered into possession of the property, Lord Dunsinane set to work with great zeal in improving the estate. The tradition of the people in the locality is, that all the available income from the property was devoted to its improvement, his salary as a Judge being sufficient for his maintenance. During his time, the whole appearance of the district was changed for the better. Substantial buildings were erected,—the carpenter work consisting of the celebrated Braemar wood, the trees upon the property not being sufficiently matured for the purpose. Until he became the proprietor of Dunsinane, there were no macadamised roads in the parish, and a very good story is told regarding the circumstances under which this improvement was effected. His Lordship, while on circuit at Inveraray, was attracted by a peculiarity in the construction of some roads leading to the Castle, and, approaching one of the workmen, asked to be favoured with a few of the stones he was breaking. This seemed to the workman rather a strange request coming from such a personage, but his astonishment must have been considerably heightened when his Lordship took out a silk handkerchief and rolled up the pieces of road metal as if they had been the most precious stones, giving the stone-breaker, at the same time, a gratuity of half a crown. His Lordship then gave the stones to his coachman, John Whittet, who died within the last twenty years, and who used to relate this story, telling him to take the greatest care of them, and convey them to Dunsinane House. John, fancying that his master had made some very important discovery,—that he had traced the presence of gold, or something of that sort,—watched over the precious parcel with the greatest faithfulness, and deposited it safely within the walls of Dunsinane House. His Lordship afterwards caused the stones to be laid in a corner near the entrance to the mansion-house, and gave instructions to some of his men to have stones broken exactly similar to the specimens he had provided, telling them, at the same time, that if they were ever in any doubt as to the size and shape of the stones, they were to take a walk up to the mansion-house, and have another look at them. After a large quantity of stones had been broken as his Lordship directed, he caused them to be utilised in the same way as he had seen in the neighbourhood of Inveraray Castle.

While Lord Dunsinane took a great interest in the improvement of the property generally, his energies were especially directed to the planting of the less profitable portions of the land, and embellishing the more picturesque spots with suitable trees. In fact, arboriculture was with him quite a hobby. Owing to the many calls upon his time, and the engrossing nature of his judicial duties, his visits to Dunsinane were seldom very protracted, and when it was impossible for him to look round the property in daylight, he was known to proceed at night, with a lantern in his hand, to examine how some favourite trees were thriving. Throughout the whole of his lifetime, he was actively engaged in laying down plantations all over the property. In 1780, he continued the work which had been commenced by the first Baronet, by planting about 300 acres of Scots fir. These were cut down between the years 1820 and 1840, and the timber proved to be of very fine quality, and was much sought after by the wood merchants of Perth and Dundee. About the same time, Collace Hill was first planted by Lord Dunsinane, larch being the principal variety. Indeed, the larch appears to have been a great favourite with his Lordship, as we gather from an old document which has come into our possession, dated October 1782, and containing eighteen separate directions to the forester, Hugh Fraser, regarding the management of the woods. In these directions the larch is frequently referred to, and the forester is instructed to obtain as many of them as he could. His Lordship appears to have obtained a great many of the plants in gift from his friends. In the first direction, Hugh Fraser is told that "Mr Donaldson, at Elcho, is to give me some hundreds of ashes fit for planting out. You must go and take them up at a proper time—go over the hill the short way, and pass at Mr Donaldson's boat opposite to the Castle of Elcho. The trees may likewise be brought over by that boat. At Kinfauns, you may inquire at the gardener about any tree seeds that can be got there, *larix* cones, &c. Lord Gray was to let me have such as I wanted." All of the eighteen directions are equally as explicit as the one we have quoted, and many of them are quite as interesting. In the next paragraph we are told that Mr Smythe of Methven had a number of plants for him, and Fraser is instructed to have them planted in the nursery as soon as he gets them. We are afterwards told that "Lintrose is to give me a few plants of the scarlet oak, and I will point out two or three places in the nursery where I would have them planted. . . . . . You may go east to Lintrose and get them as soon as the season is fit for planting them, which I suppose will be directly. Lintrose (to whom make my compliments) will give you any tree seeds that you may want. I suppose you can get from him plenty of *larix* cones and of beech mast, both which I wish to have, especially plenty of *larixes*." In addition to the large number of seeds and young trees which his Lordship appears to have got in gift from his friends, he also made extensive purchases at the Perth Nurseries. In the third of the directions referred to, Fraser is instructed to "get from the nursery at Bridgend one thousand elm seedlings, and plant them in my nursery. The price is only 5s." At another place, he is instructed to plant "the Northumberland fence at the Hill Park" with "ashes and elms—one hundred elms of six feet high, at 5s. per hundred, may be got from the Perth Nursery for that purpose. If these are not sufficient, it may be made two hundred, but not to exceed that number." All through the document instructions are given to Fraser to have every available spot planted with the trees most suitable for the soil. "Let the strip lying east from the Todhill Park be fully planted up with oaks, a few *larixes* and beeches also may be mixed with them. Let some oaks also be planted at the head of that strip near the road, although it is at present pretty much filled up with other trees, yet, as I believe there are no oaks amongst them, I would have some put in. If we find them thrive, we can make room for them by removing some of the other trees. Plant two-year-old *larixes* on the muir ground at top of the strip." Again, "in the east wood, where the natural firs are, there are a great many blanks near the roadside The *larixes* which have been planted there are thriving remarkably well; some more may be put in where there are blanks—also some oaks and beeches." The seventh direction is to the effect that "on the face of the muir where the two-year-old *larixes* have thriven so well, a great many more ought to be planted, and likewise a good mixture of oaks and beeches." Many other directions are given regarding the management of the woods, and about half a dozen instructions are given in the same document as to general improvements in different parts of the property. These directions bear out, in a very strong light, the great and intelligent interest Lord Dunsinane took in the promotion of forestry, and give us some idea of the extent of the planting he accomplished. His Lordship retired from the Bench in 1809, but he only enjoyed his retirement for a very short time, having died at Dunsinane House in 1812, the title becoming extinct at his death. According to all accounts, Lord Dunsinane was not only an excellent landlord,—judiciously expending large sums of money in the improvement of his property,—but he was likewise a most distinguished Judge. We are also told in the Statistical Account of the parish, that he was a good man in the highest sense of the term :—" To the welfare of the parish, and more especially to their religious instruction, he was equally attentive. He feared God and honoured the King. He remembered the Sabbath day, and was never absent from church except from necessity. This example was highly useful in his own time. Still more useful must such conduct be in our own times, to put to shame, if possible, increasing impiety, and to stem the overflowing tide of iniquity. Lord Dunsinane died 22nd March 1812, being upwards of eighty years old, and was buried within the walls of the old church of Collace, now converted into a mausoleum for the Dunsinane family; and to which there is access by a large arched gateway of uncommon beauty, and of the rare and ancient order of Saxon architecture."

Lord Dunsinane was succeeded in the property by a nephew, James Mellis Nairne, who followed the noble example of his uncle both as an improving landlord, as a county gentleman, and as a Christian philanthropist. In his day, the greater portion of the wood upon the property reached maturity, and was cut down between the years 1820 and 1840,—the superior quality of the timber, as already mentioned, securing for it a ready market in Perth and Dundee. Although he was largely engaged in cutting down the ripe timber which had been planted by his predecessors, he was far from forgetful of his duty to those who were to come after him, the ground being replanted at different periods, and additional plantations laid down. Amongst the first of his efforts in this direction was the planting of Dunsinane Hill. It was originally his intention to plant the hill to the summit, but on reconsideration he resolved to leave the upper portion of the hill in pasture, so as to preserve intact the classic ground immortalised by the dramatic genius of Shakespeare. The planting of Dunsinane Hill was executed about 1824, Scots fir and larch being the principal varieties. A larger portion would have been planted, apart from the site of Macbeth's stronghold, but as the ground on the west side was found to be unsuitable, it was not proceeded with further.

James Mellis Nairne was succeeded by his brother John, who has now, in turn, been succeeded by his son William. The work of planting the property had been so far accomplished before the present and the late proprietor came into possession, that they have found comparatively little to do in this branch of improvement. Still the work of improvement is going on as occasion arises, not only in the planting of waste ground, but in the increasing of the value of the property generally. During the minority of the present proprietor, Mr William Nairne, the Hill of Collace, consisting of about 112 acres, was replanted with Scots fir and larch, and it is the intention of Mr Nairne to continue to plant all the available land with both hard and soft woods.

There are no trees in the district that can be described as particularly remarkable, but everywhere we see the beneficial result of judicious and systematic planting. The greatest care is taken that only the varieties best suited to the soil and climate are planted; and, on the whole, the woods are in splendid health and condition.

VIII.—SCONE, LYNEDOCH, LOGIEALMOND, &c.

The Former Glory of Scone—The Stone of Destiny—The Ancient Abbey—The Mansfield Family—The Modern Palace—Queen Mary's Sycamore—Trees Planted by King James VI.—Other Remarkable Trees in the Policies—David Douglas—The "Boot Hill"—Appearance of the Country a Century Ago—A Landscape Forester's Blunder—The Plantations—Voracity of the *Hylobius abietis*—The Steading at Balboughty—Preservation of Wooden Posts—*Douglasii* Plantations—An Experimental Plantation—Taymount—Lynedoch—Logiealmond.

WE can hardly think of Scone without recalling the many associations, historical, antiquarian, and artistic, which cluster around it. At one time the Windsor of Scotland, its name is redolent of the far bygone days when royalty held high state within the walls of its ancient palace. The possession of the famous Stone of Destiny rendered it necessary for all the early Scottish Princes to repair to Scone for coronation, and until the stone was seized and carried to Westminster Abbey in 1296, by Edward I. of England, a halo of royal grandeur hung around the place. Scone is believed to have been a royal residence before the end of the Pictish kingdom in 843. At a period so remote as this, we must rely upon tradition to a large extent for any knowledge that has come down to us regarding its royal origin. It is said, for instance, that the Stone of Destiny was originally transferred from Dunstaffnage by Kenneth Macalpine before the end of the Pictish kingdom, he having brought it to "The Royal City of Scone," because in the neighbourhood he had fought and won the last decisive conflict with the Picts. Tradition, however, is not any clearer upon this point than it is with regard to the history of the Stone of Destiny itself,—other historians asserting that the battle of Scone, which appears to have been a very important one, was fought, not between the Scots and the Picts, but between the Danes and Norwegians and the Picts. According to tradition, the Stone of Destiny has had a wonderfully chequered career. There is a curious legend tracing its history up to the veritable pillow upon which the patriarch Jacob rested his head at Luz when he was favoured with his beautiful vision. In the mythical account, we next trace the stone to the possession of Gathelus, the son of Cecrops, King of Athens, who entered into the service of one of the Pharaohs, and married his daughter Scota, from whom Scotland is said to derive its name. He brought it from Syria to Egypt, and, to escape an impending plague, he is said to have sailed from the Nile, in accordance with the advice of Moses, with his wife and the precious stone, and landed on the coast of Spain. Gathelus afterwards sent the stone to Ireland under the care of his son, who invaded the island, and by whom it was set up on Tara Hill, where it was used as the Irish Coronation Stone for many ages. We next find it brought to Scotland by Fergus, the son of Eric, who led the Dalriadic Scots to the shores of Argyle, and who considered the stone necessary to his coronation at Dunstaffnage. Here, we are told, it remained until 834, when it was removed to Scone by Kenneth Macalpine, as already mentioned, and where it remained till it was transferred to Westminster Abbey by Edward. Matter-of-fact people, however, refuse to believe this very interesting legend, and are inclined to suppose that the stone was quarried in the neighbourhood of the ancient Palace of Scone, where, from its use as a Coronation Chair, it acquired its sacred character as influencing the destinies of the Scottish nation—

> "Unless the Fates are faithless grown,
> And prophet's voice be vain,
> Where'er is found this stone,
> The Scottish race shall reign."

Scone is not only distinguished as an ancient seat of royalty, at which most of the long line of Scottish Kings from Kenneth III. to Charles II. were crowned, but it has the honour of being the meeting-place of no fewer than ten Parliaments, from 1284 to 1401 inclusive, and there many important enactments were passed. It is also noted as a seat of the Culdees, and as the site of an abbey. The abbey was founded in 1114, by Alexander I., as a memorial of his gratitude for a narrow escape from a band of traitors at Liff, who had conspired to murder him. The abbey was dedicated to the Holy Trinity and St Michael, and was endowed with the lands of Liff and Invergowrie, which the Lord of Gowrie had gifted to his Majesty at his baptism. The abbey was sacked and burned in 1539, and what remained after this sacrilegious act fell about 1624. At this period, David, first Viscount Stormont, built a new church at Moathill, and which is described as having been a very elegant fabric. The main portion of this building was thrown down towards the end of last century, but the aisle still remains, and is an object of additional interest as the mausoleum of the Mansfield family. This family, as is well known, occupies a prominent part in the annals of our country. It traces its lineage to Sir David Murray of Arngosk and Balvaird, elder son of Sir Andrew, third son of Sir William Murray of Tullibardine. The second son of Sir David was Cupbearer, Master of the Horse, and Captain of the Guards to James VI., by whom he was knighted. He was also Comptroller of the royal revenues, and accompanied the King to England, obtaining from him the Barony of Ruthven, and the lands belonging to the Abbey of Scone, which were erected into the temporal lordship of Scone, with a seat and vote in Parliament, in which he was invested on 7th April 1605. He made a settlement of his estates and title, ratified by charter under the Great Seal, 4th October 1616, with various remainders, and was created Viscount Stormont, Perthshire, on 16th August 1621, with remainder in default of issue male to his heirs male and of entail, bearing the name and arms of Murray, &c. On 26th October 1625, he made a settlement of the lordship of Scone and other estates, with successive remainder, to Sir Mungo Murray of Dumcairn (Master of Stormont, who had married his niece), John, Earl of Annandale, Andrew Murray, minister of Abdie, &c., and their issue male respectively. Sir Andrew Murray, the minister of Abdie, succeeded, on the death of the first Viscount, to the baronies of Arngosk and Kippa, and was knighted at the King's coronation in 1633. He was a member of the General Assembly at Glasgow in 1638, and was created Lord Balvaird in 1641, with remainder to his heirs male. David, the second Lord Balvaird, was fined £1500 by Cromwell's Act of Grace and Pardon, 1654. He ultimately became fourth Viscount Stormont and Lord Scone. His only son, David, fifth Viscount, opposed the union, and was, with his eldest son, one of those summoned to surrender themselves at the breaking out of the rebellion of 1715. The fifth Viscount had other two sons, both of whom occupied prominent positions in the country. The second son, James, an advocate, was elected M.P. for Elgin burgh. The third son, William, was a King's Scholar, Westminster School, and, after distinguishing himself at the bar, became M.P. for Boroughbridge. He was made Attorney-General in 1754, was Chief-Justice from 1756 to 1788, and refused the seals of Lord Chancellor in 1757, 1770, and 1771. In November 1756 he was created Baron of Mansfield, Notts, in England; and in October 1776 he was created Earl of Mansfield. For several years he occupied the position of Chancellor of the Exchequer. David, seventh Viscount Stormont, succeeded as second Earl of Mansfield, on the death of his uncle, William, who died on 20th March 1793, and was buried with full honours in Westminster Abbey. The second Earl of Mansfield was Justice-General of Scotland, Joint-Clerk of the Court of King's Bench, was Ambassador-Extraordinary to the Courts of Vienna and Versailles, a Secretary of State, President of the Council, and a Representative Peer. He died in 1796, and was buried in Westminster Abbey. Sir William David Murray, K.G., the fourth and present Earl, is also full of honours. He was M.P. for Aldborough, 1830; Woodstock, 1831; Norwich, 1832–7; and Perthshire, 1837–40. He was a Lord of the Treasury in the years 1834–35. Besides the other dignities to which he has succeeded, he is Hereditary Keeper of the Palace of Scone. He has been Lord-Lieutenant of Clackmannanshire since 1852. He was selected as Her Majesty's Commissioner to the General Assembly of the Church of Scotland on several occasions, and the magnificence of his levees are still remembered as something exceptional.

The ancient glory of Scone is perpetuated by the magnificence of its modern Palace, which, in the hands of its noble proprietor, is one of the finest mansions in Scotland, flying the Mansfield flag from its battlements as proudly as ever it flew the Royal

102   WOODS, FORESTS, AND ESTATES OF PERTHSHIRE.

Standard. Occupying a commanding site within the sound of the ceaseless rush of the lordly Tay, the Palace is surrounded by a wide stretch of exquisitely beautiful sylvan scenery. As far as the eye can reach, forests of every variety of timber suitable to the climate wave to the passing breeze, while here and there a huge arboreal giant raises his head proudly to the sky. While the interior of the Palace can boast of its priceless treasures of art, its historical relics, and its mementos of royalty,

QUEEN MARY'S SYCAMORE.

the policies are not without its monuments of royal visits. On a beautifully-sloping bank at the south-west front of the Palace is a sycamore planted by the beautiful, but unfortunate, Queen Mary. Although the west fork is broken off, the tree is still a beautiful one, and stands 63 feet high, girthing 13 feet 1 inch at 5 feet from the ground. Nearer the river, and in a hollow, is an oak planted by James VI. It is also a magnificent tree, with a spread of branches of 75 feet, a height of 55 feet, a girth of 15 feet at the base, 14 feet 1 inch at 3 feet from the ground, and

13 feet 3 inches at 5 feet from the ground. On the terrace to the east of the Palace is another tree, a sycamore, also planted by James VI. Standing 80 feet high, with a noble head, it girths 12 feet at 4 feet from the ground. Among the other notable trees in the park are an American scarlet oak in splendid foliage, with a girth of 8 feet 2 inches at 5 feet; a Turkey oak, with thick umbrageous head, and a girth of 8 feet 7 inches at 5 feet; a Douglas fir, raised from the first seed sent to this country in 1827, from British Columbia, 75 feet in height and 7 feet in girth; a Wych elm, 95 feet high and 6 feet 7 inches in circumference. Near this place is the site of the ancient Abbey, which enclosed the famous Coronation Stone, now removed to Westminster. North of the old Scone burying ground, in which are some stones of the early part of the fifteenth century, including that of Alexander Mar, the sixteenth Abbot of Scone,—who flourished when the Battle of Flodden was fought,—is an oak planted in 1809. With 40 feet of a straight stem, this oak stands 70 feet in height. At the root the girth is 10 feet 2 inches, and at 5 feet from the ground, 8 feet 2 inches. The pinetum has some fine specimens of *Abies Menziesii, Picea Nordmanniana, Wellingtonia gigantea, Araucaria imbricata, Picea nobilis,* and *Pinus monticola,*—the latter being about fifty-one years of age, and believed to be about the best specimen in Scotland. In the flower garden there is a splendid Douglas fir, planted in 1834. This tree is 75 feet in height, and is 7 feet in girth at 5 feet from the ground. It was originally planted in a place which turned out to be inconvenient, and it was removed to its present site about thirty years ago. When looking at this tree we could not but think of the sad fate which overtook the gentleman with whose name the species is inseparably associated. David Douglas was born at Scone in 1798, and was the son of a labouring man. He received his education at the Parish School of Kinnoull, after which he served his apprenticeship as a gardener in Scone Gardens. In 1818, being then at Valleyfield, he had opportunities of viewing the garden of Sir Robert Preston, which contained a choice collection of exotic plants, and, through the kindness of the head gardener, he obtained access to Sir Robert's

botanical library. He was afterwards employed in the Glasgow Botanic Garden, where his botanical knowledge gained for him the favourable notice of Sir William Hooker, whom he accompanied in several of his excursions, including the one through the Western Highlands to collect materials for his *Flora Scotica.* He was subsequently recommended to the Horticultural Society of London by Sir William, and was sent several times to America to examine the plants growing in the neighbourhood of the Columbia River. His first visit to America was in the year 1823, in which year he secured many valuable plants, and greatly increased the Society's collection of fruit trees. He returned home in the autumn of the same year, but was sent out again in July 1824, for the purpose of exploring the botanical riches of the country adjoining the Columbia River, and southwards towards California. When the vessel touched at Rio de Janeiro, he collected many rare orchideous plants; shot many curious birds in his voyage round Cape Horn; sowed a collection of garden seeds in the Island of Juan Fernandez; arriving at Fort Vancouver, on the Columbia, on the 7th April 1825. During this visit, he traversed the country across the Rocky Mountains to Hudson's Bay, and returned to England in the autumn of 1827. Through the influence of Mr Joseph Sabine, Secretary of the Horticultural Society of London, he was introduced to the Literary and Scientific Society of London; and he was elected, free of expense, a member of the Linnæan, Geological, and Zoological Societies, to which he contributed some valuable papers. After remaining for a couple of years in London, he again sailed for America, to continue his favourite pursuit, in the autumn of 1829. He afterwards visited the Sandwich Islands, where he met with his death under very shocking circumstances, having fallen into a pit made by the natives to ensnare wild animals. He was attacked by a bull already entrapped, was dreadfully mutilated, and eventually killed. The intelligence of his dreadful end, which took place on 12th July 1834, in the thirty-sixth year of his age, created a profound sensation in the country. A very neat monument to his memory has been erected in his native village, where his talents and services endeared him to his fellow-townsmen.

Amongst the other objects of interest in the policies, mention may be made of the "Boot Hill," now the family burying-ground, and planted with several sycamores, elms, and beeches. How this mound came to get the name of the "Boot Hill," is somewhat of a puzzle, and is variously accounted for. When the ancient Kings of Scotland were crowned at the Royal Palace of Scone, all the nobles were present. It is traditionally said that in order to maintain the independence of the Barons, each brought with him from his own property a small quantity of earth, which he placed in his boots on the Coronation Day, so that he might have the gratification of saying that he stood on his own ground during the ceremony, and on leaving the place each emptied the contents of his boot on the spot, and thus formed the mound. However poetical this explanation may be, we fear it will have to be abandoned in favour of another which is more prosaic. The Highlanders called it "Yom-a-mhoid,"—"The Hill where Justice is Administered," and it is highly probable that "Boot Hill" may be a corruption of "Moothill" or "Moathill," and may signify the hill of meeting, the seat of judicial or baronial assembly. It is said that the conventions of the nobles were in ancient times held on this eminence, and that from it Kenneth promulgated his edicts called the Macalpine Laws. A little to the north-west of this spot, and embowered by lordly trees, is a fine old cross—all that remains of the ancient village of Scone. Near the Palace grounds, and skirting the Stormontfield Road, we noticed what is believed to be the only avenue of purple beeches to be seen anywhere. They were planted about sixteen years ago, and from the length of the avenue, and the symmetry and luxuriance of the individual trees, a very pretty effect is produced.

Leaving the policies for the plantations, we have ample opportunities of observing the advantages of judicious planting. A century ago there was almost no wood whatever either at Scone, Lynedoch, Logiealmond, Taymount, or Innernytie,—Lord Mansfield's estates in Perthshire,—with the exception of a number of trees in the home grounds. There were, at all events, none of the plantations which now adorn the landscape. At that time most of the property was simply moorland, with a great extent of waste land partly occupied by barren woodbrush such as hazel, birch, elder, &c. At one period Scone was said to be well wooded, but the blunder of a landscape forester denuded it of its arboreal beauty. During the absence of the Earl who then held the property, and who was at the time an Ambassador abroad, a Mr White was employed to "embellish" the estate, but his idea of embellishment was closely allied to Vandalism, as he executed his commission by cutting most of the old oak timber on the property. The "landscape forester" must have carried out his work of destruction most completely, as at the present day there is no oak timber beyond 70 years of age, with the exception of a few trees in the neighbourhood of the Palace. The late Lord Mansfield did what he could to repair the damage by planting oak very extensively. The trees then planted have thriven so well that few plantations of the same kind can compare with them for rapidity of growth. Some of these trees, which are only about 70 years old, already contain about 100 cubic feet of timber. When Mr William M'Corquodale, his Lordship's head forester, came to the estate forty-five years ago, he immediately turned his attention to the training of these oaks. He discovered amongst them a mixture of other hardwood trees. These were all taken out, and the plantations converted into pure oak. Fir plantations occupy a large area of the woods at Scone, Lord Mansfield being the proprietor of perhaps the most extensive old Scots fir plantations in Perthshire. Many of them are from 80 to 100 years of age, and the quality of the timber is magnificent. At Moorward Wood there are still a number of Scots fir trees close upon 300 years of age. Amongst these is a very remarkable tree, appropriately called "The King of the Forest," on account of its extraordinary size. It stands in the centre of a section of the old wood which was recently cut down, and is "attended" by about half a dozen trees of large size and great beauty,—worthy "attendants" upon a "King,"—to protect it from the fury of the elements. The "King" stands about 80 feet in height, has a girth of 16 feet at 3 feet from the ground, and 15 feet 4 inches at 5 feet and contains about 400 cubic feet of timber. There is no Scots fir of similar size south of Braemar, and it is doubtful if its equal can be found even there. Moorward extends to upwards of 400 acres. Extensive sections have, however, been cut out, and replanted with Scots fir. About six years ago 15 acres of this plantation were sold for £1820, being at the rate of £132 per acre—the wood being eighty years old—and this, of course, without including a handsome revenue got for the thinnings.

Colen plantation extends to 360 acres, chiefly Scots fir, with a mixture of oak. The Scots fir is from 90 to 100 years of age, and, both in point of size and beauty, there are no Scots firs, age considered, to equal them, except, perhaps, at Braemar. There are several smaller plantations of Scots fir, and, both in size and quality, the trees are equal to those in the larger woods.

At Drumshogle Wood, as well as in other plantations, Mr M'Corquodale has adopted a very effectual mode of preventing the ravages of rabbits amongst the young trees, by placing boards, firmly fastened with wire, round the stems, the boards being removed before the tree becomes too large, and replaced by wire-fencing. Drumshogle was planted about sixty years ago, and some men still living remember of cutting corn upon it. It was originally planted with all kinds of hardwood, and when Mr M'Corquodale undertook the charge, he thinned it out, leaving very few trees except oak, intermixed with spruce and larch widely apart. The oaks now average 70 feet in height and 5 feet in girth; while the few spruce that remain are from 80 to 90 feet high, and from 6 to 7 feet in girth. The strong rambling side-shoots of the oaks were foreshortened, and the tops balanced, when they were about thirty years old. In the course of one of our rambles we came across a grand weeping beech, on the right-hand side going out the Old Scone Road at Quarry-Mill. It was planted about thirty-seven years ago, and is now 45 feet in height. The plant was got from the Perth Nursery, and is acknowledged to be the finest of the kind to be seen anywhere. A little farther on, there is a magnificent black Italian poplar, from 95 to 100 feet in height, and containing about 150 cubic feet of timber.

At Burnside Wood there is a 20-acre crop of Scots fir, after an old crop of the same had been removed, which gives a complete contradiction to the belief that Scots fir crops do not thrive in succession. The soil, it may be mentioned, is of a moorish character, on hard till. The wood is thirty years old, and there was no preparation for it except thorough draining. Mr M'Corquodale explained that the land was drained before the ground was replanted, and the villagers at Scone were permitted to take out the roots of the old trees, which they did most effectually. The year after the young trees were planted a number of them died, but they were replaced the following year. During the same summer Mr M'Corquodale was surprised to find that some of the plants which had formerly been doing remarkably well were dying out suddenly, and he was accordingly compelled to institute a search, and, for the first time in his experience, he discovered the destructive beetle, *Hylobius abietis*. For the purpose of experiment, he took about a dozen of them, and placed them in an inverted tumbler, and fed them on the stems of Scots fir. From this experiment he became acquainted with their extraordinary voracity. By this timely discovery he was enabled to save the plantation, and the beetle was effectually destroyed,—the plan adopted being to place a layer of earth smoothly beaten down round the base of each tree, covered by a quarter of an inch of the flour of hot lime. The diseased plants were all taken out the following spring, and their places supplied, and Mr M'Corquodale has never seen a bad plant since. When the plantation was first thinned it realised £73. Speaking of the *Hylobius abietis*, we once heard a very good story of its tenacity of life. In order, as was thought, effectually to destroy them, a gentleman placed a number in a bottle containing whisky, and kept them there for twelve hours, and then starved them for five or six days. At the expiry of that time he examined them, and found them *dead—drunk*. They soon recovered from their inebriation, and were as lively as ever. This showed that they were essentially Scotch in their tastes, so that it is but natural that they should have a partiality for the native pine.

One of the most interesting sights in this neighbourhood is the Earl of Mansfield's beautiful steading of Balboughty, than which no better-designed or better-appointed and ordered steading can be found in Perthshire, or perhaps anywhere. The steading consists of a quadrangle of elegant and substantial buildings, erected between 1858 and 1861, from designs prepared by the Earl, who takes the warmest interest in all the details and working of the farm. The main entrance is surmounted by a handsome tower, built in 1861, and with a clock that at once attracts attention from the peculiarity of its dials. On the front face the hours are marked by the letters of the Mansfield motto, *Spero meliora;* on the north face is the Stormont motto, *Uni æquus virtuti;* on the south face is the national motto, *Nemo me impune lacessit;* and on the west side are the ordinary Roman figures. Passing under the tower, we enter the centre block, comprising cattle courts, pig-sties, &c. The floor is laid with flag-stones, and the different divisions are separated by stone partitions, mounted with iron, giving the whole a substantial and tidy appearance. The extraordinary cleanliness, and the absence of anything like disagreeable smell in the byres, at once attract attention. The freedom from smell is obtained by superior ventilation, assisted by disinfectants. The building fitted up with feeding-boxes is 140 feet long by 32 feet wide, the passage down the centre being 8 feet wide. The sarking is all of yellow pine, varnished, and the supports of the roof are of iron, the whole having a light and airy appearance. Extending down the entire centre is an underground channel, with air-shafts on either side of the exterior, and fitted with iron gratings, so that a continuous draught comes in from the outside. In addition to this, ventilators, which can be shut or opened at pleasure, are placed in the side walls; and at the apex of the roof, throughout the entire length, is another ventilator to allow the hot air to escape. The building is drained from the centre of each box, and the whole drainage of the steading goes into an underground liquid manure tank, and is utilised in the spring.

Leaving what is in every respect a model steading, we resume our journey through the woods. We had not proceeded far till Mr M'Corquodale pointed out a very interesting experiment, having for its object the preservation of wood posts. After the post is put into the ground, a groove 2 inches wide and 8 inches deep is cut out round the post, which is firmly packed in with gravel, raised up 3 inches above the surface in a conical form against the post; after which tar is poured on in its natural state, and gravel thrown over it again. This process was only tried a few years ago, but it gives every promise of being most successful in keeping out wind and water, as at present it is as hard as concrete. A part of this fence is on posts made from Douglas fir-trees twenty-five years of age, and its stability is very remarkable. In the same district we noticed three oaks of royal mein, the first two being *Quercus sessiliflora*, and the other a *pedunculata*. The former were planted in 1808, the atter a year later; and, for "plantation" trees, are very extraordinary for their age. The first has a clean stem of 56 feet; the height is over 88 feet; it girths 5 feet 7 inches at 5 feet from the ground; and has 76 cubic feet of timber. The second is about the same height, is 7 feet in girth, and has 95 cubic feet of timber. The *pedunculata* has a stem of 57 feet, girths 6 feet 9 inches, and has 114 cubic feet of timber. All of these trees are quite near the two-milestone from Perth. Here, too, we noticed the old Coupar-Angus Road, now covered with several ash trees, bordering on 100 feet in height, and got a glimpse of a fine lime avenue, fully 600 yards long.

Continuing our walk through the woods, we came across a very fine plantation of *Abies Douglasii*, covering about 13 acres, and planted about twenty-seven years ago. All these trees, as well as hundreds of the same class planted out singly, are making rapid progress. There are no trees upon the property which for their age, are so tall as the *Douglasii* firs. The average height of those planted twenty-five years ago is from 35 to 40 feet, and at present many of them are making 3 feet of growth in a season. The soil, it may be important to note, is moorish, on hard till. Mr M'Corqudale says that, in filling up vacancies in old plantations, the *Douglasii* can be recommended above all others. If there be head-room, light, and air to be had at all, this tree will push its way. These trees were originally planted at 9 feet apart, with larch nurses. The nurses at Taymount are now all removed; and, by and by, as every alternate *Douglasii* will be thinned out, they will be left to grow at 18 feet apart for the permanent growth. In addition to the *Douglasii* plantations at Scone and Taymount of which we have been speaking, there are also plantations of the same kind at Logiealmond and Lynedoch, all of which are making similar progress.

At Highfield, Scone, there is a very fine experimental plantation. It was chiefly planted with Scots fir, partly reared from Rannoch seed. A portion of it was laid out and planted with 32 different varieties of the newer coniferæ, larch being the only tree which was planted amongst them as nurses. Many of these pines have made extraordinary progress since they were planted twenty-five years ago, the largest being the *Abies Menziesii* and Corsican pine. Highfield extends to 15 acres, has a southern exposure, and is sheltered from the biting north winds by an old Scots fir plantation. The plantation is about 300 feet above the level of the sea, and the soil is marshy, on a hard till. The ground was well drained, and the trees are 36 feet apart. In rapidity of growth, the first place, as has been mentioned, must be assigned to the *Menziesii*, and the second place to the Corsican pine. Next in order comes *Austriaca*. There is also a very fine *Picea nobilis*, but unfortunately, for the sake of comparison with the *Menziesii*, some mischievous person cut off the top about twelve years ago, evidently for the purpose of making a walking-stick. Perhaps the slowest in growth is the *Pinus excelsa*, although it is a very compact, well-clothed, and beautiful tree. There were no *Douglasii* firs planted here. The trees were planted promiscuously, but care was taken to keep the branches as clear from each other as possible, so as to give each tree plenty of room to develop.

Coming to another portion of his Lordship's estate, we find that the same intelligent interest in arboriculture, as an important factor in estate management, is manifested. At Innernytie and Taymount there are several thriving plantations of Scots fir and larch. At Innernytie about 50 acres of Scots fir were cut down seven or eight years ago, and four years ago this disforested land was replanted with Scots fir and larch, and, contrary to experience in some other parts of the county, both the Scots fir and the larch are succeeding very well, the character of the previous crop having no detrimental effect, so far as can be observed. Amongst the hardwood trees at Innernytie is a magnificent oak, growing at the east march adjoining Ballathie on the river side. It is believed to be about 300 years old, and has a girth of 20 feet 10 inches at 5 feet above the ground with a grand bole of about 15 feet.

On the Taymount Estate there are 172 acres of a valuable crop of trees, about 60 years of age, principally larch and Scots fir. In the neighbourhood of Taymount House there are some fine old oaks. One of these oaks, on the river side, on the east march of Innernytie, girths 20 feet 10 inches at 5 feet above the ground. It has splendid ramification of limbs, and is altogether a noble tree. It is believed to be about 500 years of age. Immediately on the west side of the Highland Railway at Taymount there is a very fine plantation of purely *Douglasii* firs of nearly thirty years' standing, the larch nurses having all been removed a considerable time ago. The side of this plantation nearest the railway has twice caught fire from the sparks of passing engines. The last occasion when this unfortunate accident occurred was about twelve years ago, and at that time a large number of valuable young trees were destroyed. Their places were supplied by fresh plants, which are thriving well, and which, it is hoped, will meet with a better fate.

At Lynedoch a large acreage has been planted since the Earl of Mansfield purchased the property from the late Lord Lynedoch, about thirty years ago. The woods at Lynedoch are not only very extensive, but they are of great value. Here is a great silver fir, acknowledged to be one of the finest in Great Britain. Here, also, are two magnificent *Douglasii* firs, which rank amongst the best to be met with in this country. The silver fir is 110 feet high. It is about 150 years old, girths 13 ft. 10 in. at 3 feet from the ground, 13 ft. 3 in. at 5 feet from the ground, has a spread of 45 feet, and contains 425 cubic feet of timber. The largest *Douglasii* fir has lost its leader, but the other one is perfect in all its proportions. They were planted in 1834, and both reach an altitude of 85 feet. The one girths 9 feet 10 inches at 3 feet, and 9 feet 7 inches at 5 feet from the ground; and the other girths 9 feet 3 inches at 5 feet. The largest tree has a spread of branches of 70 feet, and the other a spread of 45 feet; and the cubic feet of timber in the one is 132, and in the other 121 feet. On the west side of the estate there is a large track of larch plantation about fifty years of age, the seed of 300 acres of which is said to have been sown by the late Lord Lynedoch when on horseback. From some points of view this plantation is so extensive as to convey the impression that the

ABIES DOUGLASII AT LYNEDOCH.

whole property is one vast forest, scarcely anything but larch being visible within a very wide range. The forest is in perfect health, and when it comes to be cut down will no doubt yield a most profitable return. Only a very inadequate idea of the value of the wood on the Lynedoch property can here be given, and it is probable that during the earlier lifetime of the late Lord Lynedoch, it was even more valuable than at present, as a large tract of valuable old timber was cut down before Lord

Mansfield purchased the property. Indeed, for ten years previous to Lord Mansfield securing the property, Lord Lynedoch and his representatives were making sad havoc amongst the woods, cutting down timber to the value of about £5000 per annum during the entire decade—

> "Many hearts deplored
> The fate of those old trees, and oft with pain
> The traveller at this day will stop and gaze
> On wrongs which Nature scarcely seems to heed."

Notwithstanding this wholesale destruction, the estimated value of the wood by the exposer was as high as £52,000. The Earl of Mansfield has been at enormous expense in re-foresting the plantations which were so ruthlessly despoiled. An almost incalculable quantity of young trees, chiefly Scots fir, have been trained and planted, and now many thriving young plantations occupy the ground rendered vacant by the despoiler. The *Douglasii* is planted very extensively as solitary trees and in groups, and are making remarkably good progress.

At Logiealmond, a great many new plantations have been laid down since the property was purchased by Lord Mansfield from the late Sir William Stewart, Bart. of Murtly, about forty years ago. When the property changed hands, the west end especially had a most bleak and naked appearance. This is now to a large extent removed, several plantations having been laid down, and hedgerows extensively planted with trees. These hedgerows consist of two varieties. Some are purely of oak and others entirely of sycamore. All the trees are trained in the Home Nursery at Logiealmond until they are considered strong enough for planting out. This system is found to be much more advantageous than lifting the trees from young plantations. At the nursery they are continually under the eye of the forester and his assistants, and they can be trained in a way which is impossible in the plantation. Before being planted out, they are removed and replanted two or three times inside the nursery, and when finally removed to their permanent seats as hedgerow trees, they are well furnished with root and branch. They remain in the nursery altogether for eight or nine years, and when put out have attained a height of 10 or 12 feet. They are planted about 12 or 14 yards apart; and so successful have they proved, that out of several miles of hedgerow trees which we saw there was scarcely such a thing as a blank. All the trees are remarkably well trained, not a single strong rambling side-branch being permitted to grow, nor a contending shoot allowed to compete with the leader. The tops are shaped off so as to throw all the vigour possible into the stems. By this means their symmetry is preserved, and they are at the same time protected from the force of the wind, it being almost impossible for the wind to destroy them by splitting down the heavy limbs.

In the neighbourhood of Logie mansion-house there is a large extent of fine old timber, consisting of beech, sycamore, oak, and larch. There is also a splendid avenue of walnut trees, and several gigantic silver firs. Some of the sycamores are remarkably large. One of them contained 309 cubic feet of timber when Lord Mansfield acquired the property. The tree is still very healthy, and now contains about 350 cubic feet of good timber. Along the banks of the Almond there are some very fine Weymouth pines. At the old main approach to Logie House there is a magnificent avenue of lime and beech trees, which is perhaps the finest in Perthshire. It is fully a quarter of a mile in length, is very wide, the trees are aged and of large dimensions, while the foliage reaches almost to the ground. It presents altogether a grand sight, and one upon which the noble proprietor may look with no small pride.

In the neighbourhood of Logie House there is also a Home Nursery of about two acres, teeming with *Abies Douglasii*, *Cedrus Deodara*, *Pinus excelsa*, and several pines reared from Indian seed. There are also some ash trees from Indian seed. The latter variety has a different habit of growth from home-grown timber, being much stronger and more robust in the stem. Hundreds of oak and sycamore trees are being trained here for hedgerow trees to beautify the property.

The new plantations which have been laid down upon the Logiealmond property are all on moorland, and on soil which, in the meantime, cannot be turned to more profitable account. They are very judiciously laid out in belts and clumps, which serve the double purpose of shelter and ornament. Amongst the more important of these plantations is one of Scots fir to the east of the shooting lodge, covering a large portion of the moor. It is situated 1000 feet above the level of the sea, and was planted about fifteen years ago. It has already been partially thinned, and, wherever the soil is at all favourable, is succeeding very well. Shortly after the trees were planted their growth was very much retarded by the ravages of the blackcock and greyhen eating the leading buds—a very common occurrence on hilly land frequented by these birds. After the trees got beyond the reach of these enemies, however, they made rapid progress.

Altogether, the plantations on the properties of the Earl of Mansfield extend to about 5000 acres. Almost the whole of these vast forests occupy ground that would otherwise be unprofitable, and an eyesore in the landscape. Some idea of the extent of planting in recent years may be gathered from the fact, that for many years after Mr M'Corquodale became forester he planted from 300,000, to 500,000 trees every year. His lordship's instructions are that no ground is to be allowed to lie waste, and that every acre be planted with the crop which is most profitable. We had abundant opportunity of observing how well his instructions are followed almost over the entire property, and particularly at Scone, where there is not a single acre but is thoroughly utilised as arable ground or as woodlands. We have repeatedly driven over the property, and on several occasions we have had the pleasure of being accompanied by Mr M'Corquodale, of whose fifty years' experience we largely availed ourselves.

IX.—METHVEN.

Methven Wood a Lurking-Place of Wallace—The Battle of Methven Wood—The Castle—Walk through the Ancient Wood—An Enormous Beech—The Plantations—History of the Property—The First Extensive Planter—Remarkable Trees—The Pepperwell Oak—Ancient Church of Methven—The Bell Tree—The Black Italian Poplar at Tippermallo.

METHVEN WOOD, like Birnam and other ancient forests in Perthshire, is inseparably mixed up with the history of Scotland. Methven is mentioned in the chronicles of the country as far back as 972, when the Thane of Methven appears to have been a person of considerable importance. When sorely pressed by his foes, Wallace is said to have found a safe lurking-place in the wood of Methven. After hiding for some time in the wood, he is said to have come to Perth under the assumed name of Malcolmson, and passed himself off as a borderer who had come north in search of employment. It was to Methven Wood that he returned after reconnoitring the city, and it was from there that he emerged to attack the English on their way to strengthen Kinclaven Castle. The great Scottish patriot, when in Perth under the assumed name of Malcolmson, ascertained that it was the intention of the English to send a reinforcement of ninety horsemen to Kinclaven Castle. Immediately on gaining this intelligence, he ceased to solicit employment from the burghers, and hastened to his confederates in Methven Wood, and made preparations for waylaying the force as it proceeded northwards along the banks of the Tay. This they were successful in doing, killing sixty out of the ninety soldiers, —the commander of Kinclaven Castle, Sir James Butler, falling by the hand of Wallace himself. The defeat of the English was observed from the battlements of the castle, whither the survivors fled, but so close was the pursuit that Wallace and his comrades entered the castle along with the discomfited horsemen before the drawbridge could be raised. All the English within the castle were put to the sword, with the exception of some women and children and two priests. The Scottish patriots remained within the castle for seven days, during which they carried off a large quantity of plunder, which they hid in the neighbouring forest of Shortwood Shaw. They afterwards burned the castle to the ground; but before effecting their escape, they were attacked by a large body of English, numbering about one thousand men, under the command of Sir John Butler, son of the governor of the castle whom Wallace had slain. As Wallace had only about sixty men with him altogether, the English felt pretty certain of his capture. Their attacks, however, were met with the invincible firmness characteristic of Wallace and his friends, and the English were ultimately repulsed with heavy loss, several of the Scots having also fallen during the conflict. After a short cessation of hostilities, the English were about to make a most resolute attack upon the position of the patriots, when Wallace and his friends succeeded in effecting their escape to Cargill Wood. The English made an ineffectual search for the booty carried away from the demolished castle, and afterwards returned to Perth. The Scots returned to Shortwood Shaw a couple of days after the battle, and carried off their treasure to the almost inaccessible Wood of Methven.

Methven Wood, however, is best known in history by the battle which was fought within the wood itself between the Scots under Bruce and the English under the Earl of Pembroke, and which was fought in 1306. After having ravished Galloway, Bruce marched to Perth, which was then strongly fortified, and where the Earl of Pembroke lay with a fairly large force. On arriving at the "Fair City," and finding the English army shut up within the walls, Bruce, in accordance with the chivalrous style of the age, sent a challenge to the Earl to come out and try his fortune in single combat. Pembroke replied that the day was too far spent, but that he would test his prowess against the Scottish monarch next morning. Bruce thereupon retired and encamped in Methven Wood. But the

Scots were not allowed to enjoy repose for any length of time. Shortly after they had pitched their camp, and whilst the soldiers were cooking their suppers, or were dispersed in foraging parties, an alarm was raised that the enemy was upon them. The alarm had hardly spread over the camp when the Earl of Pembroke, whose army outnumbered the Scots by 1500 men, burst in upon the weary soldiers of Bruce almost before they had time to arm. They made, however, a most determined resistance, Bruce slaying the horse of the Earl of Pembroke at the first onset. Notwithstanding their stubborn valour, the Scots could make no impression upon the superior forces of the English, and the battle was from the first almost a rout. Bruce was thrice unhorsed, and once was so nearly taken that Sir Philip de Mowbray shouted that he had captured the new-made King, when Sir Christopher Seton felled the Englishman to the earth, and rescued his master. Some of the best and bravest of the Scottish knights fell into the hands of the enemy, but Bruce himself, his brother, Edward Bruce, the Earl of Athole, Sir James Douglas, &c., with 500 men, effected their retreat into the fastnesses of Athole, where they suffered the miseries of outlawry.

The Castle of Methven, surrounded as it is by so many historic scenes, is an edifice of great interest. It is believed that there has been a castle upon this site for a thousand years; and the present building, a fine old baronial residence, has been in existence for two centuries. Large additions have been made to it at intervals, and it is now as commodious as it is picturesque. The present laird, Mr William Smythe, the respected Convener of the County, has improved the castle very considerably, covering it with a new roof, and modernising it in several respects. He has also increased the amenity of the ancient edifice by enlarging the grounds to four times what they were when he came into possession, planting a large number of valuable trees, and building an extensive range of glass erections in the gardens adjoining. The castle stands upon a commanding eminence between the villages of Methven and Almondbank, and looks down upon a prospect that is both extensive and enchanting. Standing upon the terrace on the south front, the eye rests upon the fertile valley invaded by the Crieff and Methven Railway, wanders over the dark expansive Wood of Dupplin, with the bare, treeless Ochils in the distance. A little to the left, we obtain glimpses of the Fife Lomonds. From the east front the view stretches away to the pine-clad Hill of Kinnoull; and from the south front we obtain a magnificent sight of Ben Vorlich through a vista of singular beauty.

Leaving the Castle our steps are naturally directed to the historical Wood of Methven, the scene of bloody conflicts, and where doubtless lie the remains of many of our forefathers who died for the liberation of their country from the tyranny of Edward. The ancient Wood of Methven comprises about 200 acres, chiefly oak coppice, and is almost as dense as it would be in the days of Wallace and Bruce. It has been a wood from time immemorial, and, although there are not many trees which can be said to be of a very great age, still there are several that have survived the vicissitudes of a few centuries. The wood lies along the western bank of the Almond, and several fine views are obtained of the surrounding country, and the gurgling river, flowing onwards to the Tay. In some places, we look sheer down upon the river from a height of nearly 200 feet. No respectable person is interfered with in walking through the woods, and the proprietor has very considerately cut away some of the shrubbery which hid the finer prospects, and provided seats for the more comfortable enjoyment of the scenery. Proceeding along the walk at the top of the cliff skirting the banks of the Almond, we suddenly meet with a monster beech, ranking amongst the largest in the country. It is believed to be about 600 years old, and may have been a large tree when the Battle of Methven Wood was fought. One almost instinctively lifts his hat on coming across such a venerable monarch of the forest. Who can recount what has transpired within its protracted lifetime? How many generations of mankind have lived and died since it first sprung up in that quiet spot by the side of the classic stream on whose banks Ossian sleeps? How many storms has it weathered? Has it any history beyond its general association with the bloody struggles which took place around it? If so, it must remain a sealed book for ever. The tree strikes one all the more forcibly that it confronts him unexpectedly. The walk on the summit of the cliff is not fringed with trees remarkable for their size—coppice-wood and young trees generally predominating on both sides. We therefore walk leisurely along the footpath without the slightest expectation of seeing anything noteworthy for some little time, but a sharp turn to the right, and a few steps onwards, brings us instantly to a standstill, in mute admiration of the gigantic vegetable wonder before us. The tree has every appearance of being over 100 feet in height, overtopping by far many immense trees in the neighbourhood. At one foot from the ground, it girths 21 feet 10 inches, and at 5 feet from the ground, it girths 14 feet 9 inches. It has a splendid bole of 20 feet, and divides into five great limbs, each of which would be a large tree by itself. As yet it shows no signs of decrepitude, but seems to be endowed with perpetual youth, the whole tree being in luxuriant health and foliage. It is, unfortunately, closed in to a large extent with coppice-wood, and its huge proportions are not, therefore, shown to the best advantage ; but this obstruction may, in course of time, be removed, and the giant displayed in all his beauty. Almost opposite the beech to which we have been referring is a very fine oak, girthing 11 feet 10 inches at 1 foot from the ground, and 8 feet 9 inches at 5 feet from the ground, with a bole of 30 feet. Close by the banks of the river is a beautiful sycamore, 11 feet 6 inches at 1 foot from the ground, and 9 feet 1 inch at 5 feet from the ground. The bole is about 20 feet and the branches spread far over the river. Excellent roads and footpaths have been constructed through the ancient wood, and indeed over the whole of the property, and at every turn one sees something to admire. Here we meet with a huge giant or an interesting young tree, and there we pause to admire the far-stretching view of river, valley, and mountain that is ever opening before our eyes. Although the ancient Wood of Methven proper only comprises about 200 acres, this is by no means the whole extent of the wood upon the property. The plantations altogether consist of between 900 and 1000 acres, apart from the timber in the neighbourhood of the Castle, of which we have yet to speak, and which includes a good many trees between 200 and 300 years of age. The great proportion of the wood was planted within the past century, although planting has been systematically going on since the property passed into the hands of the Smythe family in 1664. The property was purchased in that year by Patrick Smythe of Braco from Charles, the last Duke of Lennox, who died without issue in 1672, and whose honours, of which the Methven peerage was one, fell to Charles II., his nearest male heir, the King's great-grandfather and the Duke's great-grandfather having been brothers. The great-grandson of the first proprietor and father of the present proprietor has the honour of carrying out the most important of the planting operations. This was Mr David Smythe, who was educated for the Bar, and who sat as a Lord of Session with the title of Lord Methven, from 1793 to 1806. Most of the plantations were laid down between 1772 and 1800, but they have been extended at intervals up to the present, the blanks being also filled up as they occurred. Part of the woods formerly belonged to the Lynedoch property, and was transferred to the Methven estate by excambion.

Continuing our walk through the ancient woods, we come across some very fine timber at the north end, chiefly planted by Lord Methven, and somewhere about 100 years of age. The timber is principally beech, but is well mixed with larch, oak, and Scots fir. There are, throughout the entire wood, many fine specimens of Scots fir and larch. One of the former, taken at random, girths 9 ft. 11 in. at 1 foot from the ground, and 8 ft. 4 in. at 5 feet, with a clear stem of fully 30 feet. Another close beside it girths 8 ft. 9 in. at 1 foot from the ground, and 7 ft. 10 in. at 5 feet. A larch, also taken at random, girths 10 ft. 4 in. immediately above the swell, and 8 ft. 4 in. at 5 feet from the ground. A Spanish chestnut in the back woodlands, standing upon the top of an earth fence, girths 9 feet at the narrowest part of the bole. An elm in the Cow Park girths 11 ft. 9 in. at 1 foot from the ground, and 10 ft. 9 in. at 5 feet. At the back of the garden, there is an oak girthing 10 ft. 9 in. at 1 foot from the ground, and 9 ft. 6 in. at 5 feet. There are also a large number of very fine beech trees behind the garden.

One girths 13 ft. at 1 foot from the ground, and 11 ft. 2 in. at 5 feet. Another, taken at random, girths 10 ft. 5 in. at 1 foot from the ground, and 9 ft. 2 in. at 5 feet.

It is in the policies, however, that most of the more remarkble trees are to be found. These grounds, as already mentioned, have been considerably improved and extended by the present proprietor, who has laid them out with fine taste. Three acres of the dressed grounds to the west of the Castle were partly taken in during 1868, previous to which they were grass parks. The whole of this portion of the policies has been planted with specimens of the newer coniferous trees, all of which promise to become great trees, the soil, apparently, being most suitable. In walking through the ground, the first tree to which our attention was specially attracted is a fine old ash in the garden girthing 15 ft. 9 in. at 1 foot from the ground and 13 ft. 8 in. at 5 feet, above which there is a considerable swell in the trunk. The bole is about 20 feet; after which, the tree branches out into two main limbs. The height is altogether about 80 ft. The spread of the branches is very great, but owing to the confined situation of the tree, it cannot be accurately measured. It must be a very aged tree, as in 1796, when the garden wall was built, it was found necessary to construct an arch to protect the roots. In the same neighbourhood is a fine *Abies Albertiana*, planted fifteen years ago. It is over 50 ft. in height, and is growing so rapidly that 4 ft. 6 in. have been known to be added in the course of one year. It is beautifully clothed, and is altogether a very good specimen of this favourite conifer. The rapid growth of this tree is well brought out by an adjacent *Cupressus Lawsoniana*, which was planted in 1858, but is 10 ft. less than the *Albertiana* of almost half the age. There are some fine specimens of *Abies Douglasii*, also planted in 1858, and reaching a height of about 50 ft. One of them girths 6 ft. 9 in. at 1 foot from the ground, and 4 ft. 11 in. at 4 feet, with a spread of branches of 33 ft. A little to the west of the Castle is a splendid avenue of *Cedrus Deodora*, raised from seed by Mr Whitton, the gardener, in 1868. The seeds were brought from Simla by the Dowager Lady Elgin, by whom they were presented to Mrs Smythe of Methven. The avenue, which includes over one hundred trees, is appropriately named "The Simla Walk." Many of the trees are magnificent specimens for their age, and are all in fine health. They are planted 12 feet apart. Several of the trees raised from the same seed are planted out in the woods, and are also doing well. In the immediate neighbourhood there are some very good specimens of *Wellingtonia gigantea*, planted fourteen years ago, nine seeds having been sent by a friend in California, of which seven have lived to become very promising trees. One of these girths 6 feet 1 inch at 1 foot from the ground, and 4 feet 6 inches at 5 feet, the height being about 50 feet. Close beside this *Wellingtonia* is a very old holly. It girths 6 feet 5 inches at the narrowest part,—being 1 foot from the ground,—has a great spread of branches, and is in fine health. There is no appreciable difference upon this tree within the memory of the oldest people in the locality, and it is believed to be fully 200 years of age. Amongst the other noteworthy specimens surrounding the two latter trees, mention may be made of *Picea nobilis, Picea Nordmanniana, Abies Hookerii, Abies Pattonii, Picea magnifica*, the fern-leaf beech, and tulip tree, all of which are remarkably good specimens. Still keeping by the west of the Castle, we come to a fine clump of trees several centuries old. There is a cedar of Lebanon, about 100 years old, girthing 9 feet 2 inches at 1 foot from the ground, and 5 feet 9 inches at 5 feet. There are several fine spruce trees about 180 years of age, with a growth representing 1 foot for each year. One of these girths 9 feet 4 inches at 1 foot from the ground, and 7 feet 4 inches at 5 feet; and another girths 9 feet 9 inches at 1 foot, and 8 feet 1 inch at 5 feet. There are also several remarkably good Spanish chestnuts. One of the largest of these girths 14 feet 6 inches at 1 foot from the ground, and 11 feet 10 inches at 5 feet, with a beautiful bole of 25 feet. Another girths 14 feet 2 inches at 1 foot from the ground, and 10 feet 10 inches at 5 feet, and has an equally good bole. The following trees are all in the same neighbourhood, and their measurements also deserve to be recorded :—An elm, 16 feet 8 inches at 1 foot from the ground, and 13 feet 4 inches at 5 feet; a sycamore, 11 feet 10 inches immediately above the swell of the roots, and 6 feet 3 inches at 5 feet, with a bole of 20 feet;

a horse chestnut, 11 feet above the swell, and 6 feet at 5 feet from the ground; an ash, 18 feet at 1 foot from the ground, and 13 feet 5 inches at 5 feet, with a clear stem of 30 feet, and an entire height of about 100 feet; and a beech girthing 18 feet 5 inches at 1 foot from the ground, and 13 feet 3 inches at 5 feet, with a bole of 20 feet. At the east front of the Castle there is also a very good specimen of *Abies Morinda*, although it was considerably destroyed during the winter of 1879. There is also a wonderfully good Lucombe oak.

There are still three remarkable trees to be noticed, and these are by no means the least interesting. The first we shall notice is the celebrated Pepperwell oak, in the park in front of the Castle, and so called from its proximity to a refreshing spring bearing the aromatic name by which the tree is designated. Special reference is made to this tree in the New Statistical Account of the parish, where it is described as " a tree of great picturesque beauty, and contains 700 cubic feet of wood; the trunk measures $17\frac{1}{2}$ feet in circumference at 3 feet above the ground ; and its branches cover a space of 98 feet in diameter. It has attained an increase of girth of three feet since the year 1796. In the year 1722, 100 merks Scots were offered for this tree; and tradition reports that there is a stone in the heart of it, but, like the Golenos oak, it must be cut up to ascertain this." In 1867, the tree girthed 21 feet 7 inches at 1 foot from the ground, and 19 feet at 6 feet. The tree has grown very considerably since the publication of the Statistical Account, and even since 1867. At 1 foot from the ground it now girths no less than 23 feet, and at the narrowest part, or about 5 feet from the ground, it has a girth of 19 feet 5 inches, being 2 feet more than at the date of the publication of the New Statistical Account (1837), when it was measured at 3 feet above the ground. The tree is growing by the side of a steep bank, so that the size of the bole is somewhat irregular. On the upper side the bole measures about 8 feet, but on the lower side it has a height of 10 or 12 feet. Four immense main limbs start from the bole, and a fifth was broken off some years ago. The tree is altogether about 80 feet in height, and is in every respect a noble specimen of the " brave old oak." It is known to be over 400 years old at least. The second of the trees to which we have just referred is an ash, known as the "Bell Tree," in the village churchyard, so called, we presume, from the bell of the ancient church having at one time been suspended from its branches. Its age is unknown, but it is highly probable that it was planted at the time the ancient church was built, about the year 1433, when Walter Stewart, Earl of Athole, then proprietor of the barony of Methven, formed the church of the barony into a Collegiate Church, which it continued to be down to the Reformation. Part of this old church is still standing; and up till now, although it will not continue to be so any longer, it has been used as a burying-place by the Methven family. The portion of the old edifice still standing is the aisle, on one of the walls of which there is a stone showing its connection with royalty. The stone has the Royal Lion of Scotland sculptured upon it, with the crown above, and some Saxon inscription below, but the letters are so worn by age as to be completely illegible. It is believed that this portion of the ancient building was erected either by Margaret, the Queen Dowager of James IV., and eldest daughter of Henry VII. of England, when she resided at Methven Castle, or, as is perhaps more probable, by Walter Stewart, Earl of Athole, who endowed the church with lands and tithes. Being a son of Robert II., he was also a member of the Royal Family, and entitled to place the emblems of royalty upon the building erected under his patronage. Supposing the "Bell Tree" to be planted about the time of the building of the old church, this would make it about 450 years of age, which it has every appearance of being. The New Statistical Account says that "at $3\frac{1}{2}$ feet above the surface of the ground, this tree measures 20 feet in circumference, and it contains in all 380 cubic feet of timber. Forty years ago, it exhibited a magnificent top, but wearied, as it were, of its former pliancy, it now chooses rather to break than bow, and yearly it does homage to the soil which nourishes it, by surrendering a portion of its withered branches." The tree has continued to decay since this measurement was taken, and, although it still gives birth to luxuriant foliage, it has an aged and decrepit appearance. It may, however, still live for several generations, but Ichabod is written upon its venerable head. At Tippermallo there is another remarkable tree, also in its dotage—a black Italian poplar. It is not nearly so old as the other two above referred to, having been planted in 1776. It is, however, a very large specimen of this variety, although recent storms have lessened its dimensions. In 1836, it girthed $11\frac{1}{2}$ feet at 3 feet from the ground, and had 300 cubic feet in solid contents. It now girths 20 feet 7 inches at the base; 17 feet 10 inches at 2 feet from the ground; and 15 feet 9 inches at 5 feet.

Although we have presented a pretty extensive catalogue of remarkable trees upon the Methven property, we have far from exhausted the list. It is quite possible even that we have not included the largest specimens of some of the varieties, as a great many huge trees are to be met with nearly similar in size. However, the measurements we have given may be taken as fair specimens of a countless number of trees remarkable for their size and symmetry, and which are not only a great ornament to the country round the ancient and picturesque Castle of Methven, but are of considerable pecuniary value.

X.—MONCREIFFE.

"The Glory of Scotland"—Remains of an Ancient Fort—Druidical Circle—Woods Adorning Moncreiffe Hill—The Moncreiffes of Moncreiffe—Plantations on the Estate—Large and Remarkable Trees—Largest Horse Chestnut in the Country—*Taxodium sempervirens* as a Forest Tree—The Remains of the Old Chapel.

To those who have at any period visited Moncreiffe, the very mention of the place is sufficient to conjure up delightful visions of that truly charming scene, in which

"The camp-crowned Moredun lifts its piny height,"

and enables the visitor to view what Pennant has described as "the glory of Scotland." The view from the top of Moncreiffe Hill certainly is glorious. Standing at the flagstaff on the summit, the visitor may see at a glance the whole of the lower portion of the fertile valley of Strathearn, clothed with rich pastures and smiling corn fields, embellished with thriving woods, and watered by the clear, tortuous Earn, the windings of which can be traced from its entry into lower Strathearn until its junction with the Tay. By taking a few steps to the east, the view may be continued over the broad expanse of the Firth of Tay, with part of the Carse of Gowrie, and Dundee in the distance. From the crest of the hill the eye looks straight down upon the clean and tidy village which takes its name from the substantial bridge which crosses the Earn. In the background, to the south, we have a view of Abernethy, the ancient capital of the Picts, with its curious round tower, nestling at the foot of the Ochils. By this range of hills, the view southwards is interrupted, although the Fife Lomonds may be seen peeping over from the opposite side. In ancient times, Moncreiffe Hill appears to have been regarded as an important military position, as there are to be seen the remains of what is believed to have been a stronghold of the Picts, the fosse of which is still traceable.

It is circular in form, and has a diameter of sixteen yards. In those days of primitive warfare, it must have taxed the powers of the most energetic and skilful generals to dislodge a vigilant enemy encamped upon this fortress, as no foe could appear within a considerable radius without being instantly detected, if the defenders were upon the alert. It is therefore highly probable that at one period Moncreiffe Hill formed an important link in the chain of forts that protected the Pictish capital. On the lower grounds there is a still more ancient relic of the past in the form of a fine Druidical circle, to which we shall again have occasion to refer in connection with a memorial of an interesting event in the history of the Moncreiffe family. Although not visible from the top of the hill, there are certain points from which good views can be got of Moncreiffe House, a fine old edifice, built from plans prepared by Sir William Bruce of Kinross, the designer of several important buildings, and the restorer of Holyrood Palace in 1671–78. Very much to the advantage of the people of Perth, the hill is at all times open to the public, although, we regret to say, the privilege has been so often abused that the proprietor would be justified in declining to continue it. Sir Robert Moncreiffe, however, is as anxious as his father was that visitors should enjoy the privilege of visiting the hill at all times, and the policies on Fridays; and it is sincerely to be hoped that no thoughtless parties may give such provocation as will render it necessary to deny such a boon as the liberty to enjoy a walk to the top of Moredun Hill.

The hill, we need hardly say, forms a prominent object in the landscape for a very wide distance. The back of the eminence, with its load of varied timber, and its highly-cultivated fields sloping gently to the Tay, is one of the most pleasing of the many pleasant sights visible from the "Fair City." It is the front view, however, which constitutes its great attraction. Rising to a height of several hundred feet above sea-level is a sheer precipice, straight and clean, as if sawn through by Titanic hands. This would be a striking enough object in itself, but it would be almost destitute of beauty were it not for the many-tinted trees which clothe the slopes leading up to the naked cliff. These trees were solely planted with the view of adorning the hill, and, with the exception of a few old trees on the low grounds, are all within 100 years of age. Hardwood trees have found most favour,—oak, ash, elm, beech, and sycamore predominating,—although these are intermixed with Scots fir and larch. Here and there a few purple beech trees have been planted, the foliage of which presents a beautiful contrast to the light and dark green of the other varieties. The oldest trees on the hill are principally oak, although a few larch have attained about 100 years of age. One of these, when cut, yielded 100 cubic feet of measurable timber. About twenty years ago, shortly after Mr Bissett, the present land-steward, became connected with the property, a commencement was made with the planting of the finer varieties of pine upon the hill—a process which is still being continued. The favourite species have hitherto been *Abies Douglasii* and *Wellingtonia gigantea*, which have grown with remarkable vigour. Immediately above the mansion-house are a few very fine specimens of *Pinus insignis*. In walking about the hill, we noticed a dozen very fine araucarias, which had been transplanted from the lower ground in 1861, after having suffered severely from the intense frost of the previous winter. Their removal has proved most beneficial, as they have thriven every day since they were transplanted, and are at present in luxuriant health. The soil into which they have been replanted is a light loam, in which the bracken grows to a height of 6 or 8 feet, and the situation is at a sufficient height above the level of the sea to be free from the deadly rime which falls profusely upon the lower grounds. These trees were each about 8 feet high when transplanted, and they have now attained a height of about 25 feet. It is noteworthy that not one of these twelve trees has suffered from the removal, all of them having taken kindly to their new position.

The present laird, Sir Robert Drummond Moncreiffe, belongs to the seventeenth or eighteenth generation of the family in a direct male line. The surname of Moncreiffe is of great antiquity in Perthshire, and belonged to the proprietors of the barony of Moncreiffe as early as any surnames were in use in Scotland. The genealogical accounts of the family mention Rametus Moncreiffe as its founder. He lived between 1107 and 1124, and is said to have been keeper of the wardrobe to the family of Alexander I. The first laird of importance was Mathew de Moncreiffe, who obtained a charter of the lands of Moncreiffe from Sir Roger de Mowbray, at that time the superior. This charter has no date, but as Mathew got the same lands erected into a free barony by Alexander II. in 1248, it must have been given before that time. He also got a charter of confirmation in 1251 from Alexander III. of the whole lands and barony of Moncreiffe and Balconachin. In 1495 we find Sir John Moncreiffe, the eighth baron from Mathew, obtaining a charter of Moncreiffe. He was survived by his wife, who had a charter granted in 1530 to herself and William Moncreiffe, her eldest son, of the lands of Balgonie. The second son, Hugh, was the progenitor of the Moncreiffes of Tippermaloch, although he was never himself in possession of that estate. The third son, John, was the ancestor of the present laird of Moncreiffe. About this period considerable additions were made to the family property, including the lands of Kilgraston, Wester Rhynd, Carnbee, Craigie, &c., but afterwards the affairs of the second baronet, Sir John Moncreiffe, became so embarrassed that he was compelled to relinquish the whole of the large property which had so long belonged to the family. The first estate he parted with was Carnbee, which was sold in 1657 for about 40,000 merks, to William Ord, sheriff-clerk of Perth. This sale did not retrieve his affairs, and in 1663 he sold Moncreiffe itself to his cousin, Sir Thomas Moncreiffe, one of the Clerks of the Exchequer, who was created a Baronet of Nova Scotia in 1685. Sir Thomas, in 1679, erected the present mansion-house, over the principal entrance to which there are still to be seen his arms, with the name of himself and his lady. Shortly afterwards, the direct line of the first Baronet failed, but the patent devolved on the family of Hugh Moncreiffe, the ancestor of the Tippermalloch branch. The representative of this branch was Sir John Moncreiffe of Tippermalloch, well known as the author of a small work, now very rare, entitled *Tippermalloch's Receipts,* in which there are many peculiar remedies, based on superstition, for all the ills that flesh is heir to. The Moncreiffes were early connected by marriage with some of the most noble families in the country, and have filled honourable positions in all branches of the national service. Many of them have had a strong predilection for the Church, it being rather a remarkable circumstance, as regards one of the branches, that the late Sir Henry Moncreiffe Wellwood, was the sixth in lineal succession who had been officiating ministers of the Church of Scotland. The Rev. Sir Henry Wellwood Moncreiffe, Bart., D.D., of Free St Cuthberts, Edinburgh, is a kinsman. The present laird is the fifth Baronet since 1663, and the seventh Baronet of the house of Moncreiffe, besides being, as already stated, of the seventeenth or eighteenth generation of the family in a direct male line.

We have referred to the great improvements which have been quite recently effected on Moredun-Hill, but the labours of the proprietors of Moncreiffe have not ended with the embellishment of the magnificent hill which looks down upon their ancestral home. During the past century extensive planting has been going on all over the property. Craigie Wood and St Magdalene's Wood, in the immediate vicinity of Perth,—to the attractions of which they add considerably,—are fair specimens of what the proprietors have been doing over the whole estate, wherever this has been practicable or expedient. On the Moncreiffe estate 646 acres are laid out in wood, on the Balgonie property there are 48 acres of woodlands, and at Trinity-Gask there are 360 acres, making 1054 acres under wood upon the entire property. The plantations consist principally of larch, spruce, and Scots fir, and they have been planted at places not in the meantime likely to be more profitably utilised. A large portion of these woods were planted by the late Sir David Moncreiffe, but a good deal of the work was also done by the late Sir Thomas, who took a great interest in the beautifying and improving of his property. This work is being judiciously continued by Sir Robert Moncreiffe.

Although the greater part of the planting upon this estate has been effected within the past 100 years, it is not to be supposed that there is no old timber. In the immediate neighbourhood of the mansion-house we have some magnificent specimens of most of the varieties of trees grown in this country, many of them being the finest specimens of their kind in Scotland. Indeed, we believe it will be difficult to find anywhere such a collection of remarkable trees within the same space. The most of the older trees appear to have been planted about the time of the erection of the present house in 1679. Leading from the old Edinburgh Road up to the house is a very fine beech avenue from 600 to 700 yards long—one of the trees, taken at random as a fair specimen of the whole, girthing 10 feet 3 inches at 4 feet from the ground. This avenue, it is supposed, was originally a hedge planted about the time of the building of the house, and gradually thinned out as the plants required more space. In the centre of this avenue, surrounded by a small Druidical circle, is a young oak, which at once attracts attention on account of its straight clean stem, all the more interesting from the fact that it is a memorial of the birth of the late genial laird of Moncreiffe, planted in January, 1822, as part of the rejoicings in connection with the natal day. Although the tree is rather confined, being surrounded by three large beeches, it has grown wonderfully, being fully 70 feet in height, and in excellent health. The surrounding beeches have repeatedly been pruned to prevent them injuring this memorial of an interesting event in the family history.

The most of the old trees surrounding Moncreiffe House are oak, beech, ash, and horse chestnut; while a number of lime, elm, sycamore, walnut, and birch trees were evidently planted some short time afterwards. One of the oaks in the policies girths 17 feet 7 inches at 1 foot from the ground, and 12 feet 3 inches at 5 feet from the ground,—the height of the bole being about 20 feet, the entire height about 90 feet, and the spread of branches about 75 feet. The largest of the ash trees girths 18 feet at 1 foot from the ground, and 13 feet 9 inches at 5 feet,—the bole being about 30 feet, the entire height 100 feet, and the spread of branches 75 feet. At the west end of the house there is a fine group of seven old horse chestnuts, of such peculiar formation as to give them a highly rustic and ornamental appearance. They have been planted rather close, and their beauty is, on that account somewhat contracted; but the group, as a whole, with its magnificent foliage and unique form, presents a most captivating arboreal picture. The largest tree of all is a horse chestnut. This tree—believed to be the largest of its kind in Scotland, if not in Britain—girths 20 feet 6 inches at 1 foot from the ground, and 19 feet at 5 feet from the ground. At about 10 feet from the ground it diverges into three great limbs, each of which is equal to the stems of ordinarily-sized trees. Great as this tree is, it is utterly unable to support these enormous branches, and about fifteen years ago the largest of them rooted firmly into the ground at four or five different places, adding very much to the majesty of the tree. and bringing out the spread of branches to 90 feet. Unfortunately, some of the heavier branches have, from time to time, broken off through their ponderous weight. One of the largest of the limbs was blown off during the memorable storm which destroyed the Tay Bridge, and when weighed was found to be about two tons. South of the group of horse chestnuts already referred to, and on a piece of ground supposed to be adjoining the original garden, there was until recently an equally picturesque group of walnut trees. The largest of these girthed 13 feet 7 inches at 1 foot from the ground, and 10 feet 9 inches at 5 feet. The largest of the elms girths 20 feet 6 inches at 1 foot (the same as the horse chestnut already referred to), and 14 feet 8 inches at 5 feet from the ground. The difference between the measurements at 1 foot and 5 feet is largely accounted for by a good deal of vacant space being about the roots. There is another elm which cannot be passed over without notice, as not only is it of great dimensions, but the peculiarity of its form renders it one of the most remarkable trees on the property. At some remote period it appears to have been rent through the centre, and formed a sort of archway through which a person could pass. Thus weakened, the tree was unable to bear the superincumbent weight, and toppled over. Notwithstanding this accident, the tree has continued to flourish as luxuriantly as ever, and has fixed many of its branches in the earth, very much after the manner of the banyan. Near to the front entrance of the house there is a hickory, considered to be the finest in Scotland. It girths at 1 foot from the ground 6 feet 6 in. ; and at 5 feet, 5 feet 9 in., with a splendid bole of 20 feet. In the lawn to the east of the mansion-house is a very fine specimen of *Picea Nordmanniana.* Although only planted about twenty-six years ago, it has already attained a height of 30 ft. It girths at the ground 4 ft. 2 in.; at 1 foot from the ground, 3 ft. 9 in.; and at 5 feet from the ground, 2 ft. 2 in. A special interest attaches to this tree on account of its being the first of its kind to produce ripe seed in Scotland, Mr Bissett having raised a number of young plants from the seed. It is a little remarkable, however, that the plants raised from the cones taken from this tree have at present more resemblance to the silver fir than to the parent plant. The reason of this is not very apparent, but it may probably be caused by the pollen of some silver firs situated about 50 yards distant acting upon the flower of the *P. Nordmanniana.* There is also a remarkable silver willow in the lawn, about 20 yards from the tree already mentioned. This willow girths 17 feet at 1 foot from the ground, and 19 feet at 5 feet from the ground,—the difference being due to a projecting limb. On the southern side, the branches have yielded to their weight and rooted, extending the spread of the tree very considerably —the younger portions being much more vigorous than the parent stem. In the pleasure grounds there is a beautiful cedar of Lebanon, which carried between 200 and 300 cones five years ago. It reaches a height of 66 feet; and at 3 feet from the ground it girths 11 feet, the spread of branches being about 64 feet. A very good example of the value of *Taxodium sempervirens* as a forest tree for this country is to be seen in the policies. It was planted about the same time as the *P. Nordmanniana*, and already it has a height of 43 feet, being 13 feet more than the *P. Nordmanniana.* Its girth at the ground is 6 ft. 10 in.; at 1 foot from the ground, 5 ft. 10 in.; at 3 feet, 5 ft. 4 in.; and at 5 feet, 4 ft. 11 in. The spread of branches is 22 ft. It will thus be seen that the tree has a growth of about 2 feet each year. The tree has, unfortunately, lost its leader several times, or it might have shown still more satisfactory progress. The situation is sheltered, and the ground dampish. A short distance from the two trees above mentioned there is a remarkable common spruce, beautifully feathered to the ground, and with some of its branches rooted, like many others at Moncreiffe. It rises to the height of 80 feet, girths 9 feet 4 in. at 3 feet from the ground, and has a spread of branches of 50 feet. The three last-mentioned trees stand triangularly only about 50 and 70 feet apart, and afford a very good illustration of the proximity of all the remarkable trees in the policies of Moncreiffe. In the same neighbourhood there is also one of the largest Portugal laurels in the country. It is very thick and close, with a circumference of 186 feet, and would have measured fully 200 feet had it not been cut to keep clear of a walk. Within a very short distance from this laurel there is a fine silver fir, reaching a height of about 100 feet. At 1 foot from the ground it girths 13 feet; and at 5 feet 11 feet 7 in., the spread of branches being 45 feet. A purple beech girths 9 feet at 1 foot from the ground, and 7 feet 6 in. at 5 feet, the spread of branches being 60 feet. In front of the garden there is a fine clump of 12 American cedars of different species, one of which deserves to be specially mentioned. It is a *Cupressus thyoides viridis*, and is considered the best specimen of the species in this country. It is about 35 feet in height, of very symmetrical form, and with the foliage thick and close to the trunk. The policies of Moncreiffe also furnish an illustration of the rapid growth of the *Abies Albertiania*. A specimen which was planted in 1869 has already a height of fully 30 feet. It girths at the ground, 2 feet 6 in., at 1 foot from the ground, 2 feet 3 in.; and at 3 feet from the ground, 2 feet. The foliage is very rich, and close down to the ground. The splendour of this tree is all the greater when compared with an *Abies Menziesii* close beside it, and which, although planted eight years earlier, has only a height of 25 feet. Beside these two, there is also a superb specimen of *Cupressus Lawsoniana*, planted in 1859, and now 25 feet high. Between the garden and the orchard there are several large specimens of gean trees. The situation of these trees so near to the garden and orchard is of considerable advantage, as it is found that, if there be an abundant crop of geans, the birds are not in the habit of doing so much damage to the fruits.

Before leaving Moncreiffe, there are one or two objects of

interest in the policies worthy of notice. About 300 yards to the south-east of the mansion-house are the remains of a very old chapel, now used as the family burying-ground, romantically embowered with wide-spreading and richly-foliaged trees. This chapel is of hoary antiquity, and was originally an appendage to the church at Dunbarney. The occasional discovery of bones, and various appearances, as well as the uniform voice of tradition, concur in proving that the ground surrounding it was at one time used as a cemetery. The interior has been used as a burying-place by the Moncreiffe family for many successive generations. Sir Robert Douglas, in his *Baronage of Scotland*, states that at so remote a period as 1357, Duncan Moncreiffe of that ilk and his lady were buried there. The families of the Moncreiffes of Tippermalloch and the Moncreiffes of Kinmonth were also at one time interred in the same chapel. It would be almost impossible to conceive of a more beautiful and romantic spot for the repose of the ashes of a distinguished family. The walls of the edifice are nearly entire, but not a vestige of the roof remains; and the ruin is so deeply embosomed in wood, that it presents a most picturesque appearance, while the thick foliage of the trees and evergreens by which it is surrounded, so completely hide it from view that a stranger may be quite near without being aware of its existence. The old chapel is 30 feet long, 18 feet broad, and 10 feet high. It has a north aisle, containing a vault underneath, and a small belfry at the east gable, the bell formerly belonging to which is still preserved in Moncreiffe House. Here lie the remains of the late Sir Thomas Moncreiffe, who died, very much regretted, on Saturday, 16th August 1879, at the comparatively early age of fifty-seven. The late Sir Thomas was not only a great lover of trees, but he was also an enthusiastic naturalist, being widely known as an accomplished entomologist. Beside him there also lie the remains of his eldest son, David-Maule, who died when only three years of age, as well as Sir Thomas' father, Sir David, who died in 1830. The old chapel is, therefore, a place that merits the reverence it receives from the Moncreiffe family, surrounded as it is with so many hallowed memories. A pair of quaint old stones, bearing the arms of the Moncreiffe and

Tullibardine families on either side, show the close relationship which has existed between the Athole and Moncreiffe families from a very early age. In the midst of the policies there is an elaborately-constructed summer-house, made of different species of willows most artistically and curiously arranged. In the centre there is a large table of the same construction, and showing an amount of skill and labour that are now very seldom bestowed upon such an article of garden furniture. In a rockery near Mr Bissett's house there is an interesting memento of Mr Laurence Macdonald, the Perthshire sculptor, consisting of a specimen of his first attempts at that art by which he became famous, and which was executed when he was working as an apprentice mason at Moncreiffe.

XI.—KILGRASTON.

Situation—The Mansion-House—An Apiarian Curiosity—Grants of Kilgraston—Policies—" Cromwell's Tree "—Other Remarkable Trees—Pitkeathly and its Mineral Wells—Drummonie House—Large Tulip Tree, &c.

KILGRASTON, the property of Charles T. C. Grant, is finely situated in Lower Strathearn, near the confluence of the Earn and the Tay. The mansion-house cannot be said to command a very extensive prospect, but the district within range is varied and beautiful. Taking our stand upon the roof of the mansion, we see that it is entirely surrounded by hills, with the exception of the eastern side, which is opened up by the passage of the Tay and the Earn on their way to the sea. The hills, indeed, surround the house in such a way as to suggest the form of a horse-shoe, with the opening towards the east. The hills bounding the north and west are those of Moncreiffe and Callerfountain, the Ochil range skirting the horizon to the south. Within these limits there is exposed as fine a track of fertile country as could be desired, while the valley and the picturesque hills enclosing it are richly ornamented with timber. The mansion itself is an exceedingly pretty structure, of the Greek style, plain and substantial. It was chiefly designed by the late Mr Francis Grant, grandfather of the present laird; and although the interior was entirely destroyed by fire in 1870, this has not affected the outward appearance of the original building, which is built of a red freestone found upon the property. While standing upon the top of the house, we had our attention attracted to an apiarian curiosity. The parapet along the front elevation is adorned by a hugh vase with a gigantic lion on each side. These ornaments are made of plaster, and the "little busy bee" has discovered a small aperture about the feet of one of the lions, through which it can easily gain access to the interior. The place has been found so commodious and suitable that a large colony of bees have made the inside of the lion their regular home for the last forty years at least, beyond the possibility of having their stores of honey disturbed. While the workmen were reconstructing the house after the fire, they thought it necessary to turn over this ponderous model of the king of the forest, and found the inside completely packed with honey.

The family of Grant of Kilgraston is originally of Glenlochy, in Strathspey, which was their ancient residence. The first of the lairds of Kilgraston of this name was John Grant, elder son of Peter Grant, the last of the Grant lairds of Glenlochy. He proceeded to the island of Jamaica, where he was the Chief-Justice from 1783 to 1790; and, on his returning home he purchased Kilgraston and the contiguous property of Pitkeathly. Previous to this, the property was held by the Craigies, a branch of the Craigies of Craigiehall, in the parish of Dalmeny, Linlithgowshire. It is traditionally said that the property came to the Craigies through one of them having the good fortune to exchange the wilds of Orkney for the fruitful lands of Kilgraston. The family ultimately rose to a position of considerable distinction both in law and arms, while they did much to make Kilgraston what it is. "They were remarkable," says the Statistical Account, "for the elegant improvements they made on their estates; and it is to their public spirit that the community is indebted for several avenues of trees which adorn the roads in the parish." The first of the Grants of Kilgraston died without issue in 1793, and he was succeeded by his brother Francis. Francis married Anne, daughter of Robert Oliphant of Rossie, who became Postmaster-General. Francis died in 1819, and was succeeded by his eldest son, John, who was twice married,—first, in 1820, to the Hon. Margaret Gray, second daughter of Francis, fifteenth Lord Gray; and, secondly, in 1828, to the Lady Lucy Bruce, third daughter of the Earl of Elgin and Kincardine, and by them he had a large family of sons and daughters. The most distinguished of the sons of Francis was the late Major-General Sir Jas. Hope Grant, K.C.B. He was Brigade-Major under Lord Saltoun, in the war with China;

and subsequently in India, under Lords Gough and Hardinge. He took an active part in the suppression of the Indian Mutiny, his services in that war being rewarded by a Knight-Commandership of the Order of the Bath. In 1860 he commanded the British forces during the operations in China; and in November of that year he was decorated with the Grand Cross of the Order of the Bath. Sir Francis Grant, the distinguished President of the Royal Academy, was another son.

The policies of Kilgraston are laid out with beautiful shrubberies and flower gardens, intersected with most delightful walks and drives. One of the most charming of the walks is appropriately named "Paradise Walk," from its beautiful formation and commanding position on a ridge along the southern boundary of the property. The walk is about one mile in length, and for fully half a mile there is a splendid avenue, composed chiefly of beech trees of a good size. Another very nice avenue, principally lime trees, leads from the Dunning Road to the family burying-ground, a distance of half a mile. Formerly, the avenue extended to the River Earn, half a mile farther, but when the railway cut through the lands it was found necessary to remove most of the trees, so as to square the agricultural land. Nearly all the limes in this avenue are of a good size,—one, which we measured as a fair average specimen, girthing 13 feet 4 inches at 1 foot, and 9 feet 8 inches at 5 feet. The main entrance to the policies also forms a very beautiful walk. The approach (which enters from the picturesque old village of Kintillo) is through a highly artistic arch of Craigmillar stone, the lodge harmonising in design with the gateway. Perhaps the most notable of the trees at Kilgraston is a large Spanish chestnut on the farm of Kilgraston Mains, regarding the history of which there are several traditional accounts, the generally-accepted one being that it was planted on the day on which the city of Perth capitulated to Oliver Cromwell, the 3rd August 1651. The girth above the swell of the roots, about 2 feet up, is 19 feet, and at 5 feet from the ground the girth is 14 feet 5 inches. There is a splendid bole of about 15 feet, and the tree is in other respects well balanced. About 30 years ago it met with an accident which was then feared might be fatal, but has, happily, turned out to have had quite an invigorating effect. The farmer at that time obtained liberty to remove a number of roots in the field, but the workmen, misunderstanding their instructions, thought they were to remove this great tree. They accordingly, proceeded to dig a deep trench all round the tree, and had cut off a number of the roots before they were detected. The trench was immediately ordered to be filled in, and the work of destruction stopped. The tree at this time was beginning to show symptoms of decay, but since it has suffered this underground amputation, it has entirely recovered its health, and is now thriving most vigorously. The largest of the oaks on the property grows in the North Lodge Park, the girth at 1 foot up being 16 feet 10 inches, and at 5 feet, 11 feet 5 inches. It is an exceedingly shapely tree, and has a nice bole of 10 feet. The next best oak to be met with grows in the pleasure-grounds, and has a girth of 11 feet 8 inches at 1 foot from the ground, and 11 feet at 5 feet. Beside the burying-ground, which is enclosed in a picturesque building representing a ruined chapel, there are several gigantic silver firs. The largest of these girths 11 feet at 1 foot from the ground, and 9 feet 6 inches at 5 feet; and another has a girth of 10 feet at 1 foot, and 8 feet 5 inches at 5 feet from the ground. They range from about 80 to 90 feet high, and are magnificent timber trees. Behind the garden, a very peculiar pollard ash attracts attention. There are in reality two trees, growing about 3 feet apart, but when they were saplings the wind would seem to have twisted them round each other, and, now that they have become trees, they are literally locked in each other's embrace, forming a sort of arch, and must stand or fall together. A large willow tree also presents a curious feature; it was originally a prop for a neighbouring tree, and was so placed in the ground some 60 years ago; but now it has grown into a great tree, and the original plant, much decayed, is seen protruding about 1 foot through the centre of the bole, where the branches break off. The girth of the trunk of this willow at the narrowest part is 12 feet 6 inches. A cedar of Lebanon in the same neighbourhood, which is quite close to the house, girths 9 feet at 1 foot, and 8 feet 2 inches at 5 feet from the ground. A horse-chestnut quite near the house, girths 11 feet

3 inches at 1 foot, and 10 feet 7 inches at 5 feet; and a beech in the pleasure-grounds has a girth of 16 feet 2 inches at 1 foot and 12 feet 2 inches at 5 feet. The latter is an exceedingly beautiful tree, the bole being of a good size, and the branches numerous and well arranged. There are also a number of very fine beeches in the Dovecot Park, the best of them girthing 12 feet 10 inches at 1 foot up, and 10 feet 3 inches at 5 feet. Several large elms also grow in the same park, the best one girthing 21 feet at 1 foot, and 12 feet 7 inches at 5 feet, the narrowest part of the bole. Only a few of the newer coniferous trees have been planted at Kilgraston. The most notable of these is a *Wellingtonia gigantea* near the house. It has a height of about 35 feet, and is very shapely and well furnished. The girth at the ground is 8 feet 7 inches; at 1 foot up, 7 feet 6 inches; and at 5 feet up, 4 feet 8 inches. A thriving little specimen of *Thujopsis dolobrata* may be noted on account of its having been publicly planted in 1879 by Mr C. T. C. Grant, on his being initiated into the mysteries of the Ancient Order of Foresters.

Pitkeathly, the contiguous property, which was purchased along with Kilgraston, has for many generations been highly famed for its mineral wells, and these were, about eight years ago, vastly improved by Mr Grant, so as to maintain their reputation as a favourite resort. These wells are situated only about one mile from Bridge of Earn, and the surrounding scenery is of the richest description. Indeed, the wells, which are five in number, spring up in one of the most beautiful parts of Strathearn, being immediately surrounded by the seats of Moncreiffe, Kilgraston, Freeland, Dunbarney, Invermay, &c. The celebrated Moredun, or Moncreiffe Hill, is quite at hand. Mundie Hill, on the Kilgraston estate, is perhaps not so well known, but the view from the top is no less enchanting. This hill is one of the peaks of the Ochils, and lies a little to the south of the Wells. On a clear day the traveller will be rewarded for his trouble in ascending the summit by a glimpse of Edinburgh and the Bass Rock to the south, Dundee to the east, Perth and the Grampians to the north, and Ben Lomond and the western hills to the west. Close to the Wells, there is a favourite spot which has gained celebrity through a slight mistake made by Sir Walter Scott in the first chapter of *The Fair Maid of Perth*. The "Wizard of the North" there states that a view of Perth was obtained from a spot known as the Wicks of Baiglie; but as this was impossible, it is more probable that Chrystal Croftangry had come from the south by an old road called Wallace's Road, from a tradition that it had been traversed by the Scottish champion, as this would have led him to obtain a sight of Perth. This old road to the south passes over the Ochils, and comes down from Milnathort by West Dron; and from a spot on that road, a little above West Dron, the traveller may have a view of the "Fair City." Putting aside the mistake of the great novelist—which, after all, has only been caused by his giving a wider range to the "Wicks of Baiglie" than is generally accepted—our readers will have some idea of the glory of the view from the description by the "Chronicler of the Canongate" himself. In the first chapter of *The Fair Maid of Perth*, we are told:—

"One of the most beautiful points of view which Britain, or perhaps the world can afford is—or rather we may say, was—the prospect from a spot called the Wicks of Baiglie, being a species of niche at which the traveller arrived, after a long stage from Kinross, through a waste and uninteresting country, and from which, as forming a pass over the summit of a ridgy eminence which he had gradually surmounted, he beheld, stretching beneath him, the valley of the Tay, traversed by its ample and lordly stream; the town of Perth, with its two large meadows or Inches, its steeples and its towers; the Hills of Moncreiffe and Kinnoull faintly rising into picturesque rocks, partly clothed with woods; the rich margin of the river, studded with elegant mansions; and the distant view of the huge Grampian mountains, the northern screen of this exquisite landscape. The alteration of the road—greatly, it must be owned, to the improvement of general intercourse—avoids this magnificent point of view, and the landscape is introduced more gradually and partially to the eye, though the approach must be still considered as extremely beautiful. There is yet, we believe, a footpath left open by which the station at the Wicks of Baiglie may be approached; and the traveller, by quitting his horse or equipage, and walking a few hundred yards, may still compare the real landscape with the sketch which we have attempted to give. But it is not in our power to communicate, or in his to receive, the exquisite charm which surprise gives to pleasure, when so splendid a view arises when least expected or hoped for, and which Chrystal Croftangry experienced when he beheld, for the first time, the matchless scene."

It is not known when the virtues of the Pitkeathly Wells were originally discovered, but the first authentic notice which is anywhere to be found concerning them is supplied by the Session Records of the parish, where, under the date of 20th September 1711, it is recorded that " the Session met according to appointment, and took into consideration the profanations of the Sabbath by people frequenting the medicine-well of Pitkeathly, whereupon some of the elders were desired to visit the well every Sabbath morning, and exhort the people from coming to it on the Lord's-Day, and inquire what parishes they belong to, that word may be sent to their respective ministers to discharge them ; and John Vallance is forbidden to give them passage at Dunbarney boat, and Thomas Drummond desired to spread the report that they are to be stopped by constables by authority of the Justices of the Peace." From the Statistical Account we learn there are two traditions current concerning the discovery of these mineral waters. The first tradition is that the country people were attracted to them by observing pigeons resort to the place ; but the second, and more probable, tradition is, that their medicinal virtue was first discovered by reapers who, when using them to quench their thirst at their mid-day meal, experienced their strong effects. It is of little importance, however, when the Wells were discovered ; but it is interesting to know that from a very early date they became a place of fashionable resort,—Bridge of Earn, indeed, as well as the Pitkeathly villas, having been built for the accommodation of those using the Wells,—and that they still retain their healing qualities.

Drummonie Castle, an old seat of the Lords Oliphant, is in the immediate vicinity of the Wells. This castle was built in the year 1693, and in recent years became known as Pitkeathly House. Although it had become very much decayed, the present proprietor has put it in a thorough state of repair, and restored the old title by calling it Drummonie House. Surrounding the mansion-house are several noteworthy trees. The most remarkable of these is an unusually large tulip tree, which girths 10 feet at 1 foot from the ground, and 8 feet at 5 feet. There are a number of very good oak trees near the house. The largest girths 12 feet at 1 foot from the ground, and 9 feet 5 inches at 5 feet ; and the next largest girths 10 feet at 1 foot, and 9 feet 10 inches at 5 feet. There are many, however, about the same size.

The plantations upon the property are not very extensive, the greater part of the land being of the best quality for farming. A large quantity of timber has been blown down within the past few years, but the vacant spots are being replanted, and several acres additional laid down.

XII.—INVERMAY.

The Beauties of Invermay—The Belshes of Invermay—Their Improvements—The Soldier and the Laird—The "Birks" of Invermay—The Mansion-House—The "Bell Tree"—The Old House—Notable Yew and Holly Hedges—A Remarkable Spruce Tree—The Walk by the May—Old Muckersie Church—The "Humble-Bumble"—The Plantations.

THERE are few places in the environs of Perth more popular with pleasure-seekers than the beautiful grounds of Invermay. Here all the most delightful characteristics of Highland and Lowland scenery are to be found,—the broad peaceful valley, with the sluggish waters of the Earn winding their way to the more rapid Tay,—the deep woody dell, with the impetuous May and its tributary streamlets rolling along their rocky beds,—the dense forest, and the commanding hill,—all combining to attract the visitor, and increase the amenity of the property. From an early date, the place was known as Endermay, but this designation was dropped about the beginning of the present century in favour of the present more euphonious name. The estate is now the property of Lord Clinton, but for centuries it was the home of the Belshes. This family trace their descent from Ralf Belasyse, a Norman baron, who came to England with William the Conqueror. One of the Belshes married into the family of the Stuarts of Fettercairn, and ultimately succeeded to Invermay. Several of the Belshes attained to considerable distinction, the last of the family rising to the rank of a Major-General in the army, having served with distinction in the Peninsula, and obtained the war medal with four clasps. That the various members of this family took a deep interest in the improvement of the property, is testified by its present condition. All the older timber which now adorns the estate was planted by the Belshes; and the suitability of the wood to the soil and climate, and the taste with which it was planted, show that they were fully alive to the importance of the work. The late Mr P. R. Drummond, in his entertaining work on *Perthshire in Bygone Days*, tells a very good story, which shows the carefulness with which one of these proprietors looked after his woods. On one occasion, when Mr A. H. M. Belshes was walking over the property, he encountered a soldier committing, in Mr Belshes' eyes, a flagrant crime. "The soldier, when passing the east gate, observed a young oak that he thought would make a good walking-stick; so, without scruple, he laid down his musket against the dyke, and, going into the wood, cut down the tree. While he was quietly lopping the branches from his prize, the laird came along the road. He saw at once the daringly impudent and really wicked step that the fellow had taken, and his passion, of course, rose to the occasion. No remedy to the evil deed occurred to him; indeed, there was none. But the irate laird seized the gun, and, with an oath, said, 'Take back that tree, and lay it down where you got it, or I will blow your brains out.' The soldier obeyed doggedly, and, after a round of the linguistic guns, he reluctantly gave the soldier his musket, No sooner had the man of war got possession of it, than he presented it at the laird, saying, with the same oath, 'If you do not put that stick into my hands as a gift, I will blow *your* brains out.' The laird made a virtue of necessity, and surrendered it." Sir John Forbes of Fettercairn and Pitsligo succeeded to the Belshes, and Lord Clinton, marrying the only daughter of Sir John Forbes, succeeded to the property about sixteen years ago. His Lordship does not reside here, his principal seats being Heanton Satchville, Beaford, Devonshire, and Fettercairn House, Laurencekirk, but he is well represented by Mr Mackie, his land-steward.

The main entrance to the property is from the Green at Invermay, the place being so called probably from its having at one time been a clachan, although there are now no remnants to indicate that there were ever any inhabitants to sport upon the village green. Our way lies along the river side, and we are immediately in the midst of the beauties of the place. As soon as we enter the approach, we pass, on the left hand side of the drive, some splendid specimens of *Abies Menziesii*, about ten

INVERMAY. 149

years of age, and making rapid progress, the annual growth being from 18 inches to 2 feet. The ground is of a light loamy nature, and drains freely into the May. At one time almost all the trees at Invermay were oak, intermixed with beech—the oak being not only the favourite variety, but the most thriving, the soil being admirably adapted for the cultivation of the timber which so long provided our sea-girt home with its wooden walls. The "Birks of Invermay" are often spoken of as the subject of a popular song, but we are afraid that the "birks" must be set down as a poetic fiction. There are no birch trees worth speaking of at Invermay, nor does there ever appear to have been any in sufficient quantity to touch the poetic fire. There was no reference to the birch in the original edition of the song which now bears the title of "The Birks of Invermay," the following being a copy of the first edition of Mallet's pastoral:—

> "The smiling morn, the breathing spring
> Invite the tuneful birds to sing;
> And while they warble from each spray,
> Love melts the universal lay.
> Let us, Amanda, timely wise,
> Like them improve the hour that flies;
> And in soft raptures spend the day,
> Among the shades of Endermay,
> For soon the winter of the year,
> And age, life's winter, will appear.
> At this thy living bloom must fade;
> As that will strip the verdant shade.
> Our taste of pleasure then is o'er;
> The feathered songsters love no more;
> And when they droop and we decay,
> Farewell the shades of Endermay."

The original version is much truer to Nature than the "improved" one. If the author had intended to refer to "birks," he would never have spoken of them as affording shade,—a term, however, which might be appropriately employed with reference to the widespreading oak, which seems to have abounded, instead of the birch. The alteration from the "shades" to the "birks," of Invermay appears to have been made by the Rev. Alex. Bryce of Kirknewton, who composed three additional and inferior stanzas, and who evidently never saw the place, or he would not have tampered with the more accurate picture sketched by Mallet, who is one of our Perthshire poets.

A short drive brings us to the mansion-house of Invermay, situated on a piece of rising ground overhanging the May, and commanding a magnificent prospect for a great many miles. The whole vale of Lower Strathearn lies at our feet, and to the west the Crieff and Logiealmond Hills strike the sky. Northwards the classic Hill of Birnam is easily recognised, while in the remote distance we have a shadowy view of Ben-y-gloe, and the mountains around Blair-Athole. The mansion has the verdant Ochils for a background, and is surrounded by 4 acres of pleasant policy grounds, in which clumps of rhododendrons, laurels, and common bay predominate, although vegetation of a more stately growth is not absent. The "Bell Tree" ranks amongst the most notable of the ornaments in the policies in the immediate neighbourhood of the mansion. It is a very fine old Scots fir, with a grand bole of 25 feet. At the top of the bole the tree forks into two branches, in a way which suggests the thought that the tree was specially grown for the purpose of hanging the bell which regularly called the Belshes to the dinner table to the end of their days, and which is still to be seen fixed in its pretty bifurcated belfry. The "Bell Tree" is surrounded by some good Scots firs, but none of them are so suitable for the suspension of the great bell as the one which has been selected.

The old house—or "The Tower," as it is called—is situated but a short distance from the present mansion, and is rendered somewhat picturesque by a complete covering of ivy. A great part of the building has been demolished, and the remaining portion is used as a laundry, gun-room, and lumber-room. A shield over the entrance bears the date 1633 and the initials "D.D." and "E.A." The grounds around "The Tower" appear to have been laid out with considerable taste, according to the notions which prevailed at the time. The principal feature, which can still be traced, is a long yew hedge, with a holly hedge in continuation. These hedges which were doubtless planted about the time of the building of the "The Tower," in 1633, have now grown to be great trees of their kind. They are very compact, there being only a few inches between each of the trees. The trees in the yew hedge are about 50 feet high, and have a spread of fully 40 feet. The holly hedge is a particularly pretty one. It has been cut down at one time, and afterwards allowed full liberty to spread its branches. The row is about 100 yards in length, and the trees are about 60 feet high, or 10 feet higher than the yews. At the old bowling-green below "The Tower," and close by the river side, there is a very remarkable spruce tree. It is thought to be a different variety from the ordinary kind, as it is both shorter and finer in the needle. It is a little naked about the bottom, but, otherwise, it is a grand tree. The girth at 1 foot from the ground is 15 feet, and at 5 feet from the ground, 13 feet 6 inches. There is a clean bole of 14 feet,—the entire height of the tree, which is about 200 years old, being about 80 feet.

Perhaps the most interesting walk on the property is that which clings close to the river. The walk is not particularly remarkable on account of the size of the timber which adorns it, but for its splendid sylvan beauty, and the charming peeps which are got of the river as it dashes over the rocks, and cuts its way through the narrow gorges. There are, however, several trees of large dimensions. We have already given the measurement of the largest of the spruce trees, but there are several others of this species which girth from 12 to 13 feet at 5 feet from the ground. Silver firs are to be met with girthing from 14 to 15 feet at 1 foot from the ground, and from 12 to 13 feet at 5 feet from the ground. There are several very fine oaks on the banks of the May, girthing from 12 to 13 feet at 5 feet from the ground. A few sycamores girth from 12 feet 6 inches, to 16 feet at 5 feet from the ground. There are a number of very fine beeches along the drive to Muckersie Chapel. The girth of the largest is 16 feet 6 inches at 1 foot from the ground, and 14 feet 6 inches at 5 feet, the others girthing from 12 to 16 feet at 1 foot, and 11 feet to 14 feet 3 inches at 5 feet from the ground. A splendid walnut growing near the house has a girth of 14 feet 6 inches at 5 feet.

Old Muckersie Church is an interesting object on the banks of the May. It is situated upon a prominent eminence about one mile from the mansion-house, and a short distance above Muckersie Falls. This is the church of the old parish of Muckersie, before it was united to Forteviot, and is now the burial vault of the Belshes, the small piece of surrounding ground being the last resting-place of many of their servants and retainers. The church and burying-ground are completely encircled with fine Scots firs of about 150 years' standing, a few old yews being also in the neighbourhood of the building. The church is a very plain structure of moderate dimensions. Muckersie Falls, to which we have alluded, are the finest on the river. When the river is full, the fall is 30 feet wide, the water falling in a great body over a rocky precipice of about the same depth. The "Humble-Bumble" is the most romantic of the waterfalls, its name being derived from the peculiar gurgling sound of the rushing water. The scene here bears a close resemblance to some parts of the Braan, at the Rumbling Bridge, but on a smaller scale. The River flows through a deep gully scooped out of the solid rock, which in some parts rises to a height of 60 or 80 feet from the surface of the water. A few birch trees grow on the upper lands in this neighbourhood, but at some distance from the banks of the river. The predominating trees are ash and elm, which cling to the crevices in the rock, and have but a scanty footing. The visitor can here descend almost to the water's edge by a series of winding stairs, and from a small rustic bridge look down upon the surging torrent. The trees and banks are overgrown with moss, and ferns grow profusely at inaccessible places, and impart to the scene a most fairy-like aspect. The gully through which the river here passes is about a quarter of a mile in length, and in some parts no more than 3 or 4 feet in width; while the rocks on both sides rise to a height of 80 feet, almost making one giddy to gaze upon their precipitous and mossy sides. Trees have taken possession of the crevices wherever they could obtain a hold,—the prevailing varieties being oak, ash, elm, and sycamore,—lending an unspeakable grandeur to the scene.

Although the plantations are not very extensive, they contain some very fine timber. The Roe Glen Wood consists of 150 acres, principally larch and spruce of 70 years' growth, and being

INVERMAY.

fully ripe, is being cut down. At Hog's Wood there are 40 acres of ash, Scots fir, and larch of about the same age. There are about 100 acres of policy woods, which chiefly consist of hardwood trees. At Hawkhill Wood there are 24 acres of timber, 18 acres being a mixture of various kinds, the remaining 6 acres being entirely of Scots fir of 150 years' growth, and known as the Crow Wood, from the extensive rookery which has been established in its midst. In addition to these plantations there are numerous strips and clumps planted for ornament or shelter. About 70 acres have been planted within the last eight years, principally spruce, larch, and Scots fir, with a sprinkling of hardwood trees, all of which are thriving very well.

Only a few of the newer conifers have been planted, and these chiefly along the east approach, to take the place of blowndown beeches. The westerly wind blows along this approach with great fury, and these trees, consisting chiefly of *Picea Frazerii*, have not thriven very well. A few specimens of *Wellingtonia gigantea* have also been planted, but neither have they succeeded, owing to the exposed situation. The situation is back-lying, and the soil rests on a stiff, tilly bottom. All the hardwood trees, however, and especially the oaks, thrive exceedingly well, and the beeches are remarkable for the cleanness of their stems.

XIII.—FREELAND.

The Original Design of the Policies—The Village of Forgandenny—The Main Approach—The Old Home of the Ruthvens—The Den—Remarkable Trees—The Mansion-House—The Most Remarkable Elder Tree in Perthshire—The Woods—Natural-Sown Silver Firs—A Venerable Ash—Trees with an Unknown History.

THE ESTATE OF FREELAND, once the property of a branch of the historic family of Ruthven, and now in the possession of Mr Collingwood Lindsay Wood, presents one of the best illustrations we have met with of the great changes which have taken place in the surroundings of noble ancestral dwellings within the past century or so. We do not allude so much to the improvements which have everywhere been effected in accordance with modern tastes and the increase of knowledge, as to the altered appearance of the country generally,—the changes wrought by the hand of Time, as well as by the hand of man,—the removal of old landmarks and the substitution of new ones, and many other things which at one time may have been considered, like the laws of the Medes and Persians, incapable of change. By whichever way we approach Freeland, we can hardly avoid being impressed with this fact. At first, the main approach to the mansion was not, as at present, from the Dunning Road, but straight through the village of Forgandenny, with its picturesque cottages and trim garden plots. The whole of the village of Forgandenny is on the Freeland Estate, and is snugly embosomed amid tall and stately trees. Indeed, when the woods are in full foliage, and the gardens gay with flowers, it would be difficult to find a sweeter spot. Although the main approach to the mansion-house was diverted many years ago, the village has in no way suffered in consequence; and since Mr C. L. Wood became the proprietor, the appearance of the pleasant little hamlet has been greatly improved. New cottages have been built, and old ones have been completely renovated. It may here be mentioned that while workmen were making excavations for the foundation of two new cottages, they came upon an earthen jar about 18 inches underground, containing a large number of old silver coins of the reign of James VI. Some of the houses in the village are very old. The old schoolhouse, which was built in 1748, is still inhabited. It consists of a modest "but and ben," and contrasts strangely with the commodious and beautiful houses erected by School Boards in our own day. In front of this interesting building stand two splendid specimens of the common yew, supposed to be the same age as the house they now so completely overshadow. Another house in the village bears the date 1737. On either side of the avenue leading to the Parish Church there is a fine row of *Cedrus Deodara*, planted in 1853 by the Dowager Lady Ruthven, widow of James, the sixth Baron. They were reared in Freeland Garden from seed brought from India by the Hon. William Melville, brother of the late Earl of Leven. The church is a long narrow, old-fashioned building, and is supposed to be very old. Part of the east wall is older than the other portions, and is believed to be the remains of a former chapel. The churchyard contains some curious tombstones, the most interesting, perhaps, being that known as the "Covenanter's Stone," which bears this inscription:—

"Here lies Andrew Brodie, wright in Forgandenny, who at the break of a meeting, Oct. 1678, was shot by a party of Highlandmen commanded by Ballechin, at a cave's mouth, flying thither for his life, and that for his adherence to the Word of God and Scotland's Covenanted Work of Reformation."

On the left hand side of the door leading to Freeland Gallery, there is another curious stone bearing the date 1369, with this verse:—

"All men think on zovr dying day
Zit joy to die to live for ay."

In 1774, another stone was placed on the right hand side bearing the same lines, with the spelling modernised as follows:—

"All men think on your dying day,
'Tis joy to die, to live for ay."

Both the church and surrounding burial-ground occupy a very

romantic situation. Ecclesiastical events constitute an important part of the history of Forgandenny. Mr John Orme, we are told, one of the ministers of Forgandenny, was deposed shortly after the accession of Charles II. for refusing to conform to Episcopacy. Local tradition says that the Lord Ruthven of that day built a place of worship for the ejected minister within the policies of Freeland. There are persons still living who distinctly remember of a chapel, all properly fitted with pews, standing at the west end of the lawn in front of Freeland House. It was taken down about fifty years ago. There is little doubt but that must have been the building in which Mr Orme preached after his deposition. The report of such a church is further confirmed by Wodrow, who states that in 1662 Lord Ruthven was fined £4800 for attending conventicles and harbouring outed ministers.

The present approach from the Dunning Road has itself witnessed a good deal of change. From an old plan in the possession of the present proprietor, bearing no date, we learn that the ample space between the two rows of lime trees was originally laid out as a flower garden in the elaborate Dutch style,—a circumstance which fully accounts for the extraordinary width of the avenue. This avenue was planted directly in a line south of the old house, so that the Dutch garden would be seen from the windows. The whole of the policies, indeed, were at one time laid out in accordance with the Dutch system so much in favour in the beginning of last century, and of which Dupplin was the best example in Perthshire, so far, at least, as the remains of the design can be traced. At Freeland, however, we are in a better position to comprehend the original design, as a plan, dated 1783, and also preserved by Mr C. L. Wood, shows the scheme in its entirety. The design is at once simple and beautiful. The whole of the policies are, first of all, laid out in one large square, the square being enclosed on all sides by an avenue of trees. The outer square is then subdivided into twelve smaller squares of about 12 acres each, the boundaries being likewise avenues of trees. At all the points of intersection, the avenues are made to form roundels, which at once give character and variety to the design. The centre avenue, running from north to south, is very much wider than the others, and has three rows of trees on either side,—the points of intersection forming magnificent roundels. This avenue generally goes by the name of the Broad Avenue, but in the old plan to which we refer it is designated the Great Cross Avenue. Without the aid of this plan, it would now be impossible to trace the remains of the original design; but, with its assistance, we can easily follow the line of the numerous avenues which at one time intersected the policies, and can recognise the position of the roundels at various points from some huge solitary tree, the last of a once perfect circle, growing off the direct line of the avenue.

On entering the policies by the main approach from the Dunning Road, our attention is at once diverted from the great lime trees, to the unusual width of the avenue. It is exactly 85 feet in width, and 165 yards in length. The trees are all close upon 100 feet in height,—one which we tested with a dendrometer being 97 feet high. The heaviest of these trees is of the American variety. The girth above the swell of the roots, being 2 feet from the ground, is 12 feet; and at 5 feet from the ground the girth is 11 feet 1 inch, the bole being 12 feet. It was the centre of this avenue that was formerly laid out as a Dutch flower garden,—the kitchen garden being at this period intended to be on the east side of the avenue, as shown by the original plan. At the end of the avenue we obtain access to the original kitchen garden and orchard (shown on the original plan as an orchard only) through an arch of Portugal laurel; and on the south side of this floral opening there is a fine specimen of *Robinia pseud-acacia*, and a very large tree to be grown in Scotland, the stem girthing 4 feet 5 inches. It is about 50 feet high, but it inclines considerably towards the east, owing to its having been overborne by the surrounding shrubbery. On the opposite side of the avenue from this tree is a particularly handsome variegated sycamore. In exploring this portion of the policies, we came upon the beautifully laid-out family burying-ground, the sacred spot being overshadowed by luxuriant shrubbery and tall trees, the most noteworthy of the latter being a beautiful shining-leaved Spanish chestnut (*Castanea vesca glabra*).

Leaving the hallowed spot which reminds us of our common end, we have only to proceed a few yards to meet with the picturesque ruins of the old home of the Ruthvens. The Ruthvens were a very ancient family, and, although they have now no property in Perthshire, they at one time held a very large tract of land in different parts of the county. Their origin is traceable to Thor, a person of Saxon or Danish blood, who settled in Scotland under David I., and whose descendant was the Hon. Alex. Ruthven of Freeland, younger son of William, second Lord Ruthven, ancestor of the Earls of Gowrie. He died in 1599, and was succeeded by his eldest son, William Ruthven of Freeland, who married Elizabeth, eldest daughter of Sir William Moncreiffe of Moncreiffe. He succumbed to the plague of 1608, and was succeeded by his only son, Sir Thomas Ruthven of Freeland, who was elevated to the Peerage in 1651, under the title of Lord Ruthven of Freeland. His Lordship married Isabella, daughter of Lord Balfour of Burleigh, by whom he left a son and three daughters. The son, David, who was a Lord of the Treasury, died without issue in 1701, when the barony devolved upon his niece, the Hon. Isabella Ruthven, as first Baroness. The barony has since devolved upon several of the female members of the family, the last Baroness dying in 1864, and was succeeded by her grandson, Walter James, the eighth and present Baron Ruthven, from whom the property was purchased by Mr C. L. Wood. The widow of the sixth Baron is still living at Winton Castle, parish of Pencaitland, East Lothian, to which she removed after her husband's death in July 1853. The ruins of the old home of the Ruthvens, which lie at the north end of the lime avenue we have already described, show that the building must have been a most substantial one. It was burned on the 15th March 1750, and along with it was consumed the patent containing the precise specification of the honours of the house of Ruthven. There does not appear to be any record to show how the house came to be destroyed by fire, whether its destruction was accidental or the result of a raid. The building has evidently been a stronghold, and erected at this spot probably on account of the proximity of a splendid spring, known as the Lady's Well, and which may have originally been enclosed within the north-east corner of the house. The water from the Lady's Well was, in olden times, supposed to cure certain diseases, but we have not heard what these diseases were, and people used to come long distances to drink and carry away the health-giving water. A bright little stream also flows along the northern side of the old house, and passes through a den of great beauty. All that remains of the upper part of the building is the front entrance, which consists of two upright pillars of polished ashlar, and a massive but comparatively plain lintel. The main portion of the building appears to have been of rubble, the walls being of great thickness. At the door, the wall measures 3 feet 9 inches thick, the width of the doorway being 3 feet 6 inches, and the height, 7 feet 3 inches. On the eastern side of the ruin, a portion of the entrance into a dismal-looking dungeon is still open, the lower part being filled up with loose earth. On proceeding through this uninviting aperture, we can see that the house has been built on strong arches, and must have been well-suited to the troublous period at which it was built. On the east side of the main entrance,—the only one, indeed, which now remains,—and within the very portals of the door, grows a large elm tree, richly clothed with ivy from base to summit. This tree must have sprung into existence shortly after the destruction of the house, as it has every appearance of being about 130 years old. Including the ivy, the girth at 5 feet from the ground is 5 feet 9 inches, the bare bole under the ivy being exactly 5 feet.

Leaving the beautiful ruins of the old house, we proceed down the Den, and are soon lost in admiration of the many fine trees which line its sides. The stream which flows through the little dale is of inconsiderable size, but the water, bright and clear as from a spring, leaps from stone to stone, and produces the most harmonious music as it rushes along to swell the volume of the Earn. On our left hand, and immediately opposite the stables, is a beautifully-furnished specimen of *Abies Canadensis*, 40 feet in height. On each side of the Den are several varieties of fine old timber. The first to attract attention is a venerable elm in the last stage of decay, firmly held in the fond embrace of the ever-verdant ivy. Including its epiphytic parasite, which adds considerably to the magnificence of the ruin, the girth of this tree is 11 feet 6 inches at 5 feet from the ground. Immediately opposite this elm, on the east side of the burn, is a grand silver fir, girthing 13 feet above the swell of the roots, and 11 feet 7 inches at 5 feet from the ground. Continuing our walk towards the north, we pass several fine old yews, whose branches meet overhead and form a Gothic-like arch across the path. Beautiful rustic bridges span the burn at intervals, and add much to the charm, as well as convenience, of the Den. At the north end of the Den there is a fine avenue of beech trees, every tree being about 100 feet in height. One of those round which we tried the tape-line girths 12 feet 6 inches above the swell of the roots, and 10 feet 7 inches at 5 feet from the ground,—the bole being fully 30 feet. Below the icehouse in this avenue, and just within the borders of a thriving young plantation, stands a beech which girths 16 feet 6 inches above the swell of the roots, and 14 feet 6 inches at 5 feet from the ground. The height of the bole is 33 feet, the total height of the tree 100 feet, and the spread of branches 88 feet. An ash in this neighbourhood girths 11 feet 10 inches at 1 foot from the ground, and 10 feet 4 inches at 5 feet,—the height of the tree being fully 100 feet. Immediately to the north of the house is a grand beech, girthing 15 feet 1 inch at the narrowest part of the bole, which is altogether about 12 feet. Above the part at which this measurement was taken, the bole swells very considerably, and the tree afterwards breaks out into eight great branches. The entire height is about 100 feet, and the spread of branches 106 feet. A few yards from this tree is a rather remarkable Spanish chestnut, no less than six trunks springing from the main stem, which at one time had been cut above the root. The girth of the whole is 25 feet 2 inches, the largest of the six stems girthing 7 feet at 5 feet from the ground. A little farther in the direction of the mansion is a handsome larch, and a few yards behind the house is a variegated English oak (*Quercus sessiliflora variegata*) of great size. Above the spring of the roots, about 2 feet from the ground, the girth is 11 feet 10 inches, and at 5 feet from the ground, the girth is 10 feet 6 inches, the spread of branches being 88 feet and the height 84 feet.

The mansion-house of Freeland, which we have now reached,

FREELAND HOUSE.

is partly old and partly new, and presents a magnificent appearance. The older portion appears to have been erected shortly after the destruction of the former house by fire, and is a plain, substantial structure. Since the original portion was erected, it has received no less than four additions,—the first about fifty years ago,—all of them being made to harmonise with each other. At one time the front of the old portion was of whitewashed rubble, but some years ago it was completely modernised by the erection of a polished ashlar facing. Several important additions have been made since Mr Wood came into possession of the property, the most extensive of these being the erection of a large and handsome wing at the west end, to be used as a hall or chapel. The mansion-house stands 160 feet above the level of the sea. The highest point upon the estate —" Castle Law," the site of a Pictish fort—is 900 feet above sea-level, and from here a magnificent view can be obtained, stretching from the sea below Dundee on the one side to Ben Lomond and Ben Venue on the other.

A splendid beech avenue, 25 feet wide, starts at the new western wing of the mansion, and proceeds northwards for 70 yards. All the trees range from 110 feet to 120 feet high, and the girth of the largest specimen is 13 feet at 1 foot from the ground, and 11 feet 3 inches at 5 feet. Another girths 12 feet above the swell of the roots, and 10 feet at 5 feet above the ground. A third girths 10 feet 9 inches above the swell of the roots, and 9 feet 4 inches at 5 feet from the ground. Near the north end of the avenue there is a very beautiful beech girthing 9 feet 9 inches above the swell of the roots, and 8 feet 10 inches at 5 feet from the ground, with a clean, straight stem of 33 feet in height. To the south of this avenue there are three very fine specimens of hornbeam; and at the end of the avenue are two fine old spruce trees about 110 feet high, the girth of the largest being 10 feet 6 inches at 5 feet from the ground. The limbs grow horizontally for several feet, and then suddenly take a direct upward tendency, and have become miniature trees. About 120 yards north of Freeland House there is a very curious oak, having a conical excrescence thrown out from the base of the trunk, the periphery of which is perfectly circular from the level of the ground. At the ground level it girths 14 feet 2 inches; at 1 foot up, 12 feet 8 inches; at 2 feet, 10 feet

1 inch; at 3 feet, 7 feet 9 inches; at 5 feet, 5 feet 7 inches; and at 6 feet, 5 feet 4 inches, after which it takes two spiral turns, from west to east, against the course of the sun, and then assumes the normal form, the total height being 63 feet. This tree at once draws the observer's attention by its grotesque appearance, and is well seen on the left-hand side of the back road going to Freeland House from the west. What is no doubt the most remarkable elder tree in Perthshire is to be found at Freeland, growing within a few yards from the house of Mr Duff, the land-steward, and within the sight of passers along the Perth and Dunning Road. The growth is characterised by a peculiar spiral twist, exactly the same as may be seen in a gun barrel. At 2 feet from the ground the girth is 5 feet, and at 5 feet the girth is 5 feet 3 inches. It has a bole of 7 feet, after which the tree branches into two limbs. The limbs were partially broken many years ago, and although the tree is completely denuded of bark, sprigs still continue to sprout from the extremities. One of the branches girths 3 feet. Its age must be exceedingly great, and in its younger days it may have yielded many quarts of that delicacy which the old song tells us Mrs Jean was making when the Laird o' Cockpen presented himself on his matrimonial mission.

A very short walk takes us from the policies to the heart of the woods. The total extent of the estate is 3000 acres, and of this about 270 acres are planted with wood. The plantations are full of magnificent specimens of all the common varieties. In Kinnaird Wood, on the south side of Dunning Road, some natural-sown silver fir seedlings were found four years ago, the first we have heard of in Perthshire. There is a large quantity of splendid Scots fir in this wood. The following are amongst the largest specimens:—10 feet above the swell of the roots, and 8 feet 11 inches at 5 feet from the ground, with a clear stem of 26 feet, and a total height of 83 feet; 9 feet 7 inches above the swell of the roots, and 9 feet 3 inches at 5 feet from the ground, with a clear stem of 40 feet 6 inches, and a total height of 75 feet; 9 feet 9 inches at 2 feet, and 9 feet at 5 feet from the ground, with a bole of 30 feet; and another 9 feet 3 inches at 1 foot, and 8 feet 8 inches at 5 feet, with a bole of 50 feet.

At the old kennels at Kinnaird there is a venerable ash, quite hollow in the centre. It is rather curious that the sites of old mansions and homesteads are more frequently marked by aged ash trees than by any other variety. The ash is said to have been the first barren, or non-fruit-bearing, tree planted by our ancestors round their dwellings, and in many cases these venerable specimens are the only indications that the spot was once occupied by a happy homestead or a lordly mansion. Sometimes, however, the oak is found to take the place of the ash; and the observant eye of Perthshire's greatest poet, and Scotland's second Burns, Robert Nicoll, has noted this circumstance in one of his most homely pieces :—

> "Ae auld aik tree, or maybe twa,
> Amon' the wavin' corn,
> Is a' the mark that time has left.
> O' the toun where I was born."

The shell of the old ash at Kinnaird girths 13 feet 6 inches at the narrowest part, the bole being 25 feet. The Law, in front of Freeland House, contains several grand silver firs. One of these girths 15 feet 3 inches above the swell of the roots, and 12 feet 5 inches at 5 feet from the ground. There is a bole of 74 feet, and a considerable portion of the top has been broken off. Another of the same species girths 14 feet 6 inches above the swell of the roots, and 12 feet at 5 feet from the ground, with a bole of about 100 feet, and an entire height of 110 feet. A very fine larch also grows here, directly in front of the mansion. Clear of the swell of the roots, the girth is 13 feet 6 inches, and at 5 feet from the ground the girth is 11 feet, —the bole being 21 feet, and the entire height of the tree 92 feet. Like one of the spruce trees previously noticed, the limbs grow horizontally for a few feet and then turn upwards, giving the tree an unique appearance. Near this place there is a very ancient arbour, evidently constructed in connection with the original mansion. The walls are of extraordinary thickness, and the arch is beautifully formed. Passing along the east approach we come to Leslie Hill Wood, where there is a quantity of fine larch, Scots fir, and spruce in a thriving condition. On the way from Leslie Hill to the river side we pass through an avenue chiefly of oak. One of the oaks at the end of the avenue girths 13 feet 6 inches at 2 feet from the ground, and 10 feet 6 inches at 5 feet from the ground, with 18 feet of bole—an entire height of 50 feet, and a fine umbrageous head. This tree was probably one of a roundel. On the south side of a strip of plantation near Boatmiln Farm, called, "The Ass Park," there was once a row of oak trees of large size, a few of which are still standing. The tree nearest to the Station Road girths 13 feet above the swell, and 11 feet at 5 feet from the ground, with a spread of branches of 86 feet, a clean stem of 13 feet, and a total height of 83 feet. A few others reach almost the same dimensions. In this strip there are also several large Spanish and horse chestnuts, one of the Spanish variety girthing 12 feet 6 inches above the swell of the roots, and 10 feet 3 inches at 5 feet, with a total height of 64 feet. The Crow Wood has recently been planted with spruce, larch, and silver firs, but there is also some fine old timber intermixed. A Scots fir girths 11 feet 8 inches at 1 foot, and 11 feet 6 inches at 5 feet from the ground, with a bole of about 16 feet, from which spring three large branches—a fourth having been broken off by a storm some years ago. It has a peculiarly twisted and gnarled growth, and its general outline closely resembles a huge umbrella. It is, however, a noble specimen of the true Scots fir. Proceeding westwards, we pass a quantity of large silver firs, with stems ranging from 50 to 70 feet, as well as many fine specimens of oak, sycamore, and beech trees. Our way now lies through several avenues, including the Broad or Great Cross Avenue already referred to, all of which contain grand trees. The Great Cross Avenue principally consists of oaks, all of which are of large size. One of these, opposite Boatmiln Road, girths 11 feet 9 inches above the swell of the roots, and 9 feet 9 inches at 5 feet from the ground, with a clean bole of 15 feet, a spread of branches of 80 feet, and a total height of 75 feet. A few very good Scots firs also grow here, one on the east side of the avenue girthing 9 feet 6 inches above the swell of the roots, and 8 feet 6 inches at 5 feet from the ground, with a height of 64 feet. Another Scots fir on the west side of the avenue girths

8 feet 8 inches above the swell of the roots, and 8 feet at 5 feet, the total height also being 64 feet. In the avenue leading to Gallow Muir, and at a short distance from the Kennels, we pass two sycamores, regarding which it requires no great stretch of the imagination to fancy that they must have an interesting history if it were known. They are the only trees of this kind in the neighbourhood, and are placed exactly opposite each other on the roadside, and were, in all likelihood, planted as a memorial of some important event, probably a marriage, in the Ruthven family, many years ago.

Since Mr Wood came into possession of the property in 1873, he has planted many acres of wood, principally in small patches, for shelter and ornament. Some time ago he laid out a very nice nursery to supply the woods and policies. The stock comprises all the common varieties, as well as many of the newer introductions from California, Japan, &c., which have successfully withstood our rigorous climate. Under the superintendence of Mr Duff, who procured the bulk of the plants at the Perth Nursery, the young trees have come through the recent severe winters in wonderful condition, and are in every way most promising. Amongst the other improvements effected by Mr C. L. Wood is a new kitchen garden. This indispensable adjunct to a country house is situated some distance to the west of the mansion, and has a splendid southern exposure. The new garden is $1\frac{1}{2}$ acres in extent, is surrounded by a high brick wall, and is furnished with all the latest improvements in vineries, conservatories, and forcing-houses, all of which are turned to good account by Mr Routledge, the head gardener. The old garden, which still exists, consists of $3\frac{1}{2}$ acres. A fine beech avenue leads from the new to the old garden, the avenue being a quarter of a mile long and 16 feet wide, and the trees are of similar dimensions to those in the avenue to the west of Freeland House.

The present proprietor of Freeland, we have much pleasure in adding, is a most enthusiastic admirer of the woods, and takes a lively interest in their management. While he is doing much to extend their area by planting judiciously, he has a special place in his heart for the fine old trees which stud the property in all directions, and immeasurably enhance its beauty and its value. Many of these trees are not only large in size, but perfect in form, and no one can see them without a glow of admiration, and a feeling of gratitude that they are the property of one who appreciates their value, and understands the importance of preserving them so long as their beauty and their usefulness endure.

XIV.—ROSSIE.

The Oliphants—The Founder of Rossie—The Mansion-House—The Policies—Large Silver Firs—The Dovecot Park—Cedar of Lebanon—Sycamore Trees—A Royal Souvenir—An Old Marriage Record.

ROSSIE, situated in the parish of Forgandenny, and quite close to the village, is the seat of one of the branches of the ancient and historic family of Oliphant, the present proprietor being Mr Thomas T. Oliphant. The Oliphants occupied a conspicuous place in the history of Scotland from a very early period, the first of the name who appears authentically in the annals of the country being David de Olifard or Oliphant. He was a godson of David I., and saved that King's life after the memorable siege of Winchester in 1141. For this service the King gave him a grant of lands in Roxburghshire, and a few generations afterwards the family also became possessed, by marriage and royal favour, of extensive properties in Perthshire and Fifeshire. There were many distinguished men of the name; but the one best known in history is Sir William Oliphant (fifth in descent from David Oliphant), who, with Sir William Wallace and Sir Simon Fraser, were the only Scottish nobles who refused to swear allegiance to Edward I. at the Parliament held by that King at St Andrews in 1304. Later in the same year Sir William Oliphant defended Stirling Castle against the English army, and only yielded after many months of desperate fighting, when starvation and overwhelming numbers compelled a surrender. He died in 1329, and his tomb is still to be seen in the Parish Church of Aberdalgie. His son, Sir Walter Oliphant, was, as Craufurd says in his *Peerage*, "so gallant and brave a man that his merit preferred him to a marriage with the Lady Elizabeth Bruce, daughter of Robert I. and sister of David II., Kings of Scotland." From him descended many men of note, and eventually in 1458 Sir Laurence Oliphant (eleventh in descent from David above mentioned) was created Lord Oliphant. To him succeeded John, second Lord, both of whose sons—Colin, Master of Oliphant, and Laurence, Abbot of Incheffray—were among the "flowers of the forest" slain at Flodden on the 9th of September 1513. Colin left two sons—Laurence, afterwards third Lord, and William Oliphant of Newtoun and Thrumster. The main line of the family continued until 1751, when William, eleventh Lord, died without issue, declaring Laurence Oliphant of Gask to be his heir-male and successor to the title. He, however, and also his son, Laurence, were at that time abroad, having been attainted for the share which they had taken in the '45 ; and the title was never claimed by the family until 1839, when James Blair Oliphant of Gask, was served heir. The claim does not appear to have been prosecuted further, and a few years afterwards he died without issue, thus ending in the direct male line the senior branch of this ancient family ; but it is still ably represented by Mr Kington-Oliphant of Gask, who succeeded through his mother, a sister of the late laird. The claim of the Gask family to the title was through the William Oliphant of Newtoun and Thrumster already mentioned. He, as is shown above, was the second son of Colin, Master of Oliphant, who was killed at Flodden. William Oliphant had five sons, the eldest of whom was the direct ancestor of the Gask family. Of the second, fourth, and fifth sons nothing further is known than that they were alive in 1585, 1588, and 1587 respectively ; but the third son, Alexander, was the father of Laurence Oliphant of Newtoun, whose three sons—Laurence, Thomas, and James—were the direct ancestors of the Oliphants of Condie, Rossie, and Kinnedar respectively. It is seldom that a direct male descent of twenty-three or twenty-four generations can be shown, but we believe it is possessed by the present representatives of these three branches of a family, which for centuries held a high position in this country.

The property of Rossie Hill, or Rossie Ochil, as it is now called, was purchased in 1583 by William Oliphant of Newtoun from Andrew Blair, who married a daughter of James Rollo of Duncrub. William Oliphant's grandson, Laurence, being in some way displeased with his eldest son (Laurence, ancestor of the Oliphants of Condie), partially disinherited him, and left the lands of Rossie Ochil to his second son, Thomas, from whom they have passed by direct descent, through seven generations, to the present proprietor, also, as we have stated, a Thomas Oliphant. For two generations the family lived at Rossie Ochil, which lies amongst the hills about three miles due south of Rossie House; but in 1727, Robert Oliphant, the then laird, bought part of the charterslands of Forgandenny from Fotheringham of Powrie, and had it, along with the other land belonging to him in the parish, re-erected into the barony of Rossie Ochil. The total extent of the property is 1660 acres, Rossie Ochil, the original estate, being 1200 acres; Low Rossie, on which are the mansion-house and garden, 400 acres; and the plantations, 60 acres.

Previous to this time the house was called Forgandenny, or Forgan House, and was considerably smaller than at present, having been altered and enlarged to its present form by the next laird about the beginning of last century. It is a plain oblong block, "harled" on the outside, and containing three handsome and lofty public rooms, and a sufficiency of bedroom accommodation. The stables and other offices are ample, and there is a good old-fashioned garden, almost larger than the requirements of the house, being more than two acres within the walls, and having vineries, peach-house, and conservatory. The house is well situated, facing the south, and having lovely peeps of the Ochils to the south-east and south-west through the surrounding trees. Of these there are many fine specimens in the policies. Immediately upon entering by the north lodge, we came across a magnificent silver fir at the top of the west bank of the den of Rossie. At 2 feet from the ground the girth is twelve feet 8 inches, and at 5 feet it is 10 feet $9\frac{1}{2}$ inches. This is only the first of a series of great silver firs to be met with on almost every side. The nearest one to the above girths 14 feet above the swell of the roots, and at 5 feet from the ground the girth is 11 feet 10 inches. After a bole of 34 feet, the tree breaks into two forks, and reaches a total height of 87 feet. A rustic bridge close by leads directly into the village of Forgandenny, and by the side of it grows a very nice oak, not very remarkable for its size, but having a beautifully clean bole of 27 feet,—the girth at 2 feet from the ground being 11 feet, and at 5 feet, 9 feet 3 inches,—the entire height being 100 feet. To return to the silver firs, however, the largest of these grows at the west side of the house, the girth at 1 foot from the ground being 13 feet 8 inches, and 12 feet 8 inches at 5 feet. Immediately below the part at which this measurement is taken a large branch issues from the stem, taking an upright direction, and afterwards intertwining with another large limb which springs from the tree about 15 feet above the ground. At the height of 17 feet another large branch is thrown out from the trunk, taking a horizontal direction for about 10 feet, after which it assumes an upright habit. Another silver fir, a few yards from this, and within the flower garden to the west of the house, girths 12 feet 9 inches at one foot from the ground, and 11 feet 9 inches at 5 feet, with a bole of 17 feet. Both trees, which are about 86 feet high, are showing symptoms of decay. To the west of these there is another very fine silver fir, in a healthier state, but probably about the same age, with the branches feathering beautifully down to the ground. From this point we saw in the centre of a large park, also to the west of the house, a highly ornamental octagonal tower, built upon an artificial mound, intended as one of those dovecots, at one time so much in vogue as an adjunct of a mansion-house. This park is known as the Dovecot Park; and to the west of it are the remains of a once grand row of silver firs, which here lined the public road to Dunning, when it ran, about a century ago, through the policies of Freeland, Rossie, and Invermay. On the east side of the south approach, and close to the house, there is a very handsome and shapely cedar of Lebanon, girthing 12 feet at the narrowest part of the bole, being about 4 feet from the ground. The bole is about 12 feet, but it branches out at 6 feet, and the entire height of the tree is 64 feet. The beauty of the tree is to a considerable extent marred by a Portugal laurel, also a very old tree, the reason doubtless of its preservation. There are also a number of very good scyamore trees, between 200 and 300 years old. The largest of these girths 12 feet 10 inches clear of the swell of the roots and numerous sprouting branches, and 11 feet 6 inches at 5 feet from the ground. There is a bole of 15 feet, at the top of which the tree divides itself into two large limbs, and reaches a total height of 80 feet. Another scyamore girths 11 feet 7 inches above the swell of the roots, and 10 feet at 5 feet from the ground.

A short distance from the house there are three oaks with an interesting history, one being on the west and two on the east side of the lawn. According to the tradition of the family, these trees were grown from acorns which George III. had played with when a boy. The acorns were given to the present laird's great-grandfather, Robert Oliphant, by John, third Earl of Bute, who was tutor to Prince George (George III.), and became Prime Minister in 1762. The present laird's grandfather, Stuart Oliphant, was Lord Bute's godson, and named Stuart after him. In these circumstances, the oaks are regarded with some little veneration, and give promise of surviving in healthy vigour for many years to come. The largest of the three girths 11 feet 1 inch at 1 foot from the ground, and 10 feet 5 inches at 4 feet from the ground,—there being a good bole for about 20 feet, and an entire height of 78 feet. One of the others has a very fine clean stem of 30 feet, the girth at the narrowest part being 9 feet. Another object of some interest is a curious old stone set into the wall at the back of the stables, and recording the marriage of William Oliphant to Jean Drummond in 1659, but the stone must have been at some time brought from the older family residence in the Ochils.

Besides the larger trees we have mentioned, a number of specimens of the newer *coniferæ* have been recently planted, principally *Wellingtonia gigantea* and *Cedrus Deodara*. Having made the circuit of the property, we left by the south approach, where there is a lodge, and an ornamental gate hung on massive pillars of hewn freestone, surmounted by ornamental vases. The boundary leading up to the gate consists of a strong whinstone dyke, 6 feet high, the whole presenting a neat and substantial appearance.

XV.—CONDIE.

Situation of the Property—Destruction of the Mansion-House by Fire—Scenery of the District—Battle of Culteuchar—The Policies—Remarkable Trees—The Plantations.

THE property of Condie has, for nearly three centuries, been in the possession of another branch of the once powerful family of Oliphant, whose career was sketched in our former paper, the present representative of the family being Lieutenant-Colonel Laurence James Oliphant. Only a small portion of this property now belongs to Colonel Oliphant. In 1881, the entire property was exposed to sale, when portions were bought by various neighbouring proprietors. The Earl of Kinnoull bought that part known as the Newtoun of Condie, which includes the site of the mansion-house; Mr C. L. Wood of Freeland bought the Culteuchar part; Lord Clinton bought Benzian, which adjoins the estate of Invermay; and Mr Calder of Ardargie bought Auchtenny and Middlerigg. The remaining portions, including High Condie, and other farms, amounting to about 1500 acres, are still the property of Colonel Oliphant. The property lies in the parishes of Forgandenny, Forteviot, and Dunning, and, as originally constituted, comprised four distinct estates, Newton of Condie, Culteuchar, Benzian, and Upper Condie, Auchtenny and Middlerigg, stretching from the River Earn, about one mile west of Forgandenny Station, to the top of the Ochils, about four miles distant from Kinross,—being situated partly on the slopes of the Ochils and partly in the vale of Strathearn. The house stands, or rather stood, less than one mile west from the village of Forgandenny, and within a short distance of the Earn. It dates from the sixteenth century, and was considerably enlarged about fifty years ago, but it was burned down upon the 13th March 1866, and has never since been rebuilt. It faces the north, looking across the river on the beautiful woods surrounding Dupplin Castle, and the village of Aberdalgie, in the Parish Church of which still lie, under a well-preserved tomb of black marble, the remains of Sir William Oliphant, the gallant defender of Stirling Castle against the English in 1304. The walls of the house are still standing, and might be partly utilised for its reconstruction on the same site, which had been carefully selected, being situated on a nice piece of rising ground, overlooking a wide stretch of most pleasant country. The scenery in the immediate neighbourhood is exceedingly charming, some of the finest bits of Perthshire being on the one side; while Loch Leven, the Rumbling Bridge, and the Devon Valley are within easy reach on the other. From Culteuchar, Clow, and Middlerigg Hills, magnificent views can be had of the valleys of Upper Strathearn, the Pow, the Strathmore and Grampian Hills on the west, north, and east; the Carse of Gowrie on the east; and the Lomonds and Cleish Hills, near Kinross, on the south. It was on Culteuchar Hill that the well-known "battle" of Culteuchar was fought. This took place in 1678, when a party of "wild Highlanders" suddenly made an attack upon a conventicle, killing Andrew Brodie, wright, whose memory is perpetuated by a tombstone in the churchyard of Forgandenny, as mentioned in our description of Freeland.

The ruined mansion-house is surrounded by finely-timbered policies of oak, beech, and sycamore, and is approached from the main road, leading to Dunning, by a nice carriage drive bordered with shrubs and trees. There are two entrances to the approach from the Dunning Road,—one curving from the west and the other from the east,—the intervening space being occupied by a small plantation of mixed wood of 40 or 50 years' growth. Each side of the approach is planted with bay laurel, with a few hollies of different varieties interspersed. From the point of junction we proceed northwards along the approach, having on the west side a strip of mixed plantation, and on the east a row of oaks. Continuing along the approach, we reach a line of very handsome ornamental trees of the newer varieties, including several tall and shapely specimens of *Wellingtonia gigantea, Cedrus Deodara, Araucaria,* &c., all apparently about 25 years of age. One of the araucarias is an exceptionally good tree, the branches being complete and well formed, and the tree in luxuriant health. The height is 25 feet, and the stem at 3 feet from the ground girths 2 feet 6 inches. Another araucaria, in the immediate vicinity of the house, is 27 feet high, and 2 feet 8 inches in girth. The other ornamental trees are also in good health, and making fair progress. There are a number of fine silver firs about 80 feet high, the girths of those we measured being 12 feet 2 inches above the swell of the roots, and 10 feet at 5 feet from the ground; 10 feet 3 inches above the swell and 8 feet 6 inches at 5 feet. The hardwood trees are both numerous and of fairly large size. One of the lime trees girths 10 feet 3 inches at the narrowest part of a bole of 18 feet, the entire height being 70 feet. There are several sycamore trees worthy of notice. A very fine row lines the road leading to the home-farm from the east side of the garden, the best specimen girthing 11 feet 9 inches at 2 feet from the ground, and 10 feet 3 inches at 5 feet, with a bole of 22 feet, and an entire height of 80 feet. Another sycamore at the south end of the line girths 15 feet above the swell of the roots, and 13 feet 4 inches at 5 feet, with a bole of 8 feet, and an entire height of 60 feet. It is a much older tree than the others in the line, but it is still in vigorous health. A fine beech near the house girths 12 feet 6 inches above the swell of the roots, and 11 feet at 5 feet from the ground, with a bole of about 8 feet, an entire height of 80 feet, and a grand spread of branches of 88 feet. An ash a little to the north-west of the house girths 12 feet 4 inches at 1 foot from the ground and 10 feet 8 inches at 5 feet, with a bole—slightly curving to the north-west—of 20 feet, and an entire height of 90 feet. An oak in the same neighbourhood girths 13 feet 6 inches at 1 foot from the ground and 11 feet 6 inches at 5 feet from the ground, with a bole of 22 feet, a height of 70 feet, and a spread of branches of 80 feet. Another girths 11 feet 8 inches at 1 foot and 10 feet at 5 feet from the ground, and about the same height as the other. But the old giant,— the pride of the family,—a magnificent oak, where in days gone by the stirrup-cup was quaffed by the parting guest, was destroyed, a few years before the house was burned.

There are several large plain-leaf hollies and old yews immediately adjacent to the east end of the house. The hollies are 44 feet high, and the girth of the one we measured as a fair sample is 6 feet 9 inches at 1 foot, and 6 feet 6 inches at 5 feet from the ground. There are four yews altogether, and they are apparently about 150 or 200 years of age. Two of them run east and west at right angles to the walls of the house, the other two running north and south parallel with the walls, evidently in accordance with some old plan.

The entire estate, before it was broken up, consisted of over 3900 acres, of which 336 were woodlands. The timber in the plantations consists principally of oak and oak coppice, larch, and Scots fir, while there are also a number of hardwood trees in the hedgerows and parks, adding considerably to the amenity of the property. The young plantations present a vigorous and thriving growth, showing a suitability of soil and climate for the growth of timber. The garden extends to nearly 2 acres, and is enclosed by a stone wall, but has no glass. It is, however, well stocked with fruit trees,—both wall and standard,—and is not only very productive, but is specially famous for its strawberries.

XVI.—DUNCRUB.

History of the Rollo Family—The Old Thorn Tree at Dunning—The New Mansion-House—The Policies—The Pinetum—Memorials of a Royal Marriage—Remarkable Trees—The Old Castle of Kelty—Notable Trees at the Castle—The Plantations—The Village of Dunning.

In endeavouring to trace the history of the property of Duncrub, we are carried back to a very remote period. The noble family of Rollo—of which the Right Hon. John Rogerson Rollo, Lord Rollo in the Peerage of Scotland, and Baron Dunning of Dunning and Pitcairns in the Peerage of the United Kingdom, is the present representative—traces its lineage to Eric Rollo the Dane, who had obtained a settlement in Normandy as early as the eighth century, and from whom were descended the Dukes of Normandy, in whose line, passing over several generations, we come to William the Conqueror, who became King of England in 1066. Eric de Rollo, a scion of the same stock, accompanied the Conqueror to England in the capacity of Secretary. A portrait of him taken in his 98th year is still in the possession of the family. A descendant of Eric de Rollo came to Scotland in the time of David I., about the year 1130, and obtained from that monarch a grant of houses and land in the Lothians. From him descended John de Rollo, who settled in Perthshire, and founded the family of Duncrub, on obtaining a grant of the lands from Robert Stuart, Earl of Strathearn, afterwards Robert II. He acted as Private Secretary to Robert III., and died in the beginning of the reign of James I. In 1466 Robert Rollo obtained a grant of the lands of Petty from James III.; and on 25th August 1511, his successor, William Rollo, obtained a charter, uniting the lands of Laidcaithy and Duncrub into a free barony, from James IV., just two years before the untimely death of that gallant Prince at the fatal field of Flodden, where Robert Rollo, the son and successor of William, also fell. We next find the members of the family spelling their names with a final "k," for what reason does not appear, but the practice was continued during the fifteenth and sixteenth centuries,—the celebrated Robert Rollok, the first Principal of the University of Edinburgh with whose name the infancy of that famous seat of learning is intimately associated, being a cadet of the family. The next notable scion of the house of Rollo was Andrew Rollok, who took a prominent part in the politics and proceedings of the reign of James VI., by whom he was knighted on 26th June 1621. He received a commission from the ill-fated Charles I., as Sheriff-Principal of the county of Perth, dated Holyrood Palace, 25th September 1633. He obtained a charter of the lands of Kippen, dated 5th February 1639 ; and on the 10th January 1651, Charles II. raised him to the Peerage by the title of Baron Rollo of Duncrub, the patent of nobility bearing that the honour was conferred "in consideration of the antiquity of his family and the constant fidelity of his ancestors to the Crown." His adherence to the Royal cause brought the wrath of Cromwell upon his head, and by the Act of Grace in 1654 he was fined in the large sum, for that period, of £1000. The first Lord Rollo married Catherine Drummond, daughter of the first Lord Maderty, the ancestor of the Lords Strathallan, and by her he had five sons and four daughters. The youngest son, Sir William, said to have been "a young man of excellent parts and unblemished reputation," joined Montrose, and having been captured when his chief was surprised and defeated at Philiphaugh, he was publicly executed at the Market Cross of Edinburgh on the 28th October 1645. The first Lord Rollo died at Duncrub, and was buried in the church of Dunning on the 12th June 1659, being succeeded by his eldest son, James. Previous to his succeeding to the property as the second Lord Rollo, James had received the honour of knighthood from Charles I., in 1642. He is described as being "remarkable for his matrimonial alliances, having successively married the sisters of the two great rivals and most illustrious Scotsmen of their time,—James, Marquis of Montrose, and Archibald, Marquis of Argyle,—both of whom fell victims to the bigotry and misguided zeal of the opposite factions in that distracted period of our country's history." Andrew, the third Baron, succeeded in 1669, and married, in the following year, Margaret Balfour, daughter of the third Lord Burleigh, their initials and shield being still visible on the east gable of the church of Dunning above the doorway. His lot was cast in evil times, the most tragic affair in this period of the family history being the murder of his eldest son, the Master of Rollo, by Patrick Graham, younger of Inchbrakie. The tragedy arose out of the stormy character of the times. In 1690, the year following the Battle of Killiecrankie, a band of Highland robbers plundered the lands of Duncrub, and this led to a feud between the young lairds of Duncrub and Inchbrakie, who, if he did not take an active part in the foray, was suspected of countenancing and befriending the main actors. The young lairds met at Invermay, on 20th May, 1691. There were also present, James Edmonston of Newton, one of whose tenants was found in possession of one of the stolen cattle, and who also bore a grudge towards the Master of Rollo, and a number of others. A quarrel arose, and, at the instigation of Edmonston, who handed Inchbrakie his sword, the Master of Rollo was murdered on the way home on horseback, after supper. Inchbrakie and Newton contrived to detach the Master from the rest of the company, who, on hearing the clash of arms, turned round to find the Master fallen on his knees, with Inchbrakie standing over him. One of the company cried out—"Such a bloody murder was never seen." "I think not so," rejoined Edmonston, "I think it was fairly done," and he assisted Graham to make his escape. The laird of Newton was apprehended, tried as an accessory, and was sentenced to banishment for life, although it is said that he afterwards carried the Royal Standard at the Battle of Sheriffmuir; and even after that lived many years on his own estate in Strathearn. Graham succeeded in getting out of the country, and was outlawed, but returned to Scotland in 1720 on procuring a remission. In consequence of the murder of the Master of Rollo, the second son, Robert, succeeded, and became, in 1700, the fourth Baron Rollo. Like many of his neighbouring proprietors, the fourth Lord Rollo took an active part in the rising of 1715, and was amongst the first who hastened to the unfurling of "The Standard on the Braes o' Mar." He was appointed Colonel of the Perthshire Regiment of Horse,—his friend and neighbour, Laurence Oliphant, younger of Gask, being one of his lieutenants. At Sheriffmuir, this squadron was stationed on the extreme left of the Highland army, and had to bear the brunt of the attack, being opposed to the best cavalry of the Royal army, led by Argyle himself. Lord Rollo and his troopers fought with great stubbornness, but had the misfortune to belong to the defeated wing, by which Mar was forced to fall back upon Perth.

Here, we come across an incident of arboreal interest. During the retreat, Blackford, Auchterarder, Muthill, Crieff, and Dunning were burned by the Highlanders, lest they should afford food or shelter for the pursuing forces. This is the most memorable event in the history of Dunning, and it was commemorated by the planting of a thorn tree in the centre of the village. The tree, which is protected by a strong circular wall, has now braved the tempests of more than a century and a half, and, although the symptoms of decay are not awanting, it is wonderfully fresh and green, and, if properly cared for, may flourish for many generations to come. The history of this ancient thorn has been committed to verse by Mr John Philip, a respected parishioner, who died only a few years ago :—

> "So they three hundred Highlanders did bring
> To put in force the edict of the King.
> From Braco, Crieff, and Comrie they came,
> And other parts, to set the town on flame.
> Then Aberuthven, Muthill, and Blackford also,
> In fire and smoke up in the air did go.
> Dunning and Auchterarder shared their fate.
>
> . . . . .
>
> A thorn tree from Pitcairns Den
> For a memorial was planted then,
> And that no evil might the tree befal.
> It was protected by a circular wall ;
> And every year the night before the Fair,
> The Baron's workmen do with skilful care
> Trim and preserve its bowl-inverted form
> And single stem, which long has stood the storm.
> Long may it live to tell to future days
> What Dunning suffered from the Rebel ways."

The wall surrounding the mound is 3 feet 8 inches in height, but an old residenter remembers when it was not so high, and it is probable that there was originally no wall or mound whatever. The tree is about 16 or 17 feet in height from the top of the mound. The bole is 3 feet 3 inches,—the girth at the bottom of the bole being 5 feet 3 inches, and at the top 5 feet 2 inches. The tree is fantastically trimmed, and has rather a striking appearance.

To return to the history of the family, the fourth Baron submitted to the Government in the year following the battle which led to the burning of Dunning and other places on the route of the Highland retreat, and obtained the benefit of the Act of Grace. He survived till 1758, but does not appear to have joined in the rebellion of 1745. He was succeeded by his eldest son, Andrew, who, having adopted the profession of arms, greatly distinguished himself at the Battle of Dettingen, in 1743, and was promoted for his bravery. He became Major of the 22nd Foot in 1750; Lieutenant-Colonel in 1756; and commanded the regiment at the Battle of Louisberg in 1758. He was despatched to the assistance of General Murray in reducing Canada in 1760; took St Domingo on 6th June 1761; and at the taking of Martinique, in January following, the despatch bears that "Lord Rollo, and all the officers, deserve the highest approbation for their animated and soldier-like conduct." He died in 1762, and was buried at Leicester, where a splendid marble tomb, with warlike trophies, commemorates his victories. He married, first, Catherine Murray, a grand-daughter of the Marquis of Athole; and, second, Elizabeth Moray of Abercairny. By his first wife, he had a son, John, who was Major of Brigade in his father's regiment, and who died at Martinique on the 24th January 1762, a few months before his father, who returned to England broken in health. The fifth Baron was succeeded by his brother, John, whose only son became seventh Baron in 1783. He served in the Marines at Pondicherry and Manilla, and died at Duncrub on 14th April 1784, the year after his succession. Before he succeeded he had married a Fifeshire lady, a daughter of John Aytoun, Inchdairnie, who long survived him, dying in April 1817, and leaving a numerous family. John, the eighth Baron, served with the Foot Guards in the campaigns of 1793-4-5, and died 24th December 1846. He was succeeded by his son, William, who married, 21st Otcober 1834, Elizabeth, only daughter of Dr Rogerson of Wamphray and Duncrieff, and died 8th October 1852, leaving an only son, the present laird, John Rogerson Rollo, who is, consequently, the tenth Baron. He married, in 1857, Agnes Bruce, daughter of Lieut.-Colonel and the Hon. Mrs Trotter of Ballindean, by whom he has a large family, the Master of Rollo being the Hon. William Charles Wordsworth.

From the sketch we have given of the noble family occupying the property of Duncrub for so many generations, it will be seen that the Rollos have maintained a prominent position in the history of the country, being invariably in the heart of the stirring events in which the circumstances of the times called upon them to take part. But " peace hath her victories no less renowned than war," and we hope to be able to show that the present upholder of the family name has done, and is still doing, much to make his reign memorable as an improver of the family estates.

Amongst the first of the improvements to which the present Lord Rollo directed his attention was the erection of a new mansion-house. The old house was of great age, the date of its erection being lost in the mists of antiquity; and it was not only a most inconvenient structure, but was out of harmony with the beautiful grounds by which it was surrounded. Previous to the demolition of the old pile, Lord Rollo, in 1858, commissioned the eminent ecclesiastical architects, Messrs Habershon & Pite, of London, to design a little Episcopal Chapel near the mansion. The chapel is a very handsome structure, capable of seating 50 persons. All the windows are of stained glass, and a genuine "dim religious light" pervades the interior. The style of architecture is Early English, and the design displays great taste and accuracy. It was shortly after the chapel was opened that his lordship resolved to pull down the old House of Duncrub, and erect one more in consonance with the times. The designing of the new mansion was entrusted to the same architects who planned the chapel,

and it is built only a little distance from the site of the old
house, in a magnificent park containing some very fine trees.
It is approached by two avenues—one from the road leading
from Dunning Station to the village, and the other from
the Auchterarder Road, to the south of the mansion. The
latter is the principal approach, and the one from which the
Castle is best seen on entering, but they are both of great
beauty. Around the house extensive lawns spread in all
directions, and many pleasing views are obtained from the
windows. The principal front of the house faces the south,
but the western elevation is quite as beautiful; and seen
from a spot taking in both these points, the house presents
a most majestic appearance, and equals anything to be
found in the county. The style of the architecture is Tudor
Gothic richly ornamented, and the stone used is from a quarry
on his Lordship's estate—the windows, doors, &c., being
finished with the beautiful white stone of Dunmore. The
exterior is lofty and imposing—in some parts three storeys high
and in other places two storeys; while the kitchen entrance,
offices, &c., are carried down to a single storey, forming a very
appropriate flank to the main body of the building. The
gables are stepped, and the skyline generally is very effective
from the different elevations of the various parts of the house.
The main entrance is through a magnificent porch, 30 feet in
height, and richly carved throughout. Two huge lamps are
borne outside the porch on heavy buttresses of stone; and on
the front of the porch are medallion busts of Lord and Lady
Rollo cut in stone, both being excellent likenesses. The porch
terminates in four tall pinnacles surmounted by griffins carrying
gilt flags, having the letter "R" cut in them. Over every window
of the house on the ground floor is a baron's coronet, with
tracery carved in stone; while above the dining-room window
the full coat-of-arms of the Rollo family is carved in the most
exquisite manner. Indeed, every window and door of the
house has been taken advantage of for enrichments in carved
work. The total length of the principal front is 272 feet, and
of the west front 160 feet, including the chapel, which forms a
most graceful termination. On this side one of the bay windows

has a handsome flight of steps by which the lawn is reached. There are four fine towers, the principal one over the entrance bearing a lofty flagstaff. Both the exterior and interior of the house are finished in the most expensive and elaborate manner, —no material but the very best having been used, and every salient point richly ornamented with carving. The interior is quite as striking as the exterior. On entering through the porch we have described, attention is first attracted to animals in stone on each side of the doorway,—the one on the right bearing the word "Welcome," and the one on the left the word "Farewell." The main doorway leads into a vestibule about 13 feet square, paved with encaustic tiles, and richly ornamented with carved columns in Dunmore stone. Passing through the vestibule, a Gothic traceried door opens into the grand corridor, the most noticeable feature of the interior, being not only exceptionally spacious but of a unique design. It is 95 feet long by 15 feet 6 inches broad, and no less than 50 feet high to the roof. It is lighted all along the roof by stained glass windows, and otherwise. Two spacious galleries run entirely round it, one above the other, giving the corridor an exceedingly noble appearance. Entering the corridor, a serving passage runs off to the right, entered through an arched Gothic doorway of stained glass. Further on, on the same side, two doors give entrance to the dining-room; and a little further on is the entrance to the grand staircase. The corridor is terminated by a traceried door leading into the chancel, thence into the chapel. On the left side of the corridor are the doors leading into the drawing-rooms, library, &c. The grand staircase is reached from the corridor through two very richly-carved Gothic arches. The stair is about 6 feet broad, with bronze balustrade, and runs round three sides of the hall, which is lighted by a very large traceried window, which can be filled with stained glass. From the centre of the ceiling, which is of richly-carved wood, hangs a very handsome brass corona, lighted by some thirty jets of gas, and having, when lit, a gorgeous appearance. The grand staircase gives access to the first gallery, which runs all along the corridor, at a height of 17 feet from the floor, and is supported by carved stone corbels. The gallery is of wood, richly carved. It is 4 feet 9 inches in breadth ; and the railing —which is of bronze, with highly-polished oak handrail—is 3 feet high. A beautiful feature of this gallery is a large bay window, semi-circular in shape, fitted with a low and comfortable seat all round. Extensive suites of bedrooms are entered from this gallery, which also communicates with the upper gallery, which is 30 feet from the ground, and resembles the lower in every particular, save that it is a little narrower. Both galleries are richly covered with carved work, and are very beautiful indeed. We have said that access to the drawing-room is got from the left side of the corridor. The drawing-room itself is a most spacious apartment, over 33 feet long by 22 feet broad. The ceiling is a very rich one. There are four large windows, and the walls are decorated with an ornamental skirting of carved wood. The drawing-room is one of a suite of rooms which communicate and extend 105 feet in length, the suite comprising the drawing-room, library, second library, and Lady Rollo's boudoir. The library is a fine room, 30 feet by 20, the bookcases constructed in the wall being of carved wood. It is lighted by a large bay window, and over the fireplace is an immense mirror in a setting of rich carving. The second library is 25 feet by 21, lighted by a bay window, and a traceried one filled in with stained glass,—the roof being ornamented, and the walls lined with carved woodwork. Off this room is Lady Rollo's boudoir, 26 feet by 15, also lighted by a large bay window. On the opposite side of the corridor, there is a handsomely-fitted business-room, and the dining-room, a large and lofty apartment, 36 feet by 20. It is lined with carved woodwork, and has one large bay window, with five lights, and two other windows, with two lights each, and transom lights above. On the ground floors there is extensive accommodation of every sort, and in the upper storeys about 50 spacious bedrooms.

The policies which surround this magnificent mansion, although they do not contain much old timber, are of great beauty. About thirty years ago a double avenue, half a mile in length, was planted along the north approach,—the one from the station road,—right up to the house, and the trees have now attained such dimensions as to be both graceful and useful. The outside row consists of oaks and the inside of horse chestnuts, the combination producing a very pretty effect. A short distance south of the house there is a well-stocked pinetum, containing some noteworthy specimens. Amongst these is a splendid *Douglasii* seedling of 1836, but which has, unfortunately, lost its leader. The girth at 1 foot from the ground is 12 feet 6 inches, and at 5 feet, 10 feet 6 inches. A nice *Wellingtonia gigantea* girths 5 feet 9 inches at 1 foot from the ground. The largest silver fir girths 11 feet 4 inches at 1 foot from the ground, and 9 feet 10 inches at 5 feet from the ground,—the entire height being 80 feet. A short distance from this tree there is an early specimen of the handiwork of Laurence Macdonald, the eminent Perthshire sculptor. The figure, which was executed while Macdonald was working as a journeyman at Gask, is representative of the deity whom Milton alludes to as

> "Universal Pan,
> Knit with the Graces and the Hours in dance,
> Led on the eternal spring."

His connection with the "Hours" is illustrated by a sun-dial which he bears upon his head. Pan is not only the personification of deity as displayed in Creation, and especially of flocks and herds, but also the god of woods; so that the presence of the figure here, apart from its great local interest, has a certain appropriateness. The figure is rude, but displays a considerable advance upon some of Macdonald's other earlier productions extant in Perthshire, and gives indications of the genius for which he was afterwards distinguished. A number of newer coniferous trees of different varieties were planted up and down the pinetum upon the 10th March 1870, by the various members of the Rollo family, in commemoration of the marriage of the Marquis of Lorne and the Princess Louise. The trees thus planted are as follows:—*Cupressus Nutkaensis,* planted by the Hon. the Master of Rollo; *Cupressus Lawsoniana,* by the Hon. Agnes C. Rollo; *Cupressus Lawsoniana,* by the Hon. Eric N. Rollo; *Thuja gigantea,* by the Hon. Constance Agnes; *Abies orientalis,* by the Hon. Herbert Evelyn; and *Thujopsis dolobrata* by the Hon. Bernard Francis. All of these trees are thriving well, and are furnished with a plate bearing the name of the variety, by whom the tree was planted, and the event they commemorate.

Proceeding through the policies generally, we meet with many fine trees that are worthy of notice. At the corner of the wood directly opposite Kirklands there are several black poplars of considerable size. The largest one girths 13 feet 1 inch at 1 foot from the ground, and 11 feet 7 inches at 5 feet from the ground; another girths 11 feet 3 inches at 1 foot, and 9 feet 10 inches at 5 feet from the ground; and a third girths 11 feet 6 inches at 5 feet from the ground,—each of the trees having grand boles of about 20 feet. The largest sycamore grows at the top of the North Avenue, the girth at 1 foot from the ground being 12 feet 3 inches, and at 5 feet, 10 feet 7 inches,— the bole being a splendid one of about 35 feet. Another sycamore, opposite the peach-house, girths 10 feet 2 inches at the narrowest part of the bole; after which it swells out very considerably, and makes a beautiful park tree. An aged yew tree girths 11 feet at the narrowest part of the bole, being about 5 feet from the ground. The largest of the oaks girths 10 feet 4 inches at 1 foot from the ground, and 8 feet 4 inches at 5 feet from the ground. The largest ash tree is at the corner of Whitehill Park, in a strip near the home-farm,—the girth at 1 foot from the ground being 10 feet 9 inches, and at 5 feet, 9 feet 4 inches. The policies are laid out with great taste, all the trees being planted with due regard to effect, without the view from the mansion being in any way obstructed.

There are many objects of interest beyond the policies. The interesting old Castle of Kelty, at the mouth of Kelty Glen, for instance, is a really beautiful spot, loaded with various kinds of timber, including a few trees of more than ordinary size. The castle is a very old building, and has not been regularly occupied for about eighty years. In the beginning of the seventeenth century—and, it is said, for at least two hundred years before that—Kelty belonged to a family of the name of Bonar, which then held Forgandenny, Invermay, Kilgraston, and other territorial possessions in Perthshire, and whose representative is now Dr Horatius Bonar, Edinburgh. In 1692 the estate was purchased by John Drummond, the progenitor of whose family was William Drummond, first laird of Crieffvechter, the fifth in succession from Sir Malcolm Drummond of Stobhall, by his son James of Coldoch. William flourished about 1600, and left a son, James, father to the Rev. John Drummond, minister of Monzie, who had two sons, one of whom was Mr Drummond, who succeeded his father as minister of Monzie, and, dying without issue, left his fortune to his nephew, John Drummond, who purchased the estate of Kelty, which then became the family seat. John Drummond built the present castle, one of the lintels bearing his initials and those of his lady, with the date of the erection of the building, 1712. The building is a very substantial structure, and seems to have at one time been capable of defence. Like all ancient strongholds, many curious stories hover round this venerable pile. One of the most popular of these is to the effect that the lovely daughter of one of the lairds, having the misfortune to contract an affection for a man reckoned by her father as below her rank, was imprisoned in the castle dungeon till she died. Whether this story be true or not, there can be no doubt about the existence of the dungeon, which is as horrible a hole as could well be conceived. Beside the dungeon is a cell, with one small window, about 6 inches square, and which has evidently been the prison of the castle, the dungeon we speak of being right below. The walls of the castle are about 9 feet thick, and the whole interior presents a gloomy and forbidding aspect. Kelty continued in the possession of the Drummonds until it passed into the Airlie family by the marriage of David, the eighth Earl, in 1812, to Clementina, only child and heiress of Gavin, third son of James Drummond of Kelty. Her ladyship died in 1835, after which the late Lord Rollo acquired the estate by purchase.

The best of the trees in this part of the estate are all in the neighbourhood of the Castle. The most noteworthy of these is a very fine ash growing close to the bridge leading across Kelty Burn to the Castle. It has a girth of 10 feet 3 inches at 1 foot from the ground, and 8 feet 11 inches at 5 feet from the ground, with a magnificent head. The most of the other large trees here are sycamores, but none of them reach the dimensions of those we have described as growing in the policies at Duncrub. There are also three very fine specimens of *Pseud-Acacia*, or American locust tree, the seed of which is believed to have been brought by a member of the family who was for some time in America. There is a very beautiful walk through the glen, which is both romantic and picturesque,—the dark, moss-covered rocks, and the pretty, thickly-planted copse, presenting a fine contrast, while the lively, limpid stream underneath gives zest to the picture.

The plantations upon the property are pretty extensive, the most of them having been laid down in comparatively recent years. The total extent under wood is about 525 acres. These consist almost entirely of larch, spruce, and Scots fir, each variety having been planted with due regard to the suitability of the site. The wood is in a clean, healthy state, and has been so laid out as to enhance the beauty of the landscape.

The whole of the village of Dunning is upon the Duncrub property, and, apart from the commemorative thorn tree already referred to, there are several circumstances connected with it which impart interest to the town. When first heard of in authentic history, it formed part of the ancient Stewartry of Strathearn, and it figures in ecclesiastical records from a very early period. It is believed that the present Parish Church was built between the years 1200 and 1219. The style of architecture corresponds exactly with the style then prevailing, and is known as Early English. The patron saint of the parish was St Serf, a famous man of the fifth century. He gets the credit of having killed "a dreadful dragon, which devoured both men and cattle," and kept the district of Dunning in continual terror. The scene of this heroic deed is called the "Dragon's Den" until this day. Another account gives the matter-of-fact explanation, that this is merely a corruption of "dregging-den," and refers to "the practice of the milk cows of the district being brought together in the glen for the purpose of being dregged—that is drawn or milked." The church has undergone many alterations since it was erected, but much of the ancient building still remains in its original form, the most conspicuous object being its massive square tower. The village is neat and tidy, although not particularly picturesque, and, owing to the collapse of the handloom weaving, is not now so prosperous as it once was. A new and plentiful supply of water was introduced in 1872; and a magnificent fountain, the gift of a successful native abroad, has been erected in the village square. Lord Rollo takes a lively interest in everything that contributes to the welfare of the inhabitants; and the people have frequently, along with the agricultural tenantry, taken occasion to reciprocate the goodwill of his Lordship.

In addition to the improvements which we have referred to, the present Lord Rollo has also done much to improve the agricultural portion of his property. Several new steadings have been erected, and others put in proper repair. Fencing and draining have been carried out very extensively. Everything, indeed, appears to be done to encourage good husbandry, and to make the tenant, his servants, and his stock comfortable.

XVII.—GARVOCK.

Situation—The Græmes of Garvock—How the Property was Acquired—
The Mansion—Rose-Bush Planted by Prince Charlie—An Enormous
Walnut Tree—Other Notable Trees—An Ancient Camp—The Planta-
tions.

GARVOCK, although not a very extensive property, is a most
interesting one on various accounts. Situated at the foot of the
Ochils, and wedged in between Duncrub and Invermay, it
possesses many of the features which lend so much attraction to
these places—a commanding prospect over the length and
breadth of Lower Strathearn, with a screen of giant mountains
in the distant west and north; a large tract of fertile land
lying pleasantly towards the sun; a wealth of umbrageous
clothing; and many charming little nooks and corners to break
the monotony of the great expanse of cultivated fields and
waving plantations, which are spread out in the valley beneath.
In Garvock we have a property which has had its owners in the
same family for many centuries, the Græmes of Garvock tracing
their descent in regular succession from a famous Celtic chief,
Græmus, who flourished in the early part of the fifth century,
and who fought at Carron Water against the Romans under
Victorius. "He it was," says Lord Strathallan in his *Genealogy
of the House of Drummond*, "who broke through the old rampart called Severus' Wall, built from Abercorn to Kilpatrick at
the mouth of the Clyde, about thirty miles in length, and
beat the Roman garrisons from thence, for which notable action
it got the name of Græmes-dyke, which it retains to this day.
This was soon after the 400th year of Christ." A subsequent
Græme, along with Dunbar, the Earl of March, did great service
in preventing Scotland from falling into the hands of the Danes.
The same authority says that the Græmes came first to Strathearn by the marriage of Sir John Græme of Dundaffmuire to a

daughter of Malise, fourth Earl of Strathearn, with whom Sir John got the lands of Aberuthven, about the year 1242. Fifty years later we find three Græmes—Patrick de Græme, David de Græme, and Nichol de Græme—included in the list of those appointed to consider the claims of Bruce and Baliol at Berwick. After them came the illustrious patriot, Sir John de Græme, the companion-in-arms and greatest confidant of Wallace, and to whom Buchanan attributes this high character, "*Scotorum longe fortissimus habitus*," and of whom Blind Harry sings as being "right noble in war." He followed the Scottish hero through many perilous enterprises, and ultimately fell on the field of Falkirk in 1298. The next notable Græme of which we read is "Willielmus Dominus de Græme de Kincardine," who was one of the commissioners appointed to treat with the English in 1406, after the disastrous defeat at Homildon Hill. Six years afterwards he was sent into England to treat for the ransom of James I. He married, first, a sister of Robert, Duke of Albany, at that time Regent; and, second, the Lady Mary Stuart, second daughter of Robert III., by whom he had three sons, one of whom, William Græme of Garvock, was the ancestor in the direct line of the present proprietor of the estate, Robert Græme. He was a distinguished soldier, and the lands and barony of Garvock were granted to him by his uncle, James I., as a reward for his faithful and zealous services. From him, the property has been handed down in regular lineal descent from father to son, without any interruption, to the present time,—a circumstance which is very rarely met with in family history. Several of the succeeding lairds of Garvock have taken a prominent part in the making of our Scottish history. The third laird, Archibald Græme, fell fighting bravely on the disastrous field of Flodden. One of the sons of the fourth laird was the ancestor of the Grahams of Balgowan and Lord Lynedoch, the conqueror of Vittoria. The seventh laird, James Græme, acquired a considerable addition to the family property, by being served heir to John Graham of Balgowan, who had shortly before purchased the estate of Kippen. The ninth laird of Garvock, Robert Græme, great-great-grandfather of the present proprietor, was a devoted Jacobite, and having been "out" in the

'45, was deeply involved in the consequences of that unfortunate affair. After the collapse at Culloden he lurked about the country for some time, but subsequently returned to Garvock in the hope that he would be permitted to live at peace, but on this coming to the knowledge of the Government, a party of troopers were sent to apprehend him. On the approach of the troopers being observed, orders were given to shut the gates, and, at the urgent request of his friends, he assumed the dress of a female servant, with the view of endeavouring to escape. Being of a stalwart frame, he felt that he "did not look the character," as theatrical people would say, and rather than be captured in such an unmanly disguise he threw aside his female attire, sat down in his arm-chair, and gave orders to admit the troopers. He was accordingly apprehended, and conveyed to Perth, where he was confined for upwards of a year, until a friendly official, with the connivance of the authorities, as is believed, permitted the prisoner to escape. He succeeded in finding his way to France, where he served in the army with distinction, but lived to come home and spend the evening of his days at Garvock. Robert Græme, the present proprietor, is the fourteenth laird, and succeeded in 1859.

With a family so devoted to the Jacobite cause as the Græmes, it would have been strange had Prince Charlie not been a guest at Garvock, and left behind him some memorials of his visit. The Pretender is believed to have been a frequent visitor, and to have slept under the roof of old Garvock House, part of which is still in use. It is not known when this old portion was built, but, judging from the architecture, it must have been at a very early period. The principal wings of the mansion, however, were built in 1825 and 1827, and are thoroughly modern. Above the kitchen window is placed the family coat-of-arms, one of the first efforts of the famous Perthshire sculptor, Laurence Macdonald. Unfortunately, the arms have been cut upon soft freestone, off its natural bed, and the weather has told upon the work so much that it is almost entirely effaced. Close to the south-east of the oldest portion of the mansion is a rose-bush believed to have been planted by Prince Charlie during one of his visits to Garvock. At one time a wing of the mansion reached close up to this interesting memorial, and it is believed that it was planted right underneath the august visitor's bedroom window. The bush is of the variety known as the White Provence or Jacobite rose, and stands upon a small mound, caused by the surrounding ground having been levelled. It still flowers very freely, and has every appearance of being a long liver. The greatest care is taken to preserve it, by cutting away any dead wood which might injure its growth, and encouraging the young shoots to come forward. The oldest people about the place can recognise no difference upon the bush since they first saw it. A sword presented by Prince Charlie to the great-great-grandfather of the present laird is carefully preserved at Garvock House, along with his commission as an officer in the rebel army. The most notable object at Garvock, from an arborist's point of view, is an enormous walnut tree, one of the largest in Scotland. It is situated in the garden quite close to the mansion, and its massive and shapely proportions at once arrest attention. At 1 foot from the ground it girths 16 feet 3 inches, and at 5 feet, it girths 14 feet 4 inches. There is a clean bole of about 20 feet, after which the tree breaks into three great main branches. The entire height of the tree is about 80 feet, with a corresponding spread of branches. It is altogether a very fine specimen, and one of which the laird is justly proud. There are none of the other trees which reach to anything like this size, but there are a large number of trees of fair dimensions as ordinary timber trees. There is a very fine sycamore with a girth of 10 feet 2 inches at 1 foot from the ground, and 9 feet 3 inches at 5 feet. There are many larch trees of a good size, one of the best of which girths 9 feet 7 inches at 1 foot from the ground, and 8 feet 3 inches at 5 feet. It is fully 80 feet high, perfectly straight to the top, and in a most healthy condition. A very beautiful elm, close to the house, girths 11 feet at 1 foot from the ground, and 9 feet 9 inches at 5 feet. It is also a nice clean tree. Opposite the kitchen door, there is a remarkably fine fern-leaf beech tree. It is in luxuriant health, well shaped, and has exceedingly pretty foliage. The best oak tree upon the property stands in the park on the edge of the Broad Wood. It has a girth of 12 feet at 1 foot and 9 feet 7 inches at 5 feet from the ground, with a nice bole of 15 feet. Another oak, almost equally good, stands in an angle close by, at the side of Garvock Burn, and regarding which an interesting story is told. The particular piece of ground upon which the oak stands was added to Garvock as the result of an excambion with Mr Belshes, the laird of Invermay. The value placed upon the tree by Mr Belshes was £5, and he remarked that the laird of Garvock could either take it or want it. The latter, however, had little hesitation in including such a promising oak in his bargain. At a short distance from this spot are the remains of a Pictish camp, in a good state of preservation.

Garvock is considered to have been well-wooded from an early period, and at present all the available ground is well stocked with excellent timber. The Broad Wood is the oldest of the plantations, and consists of about 30 acres of standard oaks about 100 years old, intermixed with birch and oak coppice. All the standard trees are of good size, and several are of superior growth. In the Shaw Wood, which consists of about 24 acres, there are many fine oak, larch, and Scots fir trees of about 60 years of age. In Garvock Hill Wood there is a very thriving plantation of 40 acres, chiefly larch, about 30 years old. Nether Garvock Wood has about 30 acres of young Scots fir and larch, the latter being about one-third of the whole. A portion of this wood is about 25 years old, and the remainder about 12 years old, the whole of the timber being in a thriving condition. Alongside the Perth Road, there is a nice little plantation of 4 acres, known as the Birch Wood of Wellhill, and consisting of a mixture of oak and birch. There are, in addition, several smaller woods, and numerous clumps of thriving trees, and as the greatest care has been taken to plant these only in places where they are appropriate or useful, they contribute very largely to the beauty and value of the property.

XVIII.—GASK.

Attractions of Gask—Roman Remains—Gask Woods a Hiding-Place of Wallace—Visit of Prince Charles—Jacobite Relicts—The "Auld House"—The "Auld Pear Tree"—The New Mansion—Chequered History of the Property—Destruction of the Woods during the Rebellion—An Exiled Laird and his Woods—The Policies—Notable Beech Trees—A Magnificent Walnut—Other Remarkable Trees—The Garden—Lawrence Macdonald the Sculptor—The Plantations.

GASK has achieved for itself enduring fame in several widely different spheres. It is famous in history, famous in song, famous in art, and famous as a favourite of Nature. The Romans traversed its trackless forests—the great Scottish patriot sought shelter within its almost impenetrable woods—"Bonnie Prince Charlie" enjoyed the hospitality of its "auld laird," one of the most devoted of his subjects—the Baroness Nairne, and her niece, Caroline Oliphant, "The Flower of Strathearn," inaugurated the Jacobite minstrelsy of Scotland, and sung the beauties of their much-loved home; in Laurence Macdonald, Gask has placed a son in a prominent niche in the Temple of Art; and it has been favoured by Nature with the most gorgeous surroundings, and some of the noblest trees in Scotland. In the remote past Agricola's legions made Gask one of their stations, and have left memorials of their residence in the old Roman Road, by which the mansion is still approached from the north, and in two small camps on either side of the roadway. Coming down the stream of Time, we read of Wallace, pursued by relentless and treacherous foes, finding a safe hiding-place within its dark, untrodden woods. In visiting Perth in 1290, in the disguise of a priest, he was just on the point of being captured when he received timely warning, and ran, with a few followers, to the dense woods of Gask, whither the sleuth-hounds were sent in vain to track him, their progress being stopped by the dead body of the traitor Fawdown. Wallace, with thirteen adherents, took shelter in the deserted Gascon Ha', as Gask House was then called, and which is supposed to have stood in the centre of one of the largest of the present plantations. Here they passed a terrible night. While the party were soundly sleeping, Wallace was aroused by the blowing of horns, mingled with frightful yells, proceeding apparently from a piece of rising ground in the immediate neighbourhood. Scouts were sent out in all directions, until, failing to return, the patriot was left alone. He wandered about till morning, killing two Englishmen he encountered in the interval, and afterwards directed his course to Torwood, near to his uncle's at Dunipace. The most memorable affair, however, in the history of Gask was the visit of Prince Charles in 1745. Laurence Oliphant, the laird, and the whole family, were enthusiastic Jacobites, and the honour of a visit from their "rightfu' lawfu' King" sent all of them and their friends into ecstasies. A large number of Jacobite relics are still in existence within the mansion. These include the table at which the Prince took his breakfast; the chair upon which the Prince sat, and which was never allowed to be profaned by meaner occupants for scores of years afterwards; the Royal brogues; Ribbon of the Garter; a piece of the cloak of Charles I., cut up by Bishop Juxon for distribution; Prince Charlie's bonnet, spurs, cockade, crucifix, &c. One of the most cherished of these relics is the lock which the laird's good lady—"the leddy, too, sae genty"—clipped "wi' her ain hand," from the "lang yellow hair" of "Scotland's heir,"—the lady being Amelia, a daughter of the second Lord Nairne. Lady Nairne has exercised the poetic licence in her metrical account of how the lock of the Prince's hair came to Gask. She tells the true story in prose as follows:—

"Charlie's hairs were given to my grandmother Strowan, the day they were cut, by the man who cut them, one John Stewart, an attendant of the Prince. This is marked on the paper in her own handwriting. I have often heard her mention this John Stewart, who dressed the Prince's hair."

The ruins of "the Auld House," immortalised by the Baroness Nairne's exquisitely tender and beautiful song, are still to be seen. They nestle on the top of a gentle eminence, shrouded with wide-spreading trees, overhung with ivy, while

> " The wild rose and the jessamine
> Still hang upon the wa'."

The "wild rose" is simply the ordinary Scotch rose, the "jessamine" being the beautiful white clematis. They are both almost completely smothered up with the ivy, and can barely be traced. It cannot, however, now be said that there is

> " Still flourishing the auld pear tree
> The bairnies liked to see ;—
> And o' how often did they speir
> When ripe they a' would be !"

Although not now "flourishing," the trunk of the tree, a pretty large one, is still to be seen nailed to the wall as in the days when the bairnies looked forward to the ripened fruit with so much glee. The tree is thought to be amongst the first, if not the very first, jargonelle pear tree introduced into Scotland. The branches have been nailed to the wall in the way in which gardeners are still accustomed to fasten wall trees, although the swelling limbs have occasionally burst their bands. The house, like most of the buildings of the period, was very irregularly built, having been added to from time to time as circumstances would permit, or as the exigencies of the family rendered necessary. The prevailing characteristics are said to have been angles, crow-steps, pinnacled staircases, and long chimneys. The exposure was southern, and it looked down upon the fine champaign country which distinguishes the lower course of the Earn, until the view is interrupted by the serrated crest of the Ochils. Even in Prince Charlie's time, " the Auld House" could not have been very commodious, as is apparent from the opening verse of the song :—

> " Oh, the auld house, the auld house,
> What tho' the rooms were wee ?
> Oh, kind hearts were dwelling there,
> And bairnies fu' o' glee."

It is elsewhere described as a true " Honour's Broadstone," and it must have been with a pang of regret that the family left it at the beginning of the present century. The house which has taken its place is one of the most commodious and elegant country seats in Perthshire. Mr T. L. Kington Oliphant, the present laird, states in his work on the "Jacobite Lairds of Gask," that the new house was begun in 1801, "and was built far larger than the wants of the family required, owing to the promises made to the laird by a rich kinsman; promises never fulfilled. The war prices (the Peace of Amiens was but short) made the cost of building enormous, although the quarries were close at hand—quarries which the Earl of Strathearn had allowed the monks of Inchaffray to work so early as 1266."

The history of the property of Gask has been more than ordinarily chequered, on account of the lairds taking a prominent part in the Jacobite troubles. The Oliphants, as we have shown in a former chapter, are a most ancient family, coming to Scotland as far back as the twelfth century, when they at once rose into prominence. The lands of Gasknes, as the property was then called, were bestowed by King Robert the Bruce upon his trusty follower, Sir William Oliphant, about 1310. He then became the Lord of Gasknes and Aberdalgie, and from him sprang every Oliphant who made any figure in history after 1312. He is buried at Aberdalgie, the three crescents, the arms of his house, being still traceable on his tombstone. In 1458, Laurence Oliphant received a Peerage from King James II., and became the first and greatest of ten Lords Oliphant. It was he who founded the Greyfriars House at Perth. His son had two children killed at Flodden—Colin, the Master of Oliphant, and Laurence, the Abbot of Inchaffray, the great monastic house of Strathearn. The master of Oliphant, however, still left two sons—Laurence, the third Lord Oliphant, and William the forefather of the Gask line. When we come to the fifth Lord we are forced to exclaim, with Hamlet, "What a falling-off was there," he being described in the Gask papers of his age as "ane base and unworthy man." He sold his great estates in many different shires soon after the year 1600. Gask was alone saved from the wreck, and was made over to his cousin, Dupplin and Aberdalgie coming into the hands of the first Earl of Kinnoull. The sixth Lord received a new Peerage from Charles I. in 1633, and, having married a Crichtoun, obtained lands on the Deveron in Banffshire. The Gask branch of the family sprung from William, the brother of the third Lord Oliphant. William's grandson, Laurence Oliphant, bought Gask from the spendthrift Lord in 1625. The great-grandson of the ancestor of the Gask line, Laurence Oliphant, was a zealous Cavalier,—having been made to do penance by the Covenanters for his zeal on behalf of Charles I., —and received the honour of knighthood from Charles II., when that monarch was in Perth with the Covenanters in 1650. He added to his possessions by purchasing the lands of Williamstoun from Sir William Blair of Kinfauns, "and paid for them thirty years' purchase, when money was at ten per cent. interest." James Oliphant, born in 1660, is described as one who "deserves to rank with his great grandfather as a thrifty guardian of the Gask heritage." He was one of the four Perthshire lairds who on 9th October 1690 applied to the Scottish Parliament for an Act to compel all the heritors to drain and ditch the Pow of Inchaffray, which had been flooding the neighbouring lands. This is the only instance on record of a great agricultural improvement being effected under the authority of the Scottish Parliament. It is worthy of note, however, that part of this marsh, now amongst the most fertile land in Perthshire, was reclaimed by the monks of Inchaffray as early as the year 1218. The Jacobite risings proved most disastrous to the House of Gask. The Laird of Gask and his son, both Laurences, were on the field of Culloden,—the former with his regiment, and the latter as *aide-de-camp* to Prince Charles. After that disastrous battle they roamed about under assumed names—the father passing as Mr Whytt, and the son as Mr Brown. "The Auld House" was plundered by Cumberland's soldiers, although the officer who ordered it was afterwards tried by court-martial and deprived of his commission. The woods upon the property suffered very much at this period from the military depredations, large numbers of trees being cut down and sold. The laird and his son ultimately made their escape to France, and the property of Gask, like many more, was forfeited. A scheme, however, was matured, by which the estate was publicly purchased in 1753 by staunch friends, and the laird and his son, on the expiry of their period of exile—seventeen years—returned once more to Gask, the angry passions of the rebellion having cooled down. A letter in the family papers shows that the laird took a considerable interest in the management of the property while he was in exile. In the course of one of his letters to his wife, signed "John Whytt," he shows that the woods had a special interest for him :—

"I recommend that you'd cause look about your small plantations, and that you'll recommend to Simon [the laird of Condie] and Lawson, that they would be severe to all that shall cutt trees or brake down any branches for firing."

The age of many of the existing trees indicates that this laird and his son were the planters. The father died in 1767 at the age of 75, a year after his granddaughter, the poetess, first saw the light, she having been born at Gask, on 16th August 1766. His reinstatement at Gask made no difference in the political views of Laurence, but the times were more orderly and tolerant. He lived till 1792, and Mr Kington Oliphant, in the work already quoted, gives a faithful and, on the whole, a flattering picture of his character. Although he had his faults, —largely due to the age in which he lived and the circumstances by which he was surrounded,—yet, taking him all in all, he seems to have been "chivalry embodied in the shape of man." "In him was found a man's thoroughness, a woman's softness, a child's simplicity." He appears to have been an ardent arboriculturist. "This laird had none of his father's learning. Trees he understood better than books; many of the Gask woods were planted by him, and named after his daughters." The poetess has herself given us a very beautiful picture of her father—the succeeding laird :—

> "The auld laird, the auld laird,
> Sae canty, kind, and crouse;
> How many did he welcome to
> His ain wee dear auld house!"

After the death of the grandfather of the poetess, the Oliphants softened down their politics considerably,—the next laird, the Baroness Nairne's father, also a Laurence, enlisting under the banner of George III. He built the new house of Gask, and effected other improvements upon the property. He pulled down the "Auld Kirk" as well as the "Auld House," taking care, however, to preserve the monument of his forefather who bought Gask in 1625. "The new kirk," writes the present laird, "was built a mile from the house; the villagers, not liking the change of the burial-ground, used to force their way to the old kirkyard, that they might bury their dead near their own kin ; and this went on for at least twenty years." He died at Paris in 1819, and lies buried in Père la Chaise. The new laird, who bore his father's name, was only twenty-one when he succeeded. He died at the end of 1824, unmarried, and was succeeded by his only brother, James Oliphant, who was one of those who escorted Queen Victoria on her first visit to Edinburgh. He died in 1847, and was the eighteenth in unbroken male succession from the William Oliphant upon whom King Robert the Bruce bestowed the lands of Gasknes, and who, in 1296, was taken prisoner by Edward I. at Dunbar, and sent into England with many other Scottish captives. James Oliphant was succeeded by his nephew, in the female line, Thomas Lawrence Kington Oliphant, the present laird.

The new mansion—which, as we have said, was built in 1801—is a substantial and elegant square house, with a massive and beautiful portico, and has quite a modern appearance. It stands within a short distance from the "Auld House," and occupies a commanding situation. It is so embowered amidst overshadowing trees, however, that its presence is not apparent until the visitor is close upon it. It is now inhabited by the widow of the late proprietor, the present laird occupying Charlesfield House for some part of the year, a large and beautiful residence, erected in 1873 upon the southern slope of the Braes of Gask, a little to the north of Dalreoch Bridge, and a short distance from the mansion-house. It commands a most extensive and charming view, the greater part of the rich valley of the Earn, with some fine stretches of the river, and a wide range of the Ochils, forming part of the landscape. The residence of the laird, although occupying an open situation, is surrounded by thriving young timber, which is already affording abundant shelter. The policies at the mansion-house are richly ornamented with hardwood trees, evidently planted with a liberal hand during the lifetime of the historic "auld laird" and his father. Although the great bulk of the timber is somewhere about 100 years of age, there are a good many more venerable trees, the survivors of the military depredations to which the woods of Gask were subjected during the Jacobite troubles. Immediately on entering one of the approaches from the north, leading off the old Roman Road,—now the ordinary Perth Road,—we come across a row of splendid beech trees, under which, it is said, the soldiers of Prince Charles were regaled by the laird of Gask in 1745, while their chief was enjoying the hospitality of the "Auld House." The largest of these trees, the one nearest the road, girths 15 feet 4 inches at the narrowest part of the bole, which is immediately above the swell of the roots. The tree afterwards swells out considerably, and at 5 feet from the ground the girth is 18 feet. There is just 6 feet of a bole, the tree then breaking into three great limbs. The other trees in the row have boles of much greater size, but they are not so large in the girth, although more valuable timber trees. The largest of them girths 14 feet 1 inch at 1 foot from the ground, and 12 feet 7 inches at 5 feet; 14 feet 5 inches at 1 foot, and 12 feet at 5 feet; 14 feet 3 inches above the swell, and 12 feet 5 inches at 5 feet; and another girths 11 feet 8 inches at the narrowest part of the bole. The largest tree in the policies is a magnificent old walnut, growing in a walnut grove, a little to the west of the "Auld House." At 1 foot from the ground the girth is exactly 19 feet, and at 5 feet from the ground the girth is 17 feet 5 inches. The trunk then gradually swells until a girth of 21 feet is attained,—the tree at 8 feet from the ground breaking into a number of gigantic limbs. It is altogether a notable tree, and is still very healthy. There is another grand walnut a little to the north of the chapel, near the "Auld Kirkyard," and known under the sobriquet of "Lord Oliphant," whether on account of some family association, or because of its lordly proportions, we cannot tell. At 1 foot from the ground the girth is 17 feet 2 inches, and at 5 feet the girth is 13 feet 8 inches. There are about 25 feet of a bole, the entire height of the tree being about 100 feet. There are several ash trees of more than ordinary dimensions. The largest of these girths 16 feet 5 inches at 1 foot from the ground, and 14 feet 4 inches at 5 feet, with a bole of 10 feet, from which spring four gigantic limbs, held together by a strong chain. The tree is now in the "sere and yellow leaf," and is beginning to decay. Another ash, a beautifully clean tree, girths 14 feet 6 inches at 1 foot from the ground, and 12 feet 8 inches at 5 feet. There is a fine bole of 10 feet; after which the tree divides into two large limbs. The scyamores are both numerous and of considerable size. The largest of these has a magnificent bole of 35 feet, with a girth at 1 foot from the ground of 13 feet 4 inches, and 11 feet 7 inches at 5 feet from the ground. Another, a more valuable timber tree, has a bole of 20 feet, and a girth at 1 foot from the ground of 11 feet 10 inches, and at 5 feet of 9 feet 9 inches; and another, with a bole of 10 feet, has a girth at 1 foot from the ground of 13 feet 6 inches, and at 5 feet of 11 feet. There are a great many oaks, the largest girthing 15 feet at 1 foot from the ground, and 11 feet 9 inches at 5 feet, with a bole of 20 feet; the next largest girthing 11 feet 7 inches at 1 foot from the ground, and 10 feet four inches at 5 feet. A cedar of Lebanon may also be included in the list of the larger trees, the girth at 1 foot from the ground being 10 feet 2 inches, and 7 feet 4 inches at 5 feet from the ground.

The garden is almost equally as interesting as the policies. One of the principal features of the garden is the gorgeous display of rhododendrons, which in the early summer present a perfect blaze of the most brilliant colours. Not the least interesting object in the garden is a marble life-size statue of Apollo by Lawrence Macdonald. The statue, which was originally at Orchill, has suffered a little from exposure, but it is a capital specimen of the better class of the sculptor's early work. The figure holds a lyre, and the expressive, listening attitude of the Greek god of music and poetry is exceedingly good. Lawrence Macdonald was born at Gask in 1798, in one of a group of comfortable-looking cottages bearing the appropriate name of "Bonnie View," and situated about half a mile west from the mansion-house. His father was exceedingly poor, earning his living by violin-playing, assisted by his wife, who acted as occasional nurse to the families in the neighbourhood. Macdonald served his apprenticeship as a mason with Mr Thomas Gibson, the builder of Murray's Royal Asylum, Perth, and, as he spent a great deal of his spare time in the winter months in modelling heads, he soon attained an uncommon degree of proficiency, as is testified by the juvenile productions which now adorn the gardens at Moncreiffe and Duncrub. After doing some really good local work in the shape of coats-of-arms, &c., he proceeded to Edinburgh, where he gained a considerable reputation. Under the patronage of the Oliphants of Gask, he proceeded to Rome in 1822, and was speedily acknowledged as the leading British sculptor in Italy, all the noble Scottish and English visitors to Rome patronising his studio. He did not forget his early patrons in the days of his prosperity, and some of his finest works were gifted to the family, and are now carefully placed in Gask House. He died in 1879, but his fame is still being maintained in the Eternal City by his son, Mr Alexander Macdonald, who had, in 1872, the honour of executing a commission for the Prince and Princess of Wales. The figure of Apollo in the garden at Gask stands at the end of a broad grass walk, and is a very prominent and beautiful object. The only noteworthy trees in the garden are four very fine araucarias in the flower garden in front of the ruins of the old house. Two of these are particularly large, one of them being about 35 feet high, and finely shaped.

The plantations are pretty extensive. Several of them are loaded with fine old wood, principally Scots fir, larch, and spruce. The Wood of Cowgask contains about 100 acres of timber, chiefly Scots fir, about a century old. About 15 acres of this wood were replanted 14 years ago. At Witchknowe 15 acres of Scots fir were planted about 12 years ago, the wood being intermixed with a few young natural beeches. At Drumharvie about 50 acres of Scots fir, larch, and spruce were laid down about 20 years ago; and at Westmuir 50 acres of the same varieties of timber were planted five years ago. In almost all these plantations the trees are doing very well; and at Westmuir, where the ground is a little damp, the spruce are in an exceptionally thriving state. In all of these newer plantations the ground was carefully prepared before being planted, and thoroughly drained wherever that was necessary. Most of the available land upon the property is now completely clothed with mature or thriving young timber; and some idea of what has been done in recent years will be gathered from the fact, that during the past two decades no less than 100,000 trees have been planted at Gask.

XIX.—DUPPLIN.

The Interest Attaching to Dupplin—The Battle of Dupplin—Its Antiquities—The Kinnoull Family—Dupplin Castle—The Policies—The Queen's Visit to Dupplin—The Original Design of the Policies—The Pinetum—The Octagon—Remarkable Trees—A Mysterious *Wellingtonia*—Kinnoull Hill—The Plantations.

DUPPLIN is one of the most interesting, as well as one of the most beautiful, of the places we have to describe in connection with the woods, forests, and estates of Perthshire. The interest attaching to it is not limited to its magnificent castle, its gorgeous surroundings, or its flourishing woods and remarkable trees, but extends to its connection with the stirring events of the past, and its relics of antiquity. The most notable historical event with which it is associated is the Battle of Dupplin, which was fought within a short distance from the site of the present castle. This battle was fought in August 1332, and arose out of the generally disorganised state of the country at that period. The "disinherited barons," who combined to recover the crown from the Bruce family, had gathered round Edward Baliol, and, putting him at their head as King, resolved to try their fortunes in Scotland. The English Government being at peace with this country at the time, the barons and their leader found it impossible to cross the Border, but availed themselves of the open seaboard, landing on the shores of Fife, at the date mentioned, to the number, as is said, of 500 mounted men and 3000 foot. The place of landing was so unfavourable for the disembarking of cavalry that a small force, under a competent commander, would have put an end to the daring enterprise at one blow. The Earl of Mar, the new Regent of Scotland, was then at the head of a Scottish army ten times stronger than the invaders, but he lost his opportunity; whilst Alexander Seton, with a handful of soldiers, were cut to pieces in their attempts to drive the English from the Scottish soil. Baliol lost no time in advancing to Dunfermline, and, as he gained more confidence, he commanded his fleet to sail round the coast and anchor at the mouth of the Tay, while he himself pushed on in the direction of Perth. He proceeded as far as Forteviot, where he encamped on "Millar's Acre," with his front defended by the River Earn. The Earl of Mar was in the immediate neighbourhood with the view of intercepting the usurper. His army numbered about 30,000 men, was excellently equipped, and was commanded by the principal nobility of Scotland. The Earl of March, too, was at Auchterarder with an army nearly as numerous, threatening the English in flank. The Earl of Mar posted his army on those fine slopes which dip down from Dupplin Castle to the Earn, and which now command the attention of railway travellers from their rich garment of umbrageous trees, which present a most refreshing contrast to the somewhat dull scenery which prevails for several miles farther south. The situation seemed most perilous for the English, especially as their commander had not the reputation of being the most consummate general. But the enemy had trusted friends in the Scottish camp. Several of the nobility whose relatives had suffered in the Black Parliament, secretly favoured the "disinherited barons," and by their treachery, the English General was enabled to inflict an overwhelming defeat upon the army of Mar at the moment when his own destruction appeared inevitable. The Earl of Mar appears to have been too confident of his strength, as, notwithstanding his knowledge of the immediate presence of the enemy in front, no watch was kept, and the soldiers, in violation of all military discipline, were permitted to abandon themselves to riot and intemperance. It has been asserted by an English historian, on the authority of an ancient manuscript chronicle, that the Regent had entered into a secret correspondence with Baliol, but it is stated, on the authority of Barnes (*History of Edward III.*), that the conduct of Mar was rather the result of weakness and presumption than of treachery. The English were conducted to a ford in the Earn by Andrew Murray of Tullibardine, who served in the army of March, and who guided the English to the spot by driving a large stake into the channel of the river. Baliol crossed the river in the silence of midnight, and, marching by Gask and Dupplin, broke into the Scottish camp, and commenced a dreadful slaughter—Mar's soldiers being for the most part drunken, and heavy with sleep. An effort was made to check the first onset; and it is believed this would have been successful but for the incapacity of the Regent, who failed to improve an opportunity which occurred. The immense mass of warriors—spearmen, bowmen, horsemen, and infantry—became huddled and pressed together, multitudes of the Scottish soldiers being suffocated and trampled upon by their comrades; while the English preserved their discipline so well that within a few hours the whole of this grand Scottish army was slain, dispersed, or taken prisoners—the English loss being inconsiderable. At one part of the ground the mass of the slain was a spear's length in depth. This was a most fatal day for Scotland. There does not, indeed, occur in the whole of our Scottish annals a more disastrous defeat than took place on those beautiful slopes which have now such an artistic effect upon the landscape. It is some consolation to know that Murray of Tullibardine was speedily overtaken by the avenging Nemesis, having been made a prisoner at Perth, tried, condemned, and executed. An ancient stone cross, a little to the west of the Castle, and almost due north from the ford by which Baliol passed, marks the site of this bloody conflict. The cross has been most elaborately carved, but the figures are now almost obliterated by the destroying hand of time. Figures of horsemen and bowmen, however, are traceable, the infantry being represented in that peculiar processional order with which early Egyptian sculpture makes us so familiar. About half a mile north of the cross which marks the site of the Battle of Dupplin a large tumulus or cairn was opened, and found to contain some coffins formed of rough flat stones containing fragments of bones. About ninety years ago a stone was found near the site of the castle, having two lambs carved upon it, and was taken possession of by the then Lord Ruthven.

The family of which the Earl of Kinnoull, the proprietor of the Dupplin estates, is the representative, has a long and honourable history. Tracing its descent to Sir John de Haya, who lived in 1238, and was the ancestor of the Marquis of Tweeddale, the first of this branch of the family we read of is Peter Hay of Melginche, whose second son, Sir James Hay, K.G., accompanied James VI. to England, and was created Lord Hay, with precedence next to the Barons of the realm; and afterwards Viscount of Doncaster, and Earl of Carlisle. He was entrusted with several important missions to foreign countries. James, the second Earl of Carlisle, resided in the island of Barbadoes. On his death, in 1660, his titles became extinct, and the island of Barbadoes devolved upon the Earl of Kinnoull, who sold it to the crown in 1661. Sir George Hay, second son of Peter Hay of Melginche, was Gentleman of the Bedchamber to James VI., who granted him the Carthusian Priory or Charter House of Perth, with a seat in Parliament. He was High Chancellor of Scotland, was created Viscount of Dupplin and Lord Hay of Kinfauns, in 1627, to the heirs male of his body; and created Earl of Kinnoull, Viscount of Dupplin and Lord Hay of Kinfauns, May 1633, to his heirs male for ever. Many of the succeeding members of the family have occupied high positions in the Government of the country, and as plenipotentiaries to foreign courts. The ninth Earl assumed the name and arms of Drummond, as heir of entail of his greatgrandfather, William, Viscount Strathallan, by whom the estates of Cromlix and Innerpeffry were settled as a perpetual provision for the second branch of the Kinnoull family in 1796. The ninth Earl, with his eldest son, father of the present Earl, was appointed Lord Lyon King of Arms. The present Earl succeeded his father in 1866. He married, in 1848, Lady Emily Blanche Charlotte Somerset, daughter of Henry, seventh Duke of Beauford, and by her he has five sons and two daughters. Dupplin Castle, which occupies the site of an older one accidentally destroyed by fire in 1827, is in the Elizabethan style of architecture, and was erected at a cost of £30,000. It occupies a beautiful site on the top of the slopes of which we have been speaking, and its surroundings have been embellished by all that art could suggest and wealth command. At the south front of the Castle is a neat flower garden, with figures artistically cut in the grass. The garden is further ornamented by four very unique bronze vases of huge proportions. At the south side of this garden there is a four-feet wall beautifully clothed with ivy. To the right is a handsome balustrade stair leading to the lower flower garden, with neat bronze vases placed along the top at intervals. On the lower grounds the walk curves gently to the left round the top of a deep den, with a fine run of water, which has been diverted so as to form a pretty waterfall opposite the Castle, thus heightening the effect very considerably. The banks of the ravine have been most tastefully planted with rhododendrons, ferns, roses, bays, and yews, with here and there fine specimens of *Cupressus Lawsoniana* and *Cedrus Deodara* growing luxuriantly

"Beneath the Castle's sheltering lee."

To the east again, we have the Horse-shoe Den, forming a part of the other, and planted very much in the same manner. Here there are some of the finest specimens of *Abies Albertiana* in Scotland. They are about thirty years of age, and are consequently amongst the first planted in this country. They are at least 30 feet in height, and are beautifully furnished. Fine views of the glen are obtained from the windows of the Castle; and when the rhododendrons and other plants are in flower, the effect is most charming. At the top of the den, and stretching away to the south and east, there is another flower garden, with large clumps of many-coloured rhododendrons, each figure consisting wholly of one variety, edged with hardy mixed azaleas. These plants have not as yet attained a very large size, and here and there open spaces occur in the figures; but advantage has been taken of these to sow the single scarlet poppy, which, when in flower, produces a very striking effect. Here there are also a few very fine specimens of Swedish junipers, Irish yews, &c.; and in the centre of the garden is a small stream of water crossed by a neat rustic bridge. The banks of the stream are planted with large clumps of bays and common yews, all neatly and evenly cut back to within two feet of the ground. There are also some fine specimens of araucarias averaging about 30 feet in height; and several large figure beds of rhododendrons, each figure being planted with one sort only, and bordered with azaleas. There are also some very fine trees of different varieties here and there, of which we shall have more to say afterwards. This garden is intersected by fine gravel walks, and is always kept in splendid order.

Although the Castle appears, when viewed from the railway, as if it were shut out from the surrounding landscape, this is not really the case, as both from the windows, and from the policies and gardens magnificent views are obtained, all obstructions having been removed that were likely to detract from the amenity, while the utmost care has been taken to conceal the means by which the prospect is opened up. Straight at our feet lies the whole of Lower Strathearn, with the river winding its crooked way to join the Tay. Yonder is the busy iron thoroughfare, along which travel, in rapid succession, the vehicles of modern transit, the eye following the streaks of snowy vapour until they are buried underneath the verdant Hill of Moncreiffe. Westwards, we trace the whole range of the Ochils. Eastwards, we look towards the thickly-wooded rock which culminates in Moredun top, and gaze on the prettily-situated village of Bridge of Earn, which nestles under its shadow. All along the line in front are noble seats famed for their beauty, and, in the case of some of them, celebrated in song. Here we look upon the lordly Duncrub, there is Freeland, Condie, and Ardargie, and far away to the west are the woods and mansions of Ardoch. Wherever the eye turns it rests upon scenes of great natural grandeur, and we are irresistibly compelled to halt in mute admiration.

> "How oft upon yon eminence our pace
> Has slackened to a pause, and we have borne
> The ruffling wind, scarce conscious that it blew,
> While admiration feeding a the eye,
> And still unsated, dwelt upon the scene."

The magnificent surroundings of Dupplin Castle did not fail to attract the attention of Her Majesty when she paid her memorable visit on 6th September 1842—"the pretty view of the hills on one side, and the small waterfall close in front of the house," being specially mentioned in *Leaves from the Journal of our Life in the Highlands*. This visit is one of the notable events in the history of Dupplin. The royal party entered the policies by the eastern gate, approaching the Castle by an avenue skirted by majestic trees, with glimpses at intervals of park, landscape, and lawn, which could not fail to catch the observant and admiring eye of the Queen. Her Majesty was received by a most distinguished party, and the Royal Standard, specially sent for the occasion by the Lords of the Admiralty, floated from the turrets of the Castle, on which were also displayed the colours of the Royal Perthshire Rifles, of which the Earl of Kinnoull was Colonel. The royal party were first conducted to the Baronial Hall, and were afterwards ushered into the Library, where loyal addresses were presented from the county, the address to the Queen being presented by the Earl of Kinnoull, and the address to Prince Albert by Mr Home Drummond, M.P. for the county. Addresses were also presented to the Queen and Prince Albert on behalf of the city of Perth, the Lord Provost (the late Mr Charles Graham Sidey) having the honour of kissing her Majesty's hand. The royal party then partook of the hospitality of the Castle in the great dining-room, a large party of the nobility having the honour of sitting at the table with Her Majesty and the Prince Consort. The whole of the proceedings upon this memorable occasion reflected the greatest credit upon all concerned, and the royal visitors were exceedingly gratified at the enthusiasm which was displayed.

After perambulating the policies of Dupplin, there is no feature which will strike the visitor more than the systematic plan upon which the grounds appear to have been originally laid out. There is no record to show when the older trees were planted, but from the geometrical character of the design, they evidently belong to the period of the introduction of the Dutch landscape system, and were, therefore, probably planted during the reign of William and Mary (1689-1702). The design is somewhat similar in character to the grounds at Hampton Court, and several other equally well-known places in England. From accidents and other causes, the original plan has been very much disturbed, but enough still remains to show its artistic character. At first sight, the rich timber which clothes the slopes of the hill upon which the Castle is built appears as if it had been planted upon no particular system; but a closer examination brings out the fact, that those trees are all part of a harmonious plan, which cannot now be described in its entirety. The main feature in the design is the predominance of avenues, roundels, and octagons—the evident intention having been that the Castle should be in the centre of those figures, as it probably was in earlier times. The design is still tolerably complete, and it may be traced with some little interest. Along the side of the road leading from Forteviot Station there is a fine double oak avenue, ending in a roundel at the South Lodge. A continuation of this avenue extends from the South Lodge to the garden, all the trees being of large dimensions. Above this oak avenue is a double avenue of beech trees, all of large size, and some of them amongst the finest in Scotland. One of these to which we applied the tape measured 19 feet in girth at 1 foot from the ground, and 15 feet 6 inches at 5 feet. Another in the same avenue girths 18 feet at 1 foot, and 13 feet at 5 feet; while a third girths 14 feet at 1 foot, and 11 feet at 5 feet, with a bole of 30 feet, and an entire height of about 120 feet. This double avenue, with its two rows of grand trees on either side, for about a quarter of a mile, is really one of the most magnificent walks of the kind in Scotland, and at one time it was proposed to make the approach from Forteviot Station to the Castle through its stately portals. Nearer the Castle is another double avenue of beech trees, terminating in a roundel leading to the Parsonage. Between these two beech avenues there is a most thriving plantation of *Abies Douglasii* called "Oswald's Strip," in honour of the planter, the late Mr James Oswald, who was for thirty years gardener at Dupplin, and latterly Superintendent of Wellshill Cemetery, Perth. They synchronise with the *Douglasii* trees at Murtly, and their luxuriant growth is accounted for by the very favourable soil, which is light and friable, with free drainage underneath. To the south of the Castle is a fine lime avenue of great age, and, like the oak avenue, leads to the South Lodge. There are several other arboreal features in the same neighbourhood. At the south of the Bowling Green we noticed a few of the finest specimens of

A. Douglasii glauca in Europe; and beside these is a magnificent specimen of *Picea Cephalonica*. Here, also, is one of the finest, among the many fine silver firs in the policies. At 1 foot from the ground it girths 18 feet 10 inches, and at 5 feet the girth is 16 feet. It has a beautiful straight stem of 30 feet. In this neighbourhood, also, there is a Norway spruce of a very remarkable formation, having a numerous offspring attached to it. The parent tree girths 6 feet 6 inches at 5 feet, but the "parent" forms only a small part of the plant. The lower branches have spread to a considerable extent, and taken root, several of them having grown into pretty large trees. One of these rooted branches girths 5 feet 2 inches at 5 feet up. Another branch crawls, serpent-like, along the ground for 61 feet, but has not taken root, although one of its lateral branches has set out on its own account. The length of this great branch, and its peculiar creeping habit of growth, adds very much to the picturesque appearance of the tree. There are altogether eight thriving trees springing from the parent stem.

A nice walk leading to the Den is planted on either side with *Cupressus Lawsoniana* of large dimensions, alternating with fine specimens of old Scottish yews. At the foot of this walk a neat rustic bridge crosses the burn, and from this point one of the most magnificent views in the policies is obtained, embracing the Castle, with the silvery waterfall in front, and the Den beneath.

Leaving the Den, the burn winds through the policies, passes the old burying-ground of the Kinnoull family, and falls into the Earn about a quarter of a mile below the Castle. On the north side of the burn is another beautiful walk leading towards the garden, on each side of which there is a magnificent avenue of silver firs, which were specially brought under the notice of Her Majesty and Prince Albert by the late Earl of Kinnoull the royal party admiring them very much. To the north of the Castle, again, are some of the largest silver firs in the country and amongst the first planted in Scotland. One of these girths rather more than the fir at the Den, being 20 feet at 1 foot, and 17 feet at 5 feet. It has a grand bole of 40 feet; after which it breaks into two great branches, the entire height of the tree being about 120 feet.

The present Earl of Kinnoull is well known for his keen arboreal tastes, and his lively personal interest in arboriculture is shown in many ways. Several years ago his Lordship made a beautiful avenue approach to the castle from the Perth and Glasgow Road, through a very fine plantation, the trees in which are thriving admirably. On the western side of this new avenue, 10 acres of excellent land were laid out as a pinetum about twelve years ago. This interesting portion of the policies is finely sheltered, and intersected in all directions with grass walks 20 feet in width. Amongst the more notable of the conifers in the pinetum we may mention *Pinus Cembra, Picea Lowii, Abies Menziesii, Picea Nordmanniana, Picea nobilis, Abies Hookerii, Pinus monticola, Picea grandis, Thujopsis borealis, Abies Albertiana, Cupressus Lawsoniana, Pinus tuberculata, Juniperus Chinensis,* many fine specimens of the glaucous variety of the *Douglasii; Taxodium sempervirens, Picea Cephalonica, Thuja Warreana, Thuja Lobbii, Cedrus Deodara, Wellingtonia gigantea, Cryptomeria Japonica,* &c., &c. All these varieties of the newer conifers are growing well, and some of them are making exceptionally good progress. The largest of the trees is *Abies Douglasii*, which is over 60 feet in height. *Taxodium sempervirens* has reached over 50 feet in height. *Cedrus Deodara* and *Picea Cephalonica* are each about 45 feet high. These trees are all being reared under most favourable circumstances, and an excellent opportunity is thus presented for testing their value as forest trees.

Leaving the pinetum, and wending our way along the numerous broad grass walks, and through the seemingly endless avenues, we meet with many other objects of much interest both to the ordinary visitor and to the arborist. After proceeding a short distance along the road to Perth, we reach one of the great features in the original design. The road is skirted on either side with many magnificent specimens of the newer coniferous plants and aged hardwood trees, and before we can distract our attention from these we find ourselves in the centre of one of those grand beech octagons, which form so prominent a feature in the Dutch landscape style. This octagon is admirably designed in accordance with the ideas of the time at which it was planted, and was evidently planned so as to open up what is even yet a most charming prospect. Following this road, which leads to the East Lodge, we pass, quite near the Castle, a fine row of *Araucaria imbricata*, planted about twenty years ago, and growing evenly and well, the average height being about 20 feet, or about one foot of growth for each year. Clumps of rhododendrons have recently been planted in front of these to

THE DUPPLIN ARAUCARIA.

heighten the effect. The finest specimen of *Araucaria imbricata* to be found in the policies—if not, indeed, in the whole country—is situated in front of the conservatory. It was planted about thirty years ago by the late Mr Oswald, and is now 46½ feet in height, its fine prickly branches growing close to the ground. The stem girths 5 feet 7 inches at 1 foot from the ground, and the circumference of branches is 65 feet. It is altogether a grand specimen, and secures the admiration of all who see it.

Almost every variety of tree grown in this country has several remarkable specimens at Dupplin. It would be tedious to enumerate all of these, and we shall therefore only refer to a few more of those which struck us as being most noteworthy, and which may be taken as a type of many that are equally good. Probably the best oak upon the property is one standing near the north front of the Castle—

> "Full in the midst of his own strength he stands,
> Stretching his brawny arms, and leafy hands."

Immediately above the roots, being 2 feet from the ground, it girths 16½ feet, and 14 feet at 5 feet. It has a fairly good bole, and a spread of branches of 80 feet. In the same neighbourhood there is a very fine Spanish chestnut, girthing 23½ feet at 1 foot, and 20 feet at 5 feet, with a very good bole, but rather knotty. Another tree of the same variety girths 19 feet 9 inches at 1 foot, and 16 feet 6 inches at 5 feet, with a splendid bole of 20 feet, supporting several arms of gigantic size. There is another tree that is very largely represented in the policies, the familiar tree of which Churchill sings—

> "That pine of mountain race,
> The fir, the Scots fir, never out of place."

One of the grandest specimens is to be found within sight of the Castle, a veteran of about 200 years of age, and of huge proportions. At 1 foot from the ground it girths 13 feet, and at 5 feet the girth is 11 feet 6 inches. It has a clear bole of 15 feet, and an entire height of about 80 feet. Another is slightly larger, and quite as good a specimen, the girth at 1 foot being 13¼ feet, and at 5 feet, 11 feet 6 inches,—exactly the same as the other, the bole, however, being 20 feet, or 5 feet larger than the one previously mentioned. Several larches of large size and fine quality are to be met with all over the property. One of these girths 12 feet 3 inches at 3 feet from the ground. A spruce tree in the neighbourhood of this larch is almost exactly the same size. In the latter portion of a beautiful walk leading to the gardens there is an ancient avenue of "sable yew," bearing numerous traces of having been made serviceable in the making of bows in the barbaric days of yore, before an advancing civilisation had discovered gunpowder. One of these ancient yews girths 10 feet at 3 feet from the ground, and another has a spread of 21 yards.

The newer coniferous trees are also largely represented throughout the policies, as well as in the pinetum. Amongst the more noteworthy of these is a very nice *Taxodium sempervirens*, with a girth of 6 feet 6 inches at 1 foot, and a height of about 50 feet. *Thujopsis borealis* has been very largely planted. The variety is exceedingly fine, and there is scarcely a bad plant amongst them. To the east of the large araucaria is a fine specimen of *Cryptomeria Japonica*. At the back of the garden there are many beautiful specimens of *Cupressus Lawsoniana*. One of these, a very compact tree, finely feathered to the ground, measures 33 feet round the branches. A *Cedrus Atlantica* measures $15\frac{1}{2}$ feet at 3 feet from the ground; another measures 13 feet at 3 feet. One of the most interesting of the newer coniferous trees is a very good *Wellingtonia gigantea*. This specimen made its appearance in Dupplin very mysteriously—in fact, it is said to have, like Jack's famous bean-stalk, grown up in a single night, at a time when a *Wellingtonia* could not be got for love nor money. Its appearance was first noticed on the 9th July 1859, after one of those interested in Dupplin woods had returned from a midnight visit to Murtly. It is said that the visitor did not find matters at Murtly to proceed at first as satisfactorily as he could wish; but the "pocket pistol," which proved so potent at Dunkeld when the larches were first introduced, had the effect of making matters more pleasant all round. This was on the 8th July 1859, the day before the *Wellingtonia* made its mysterious appearance at Dupplin. How it got there nobody can tell; but it is generally supposed that, like Topsy, it "grew," although a ticket has now been attached, conveying the information that it was "planted" on the day we have mentioned. The tree now girths 9 feet 9 inches at 1 foot from the ground, and 7 feet 4 inches at 5 feet. It has not adapted itself very well to the confined locality in which it has taken root. Opposite this *Wellingtonia* is a magnificent purple beech,—not so remarkable, perhaps, for its size as for its beautiful form.

The plantations on the Dupplin property are quite as interesting as the policies, although they do not, of course, present the same variety of timber. The systematic planting of the estate was commenced upwards of a century ago, as Pennant, in describing his visit, mentions that Lord Kinnoull was at that time (1769) engaged, with great spirit, in carrying out extensive improvements both in planting and in agriculture generally. The then proprietor planted, according to Pennant, "no fewer than 80,000 trees, besides Scots firs," so as to provide "future forests for the benefit of his successors and the embellishment of his country." During the time of the late Earl, as well as since the present one came into possession, due advantage has been taken of the sites that were most suitable for extensive planting.

In any reference to the plantations belonging to the Earl of Kinnoull, the mind instinctively turns first to Kinnoull Hill, of which Perth people are so justly proud. This eminence, we need hardly stay to remark, is one of the greatest attractions of the "Fair City"; and no visitor can say that he has seen the real beauty of the situation of Perth until he has ascended Kinnoull's "craig-encircled hill," and feasted his eyes upon the gorgeous panorama spread out before him. Kinnoull Hill has attracted the attention of travellers from a very early age, not only on account of the magnificent prospect it affords of a large extent of surrounding country and its fine woods, but also on account of its geological formation. St Fond, in his *Travels through Scotland*, states that a desire to examine "the volcanic mountain of Kinnoull" determined him to pass through the town of Perth. He was highly pleased with his visit, and obtained such a collection of lava and agates that he spent a whole night in sorting and ticketing the specimens. The public are not only much indebted to the Kinnoull family for free access to the hill every lawful day, but for the construction to the summit of a pleasant walk or drive, about one mile in length. The path winds through the grateful shade of flourishing larch, spruce, and Scots fir trees, with openings here and there through which glimpses are got of the glorious scene which meets the eye when the summit is attained. The highest peak is marked by a huge stone table, and from this point the visitor looks down upon one of the grandest scenes that can be imagined. Right in front Moncreiffe Hill raises its "piny height," and spreads out its fertile fields. Through the valley at our feet flows the silvery Tay, broad and deep, and all along its course the busy fishermen may be seen capturing its finny treasures, while an occasional steamer or a heavily-laden ship imparts to the scene an air of mercantile activity. Away to the south-east we scan the Ochil Hills, the Lomonds of Fife, the ancient Castle of Elcho,—memorable from its association with Wallace,—and the villages of Newburgh and Abernethy in the foreground. About 200 yards farther to the east of the stone table there is a ruinous tower, from which an even more enchanting view can be had. The view from Kinnoull Hill has exercised the muses of many local poets, but none of them describe its glories in more glowing terms than the Rev. Dr John Anderson, who, in his *Pleasures of Home*, bursts out in pleasing verse—

> " Here rise thy Gothic glories, proud Kinfauns!
> Amid umbrageous elms and swelling lawns,
> There Elcho frowns as grimly as of yore,
> When mail-clad Wallace swam from shore to shore;
> And yonder looms the lonely spectral tower,
> Round which the clouds of grey tradition lower;
> While 'mid the vale, with Earn's wanderings bright
> The camp-crowned Moredun lifts his piny height.
> There Scone, half-buried 'mid the olden trees,
> That speak of regal crowns to every breeze.
> Lo! green Dunsinane rises o'er the plain,
> For Shakespeare famous and the murderous Thane
> And yonder Birnam stands, as then it stood,
> When onward marched its dark, portentous "wood."
> Lo! what a vision of the rolling Tay,
> Leaving his mists and mountains far away;
> Like a bold chief, he brings from Highland hills
> His mingling myriads of resounding rills,
> And ends his race beneath yon castled steep
> That rears its hoary head above the restless deep.'

The natives of Perth have so long been accustomed to admire the rich umbrageous covering of Kinnoull Hill, that it is a little difficult to imagine that it was not always so beautifully clothed, or that there was ever a time when the pine had no place upon its delightful slopes. There was a time, however, when it presented a very different aspect. In ancient times this hill, like a great portion of central Scotland, was a dense forest. The principal wood which then flourished upon Kinnoull Hill was the British oak, and there is some slight evidence to show that the trees must have been of considerable size—Mr Cant, in his notes to his edition of Adamson's *Muses Threnodie*, published in 1774, stating that it was the oak timber on Kinnoull Hill "which produced the great beams in St John's Kirk above four hundred years ago." The Hill of Kinnoull proved no exception to the destructive tendencies of the ruthless ages which brought about the disforesting of the greater part of the country; and this fine forest passed away so completely, that in 1774 Mr Cant, in the note above quoted, mentions that Kinnoull Hill "has been for many years no better than a barren common." He mentions, however, with evident pleasure, that the proprietors of the hill at that time were showing that they were fully alive to the advantage and importance of maintaining such a commanding eminence as something better than "a barren common." He says:—

"The industry and improvements of Lord Gray towards the east, of the Earl of Kinnoull in the middle, and of Doctor Threipland towards the western part, will soon change this dreary desert into a beautiful plantation of fir, oak, and other useful trees, and good arable land on the skirts. The planting on Doctor Threipland's part of the hill is in good heart and forwardness, and is a considerable ornament, not only to the estate, but also to the town of Perth, from whence it is seen from the opposite side of the river which separates it from the South Inch."

We are now, fortunately, in a position to enjoy the beauties of the work which was commenced over a century ago, and can most heartily endorse the opinion then expressed by Mr Cant, that the planting is not only a considerable ornament to the estate itself, but also to the town of Perth, as there is nothing in the neighbourhood which strikes the visitor more forcibly than the fine effect of the luxuriant timber on Kinnoull Hill,—an effect which is not a little enhanced by the many handsome villas which adorn its lower slopes. There are altogether about 100 acres of wood upon Kinnoull Hill. The other plantations upon his Lordship's estate are much similar in character, and extend in all to about 2250 acres, the wood being principally larch and Scots fir.

XX.—ABERCAIRNY.

History connected with the Property—Notable Proprietors—The Mansion—Queen's Visit—The Old Mansion—A Venerable Ash—Design of the Policies—Remarkable Trees surrounding the Policies—The Nursery—Remarkable Trees within the Policies—The Den of the Muckle Burn—Lady Fanny's Grass Walk—The Shrubbery—Inchbrakie—Historical Yew—The Village of Fowlis-Wester—The "Tree of Fowlis"—The Araucaria in the Manse Garden—Druidical Remains—The Plantations.

BOTH as regards the richness of its sylvan beauty, and its many interesting associations, Abercairny, the seat of Charles Stirling Home Drummond Moray, presents considerable attractions. It is situated three miles east of Crieff, and is a place of great antiquity, being mentioned in the *Chronicles of Scotland* as existing contemporaneously with the Abbey of Inchaffray, founded before the year 1200 by Gilbert, second Earl of Strathearn. This Abbey is situated in the parish of Maderty, and was built on ground previously sacred to the founder and his Countess, who had laid within it the dust of their first-born, and selected it as a place of sepulchre for themselves. The Abbey, with a few acres surrounding it, is now the property of the Earl of Kinnoull, but nothing remains of the ancient building but a ruined gable and a mass of fallen stones, pointing to the mutability of all things earthly. The property of Abercairny has been in the possession of the ancient and honourable family of Moray, descendants of the Earls of Strathearn, for more than five centuries. The place won distinction for itself at a very early period. Prior to 1138, Abercairny was signally ennobled in the person of Malise, the first earl, who took a conspicuous part in the Battle of the Standard, and the foundation of whose castle is still traceable at a place called Castletown in the neighbourhood. The Battle of the Standard was fought between the Scots and the English on Cotton Moor, in the neighbourhood of Northallerton, in the year 1138, the object of the Scots being to vindicate the right of Matilda, daughter of Henry I. to her father's throne, against the usurper Stephen, who had deposed her. King David I., who commanded the Scots, proposed to begin the battle with an attack by the men-at-arms and the archers, who were clothed in armour of steel. The Galwegians claimed the honour of leading the attack, as theirs by right of ancient custom; but Malise interposed, and, on remonstrating with the King, indignantly exclaimed— "Whence arises this mighty confidence in these Normans? I wear no armour, yet they who do shall not advance before me this day." The Galwegians were called Normans because most of them were English subjects of that race, who had taken refuge at the court of the Scottish King, and joined his army. To the proud words of the Earl, Alan de Percy retorted—"Rude Earl, these are proud words, but, by your life, you shall not make them good this day." The King was compelled to yield to the demands of the men of Galloway, but, notwithstanding their armour of steel, the "Normans," as Malise tauntingly called the vanguard, did not justify the confidence of the monarch, as the battle ended so disastrously for the Scots that it is said they lost 10,000 men, and Earl Malise had to be surrendered to England as one of the hostages for the maintenance of the Treaty of Peace which followed. In 1346 the title was forfeited in the person of Malise, the seventh Earl, during the minority of David Bruce, in consequence of the Earl's hostility to Edward Baliol. Sir Maurice Moray, the son of Mary, sister to the seventh Earl, who married Sir John Moray of Drumsargard, the lineal heir of Sir Andrew Moray of Bothwell, afterwards had the title restored to him by King David II.; but Sir Maurice Moray having been slain at the Battle of Durham, in 1346, and leaving no issue, the title was given to the King's nephew, Robert, High Steward of Scotland, and was ultimately annexed to the Crown by James II. in 1454. The Morays of Abercairny are thus not only the lineal representatives of the ancient Earls of Strathearn, but of the Lords of Bothwell as well. Two members of the family, Andrew Moray of Bothwell, and his eldest son, George, fell at Flodden, while George's son was slain at the Battle of Pinkie. Another member of the family, Sir David Moray of Gorthy, son of Sir Robert Moray of Abercairny, held the important office of Keeper of the Privy Purse to Prince Henry, eldest son of James VI., and was greatly distinguished for his scholarship and poetic genius. In 1611 he published a volume of poems, entitled "The Tragicall Death of Sophonisba; Coelia, containing certain Sonets and small Poems, and a Paraphrase of the civ. Psalme." The volume, which was dedicated to Prince Henry, was reprinted by the Bannatyne Club in 1823, it having become so scarce that a copy realised thirty-two guineas at a public sale. His brother John was minister of Leith, and an intimate friend of the celebrated Andrew Melville. Being a staunch Presbyterian, he suffered a good deal for his opposition to the prelatic schemes of James VI., by whom he was imprisoned for publishing a discourse containing some observations on the Bishops which offended his Majesty. In the troubles of the seventeenth century, the Morays of Abercairny attached themselves to the cause of the Stuarts. An amusing story is related by Dean Ramsay of how one of the lairds of Abercairny was prevented from joining the Rebels. "One of the lairds of Abercairny," says the Dean, "proposed to go out in '15 or '45. This was not with the will of his old serving-man, who, when Abercairny was putting on his boots, overturned a kettle of boiling water upon his legs, so as to disable him from joining his friends—saying, 'Tak' that; let them fecht wha like, stay ye at hame, and be laird o' Abercairny.'" In modern times several of the lairds of Abercairny have obtained distinction in the army. Charles Moray, who married the eldest daughter and heiress of Sir William Stirling, fourth baronet of Ardoch, attained the rank of Colonel; and their second son, William, served ten years in India, and was wounded at Waterloo. He rose to the rank of Major, and, on succeeding to the property of Ardoch on the death of his mother in 1820, assumed the additional name of Stirling. The late Henry Home-Drummond of Blair-Drummond married Christian, daughter of Colonel Moray, and their son, the present laird, succeeded his mother at Abercairny in 1864, and his brother at Blair-Drummond, &c., in 1876. He married, in

1845, Lady Anne Georgina, youngest daughter of Charles, sixth Marquis of Queensberry, and their eldest son, Colonel Henry Edward Stirling Home-Drummond Moray, had the honour of succeeding the distinguished Sir William Stirling-Maxwell, as Member for the county of Perth—a position which he occupied till the dissolution of Parliament in 1880.

The mansion of Abercairny is a spacious edifice, and was built from designs by the late eminent architect, Mr Crichton, Edinburgh. It is in a most handsome style of florid Gothic, and its elegant proportions and beautiful elevation harmonise well with the delightful scenery by which it is surrounded. The building was commenced in 1806 by the grandfather of the present laird, Colonel Charles Moray. The house was not completed when Colonel Moray died in 1810. He was succeeded by his eldest son, James, who, with much taste, had a great turn for improvements, and during his occupancy the plans of the house were frequently changed; so much so, that when he died in 1840 the house was not even then thoroughly finished. His brother, Major William Moray-Stirling, who succeeded, had very different tastes, and was most anxious to have the house completed, and the policies put in proper order. He built additions to the west and east wings of the house, and effected several important improvements upon the property generally. A very effective addition was made by the present proprietor in 1869, in the shape of a handsome tower, from the top of which the beauty of the richly-diversified landscape is seen in all its glory. The front view is to the south, and here a splendid prospect is obtained of the beautiful vale of Strathearn, and, ten miles distant, the Ochil range—some portion of which is the property of the family—closes the view in this direction. To the west, the eye rests upon the umbrageous splendour of the Knock of Crieff and Turlum, the craggy magnificence of the Aberuchills, and the rude peaks of the Grampians. To the east, the eye wanders over another section of the rich and fertile Vale of Strathearn; and northwards, there are extensive plantations and rising grounds, which strike the horizon in that direction. The various alterations which have taken place in the building of this handsome edifice have all been strictly in keeping with the original design, so that the house is as perfect in its outlines as if its erection had suffered no interruption. The mansion stands on an elevated terrace, looking over a beautiful artificial lake, to which swans and other aquatic birds lend life and interest. The Queen honoured Abercairny with a visit on the occasion of her first tour through Scotland, and greatly admired its beauty, and the loveliness of the policies and surroundings. The offices —consisting of stabling, coach-houses, dairy, &c.—are detached from the house some 300 or 400 yards, and are very complete, the design being in thorough keeping with the mansion.

The old house of Abercairny was situated a little to the south-east of the present building, and on the south side of the lake, which is such a prominent and beautiful feature in the policies. The older part of this house was baronial in character, but latterly, from the many additions and alterations which had been made upon it, the design was considerably mixed. The last portion of the old house was removed about forty years ago, its latest occupants being several families engaged upon the work of the estate. The site of the old house was, until October 1881, when it fell, indicated by the trunk of an ancient ash which grew a little to the west of the building. This tree would have been removed long ago but for its kindly association with "the auld house." Mr Drummond Moray had evidently a warm feeling towards this old monarch of the woods, as he caused the stem to be covered with ivy to preserve its outline from decay, and make its form more picturesque. There was a little life in it until lately, but before it fell the vital spark had been extinguished. This interesting tree was reckoned to be about 300 years of age, and was situated in dry loamy soil, at an altitude of about 200 feet, with a southern exposure. The length of the trunk was 20 feet, the entire height of the tree having been 90 feet. The girth of the trunk, at about 3 feet from the ground, was 18 feet 8 inches.

The policies at Abercairny have, in the course of years, undergone considerable change, but the original design has not been altogether departed from. About the beginning of the century, the grandfather of the present laird employed Mr White, the landscape gardener, to lay out the grounds with the evident intention of building the present mansion. The lake in front of the house, consisting of several acres, was part of this plan, which appears to have been rather artistic. This design, however, was partly altered by Major Moray, who succeeded in 1840. During the ten years which he occupied the property so many changes were made that it is now impossible to conceive the exact character of the original design. The flower garden, which is situated a little to the north of the mansion upon gently sloping ground, was also laid out by the grandfather of the present laird, although many improvements, including a charming statue walk, have since been introduced, tending to enhance the beauty of the grounds. The kitchen garden is the same as was attached to the old house, but has also undergone many alterations in accordance with advancing science, and under the management of Mr Brown, the present head gardener, has achieved no mean celebrity for all classes of fruits and vegetables. During the existence of the old house the policies were rather confined by the surrounding farms,—a circumstance which may explain the absence, in any quantity, of very old trees. These farms were gradually taken into the policies as opportunity occurred, and the grounds remodelled. In recent years some very desirable changes have been made upon the policies. Formerly many of the most enchanting views were shut out by trees which had nothing special to recommend them. These have from time to time been removed by Mr Drummond Moray, who has shown a most refined taste in carrying out this work. By the removal of a single tree from a particular place, views have occasionally been opened up stretching over many miles of the fairest country, and adding a a new charm to the grounds. Mr Moray has also introduced another feature into the policies which will be even more marked in after years than at present. Hitherto the parks have been chiefly ornamented with deciduous trees, which, in winter, presented a somewhat cold and cheerless aspect. This defect has been remedied by the planting of clumps of fir trees, including many of the newer conifers, at the most suitable points, so that the scene is considerably brightened during the season when the other trees are denuded of their leaves, while in summer a very pleasant contrast is presented. These fir trees are still quite young, but they are sufficiently grown to show how much they will heighten the general effect.

Although there are comparatively few very old trees, there are several of considerable size both within the policies and in the surrounding lands. The largest oak upon the property is not within the policies, but on the farm of Kintocher, within a short distance from the road leading from Abercairny Railway Station. It throws its "broad brawn arms" right across the Pow, and reflects

> "Its reverend image in the expanse below."

Its roots are planted upon a steep embankment, a single foot from the ground on the one side being equal to 5 feet from the ground on the other. Taking the trunk at 1 foot from the ground on the upper side, the girth is 16 feet, the narrowest girth being 14 feet 4 inches. Still measuring from the upper side, we have a bole of about 8 feet, at the top of which the tree breaks out into several large limbs, with a spread of branches of about 80 feet. The tree appears to be one of a row originally planted to embellish the banks of the Pow. In the same field as this oak are two very large sycamores. In the first of these the bole is short, but it is of respectable girth—the measurement at 1 foot from the ground being 13 feet 3 inches, and 12 feet exactly at 5 feet from the ground. The entire height of the tree is about 70 feet, and the spread of branches about 50 feet. The other tree is much superior, both in size and symmetry. Above the swell of the roots, about 2 feet from the ground, the girth is 17 feet 2 inches, and at 5 feet up the girth is 14 feet 2 inches. There is a fine bole of 12 feet. The tree then breaks out into three main limbs, two of which are equal to the dimensions of large-sized trees. The tree rises to a height of about 70 feet, and there is a spread of branches of about 60 feet.

Before entering the policies we reach the house of the head forester, Mr John Edward. Here a good nursery was laid out about six years ago. There are about two acres altogether, with a fine southern exposure, the nursery being effectually sheltered from the winds of the north and west by thick plantations. It is also well sheltered from the biting east wind by a strip of rising ground and clumps of trees. The soil, however, is of a poor character, being a stiff retentive till, but may gradually be improved by working. The plants chiefly reared are larch, Scots fir, and spruce, with a few silver firs, hard wood, and ornamental trees. During the short time the nursery has been in use, it has been found that the Scots firs thrive best. A portion of these have been raised from seed by Mr Edward himself, the remainder being obtained from the Perth and Crieff Nurseries. The ornamental plants include *Picea Nordmanniana, Pinus Cembra, Pinua Strobus, Picea nobilis, Picea Pinsapo, Cedrus Deodara, Picea Cephalonica, Abies Douglasii, Wellingtonia gigantea, Picea grandis, Picea bifida. Pinus parviflora, Picea Webbiana, Abies Albertiana*, and *Araucaria imbricata*. The latter plants were obtained from Mr Whitton, Methven Castle, and are said to have been raised from seed procured at Bicton, Devonshire, and believed to be the first araucaria seed successfully raised in this country. There are also a few specimens of *Pinus excelsa, Pinus monticola, Pinus aristata*, and *Pinus densiflora*. There are two specimens of the fern-leaf lime tree which Mr Drummond Moray had seen at Moncreiffe, and taken a fancy for. The hardwoods being trained in the nursery include the variegated sycamore, purple-leaf sycamore, ash-leaf maple, Corstorphine sycamore, purple flowering laburnum, purple beech, tulip trees, as well as all the commoner varieties.

Entering the policies by the Auchterarder approach, we proceed to note the more remarkable trees. Immediately on entering the grounds, we meet with three large ash trees crowning a small eminence, the largest of the three girthing 12 feet 3 inches at the narrowest part of the bole. There are several oaks of considerable size scattered throughout the policies, although none of them come up to the dimensions of the one at Kintocher. One of the largest of the policy oaks was struck with lightning about ten years ago, and now presents a rather curious appearance. The top was completely knocked away, the electric fluid passing down both sides of the tree, and splitting it up, so that we can look through the opening. The bark, too, was almost peeled off, but there was still sufficient left to carry up the nourishment. The way in which the tree has recovered from this accident is really marvellous. It has formed a fine new head for itself, while the bark is gradually closing up the wounds. After this accident the tree might have been destroyed, but for the fact that it stands at a sharp curve of the road, and is useful as a turning-point. The largest oak in the policies girths 14 feet 4 inches at 1 foot,— exactly the same as the Kintocher oak at the narrowest point,— and 12 feet 4 inches at 5 feet. Although it is not so large in girth as the Kintocher oak, it is a very much finer tree, having a splendid bole of about 45 feet, and an entire height of about 80 feet. The largest of the walnut trees girths about 12 feet at 3 feet from the ground. It is a very old tree, but, although the trunk is decaying, the top is fresh and vigorous. Mr Drummond-Moray has long taken a deep interest in the more remarkable trees both here and at Blair-Drummond, and, from his careful observations and measurements, we are able to contrast the growth of many of them since the year 1860. This walnut, which stands near the approach to the old house, is the first of those we are able to compare in this way. In 1860 the girth was 10 feet 6 inches at 3 feet from the ground; at present the girth, at 3 feet, is 12 feet, being an increase of 18 inches since 1860. In the same park as this tree there is a hornbeam with a remarkably fine spread of branches. The foliage is very dense; and, as it affords the best shelter in the park, it is the favourite resort of the cattle, the branches being lifted about 6 or 7 feet from the ground by their regularly congregating under its spreading boughs. The girth at the narrowest part of the bole is 6 feet. We have already alluded to the veteran ash which stood near the site of the old mansion, but there are several other very good trees of the same species in full vigour in various parts of the policies. There is a fine one near the spot where the old farm-steading stood. At 1 foot from the ground the girth is 14 feet 6 inches, and at 5 feet the girth is 12 feet 11 inches. Another ash in the same neighbourhood girths 14 feet above the swell of the roots at 2 feet from the ground, and 11 feet 9 inches at 5 feet. Although the latter is the smaller of the two, it is a far better tree, both as regards quantity of timber and shapeliness. The largest Spanish chestnut girths 14 feet 10 inches at 1 foot, and 12 feet 7 inches at 5 feet, with a bole of upwards of 20 feet. A horse chestnut girths 9 feet 5 inches at 1 foot, and 8 feet 7 inches at 5 feet, and has a fine spread of branches. There are several very good sycamore trees within the policies. The largest of these grows about 200 yards south-east from the house, and almost 60 yards south of the approach leading to the house from that direction. At 1 foot it girths 21 feet, and at about 2 feet from the ground it divides into two portions girthing 13 feet 5 inches and 13 feet 3 inches respectively; and, again, at $3\frac{1}{2}$ feet and 4 feet from the ground, these two parts divide each into other two parts, girthing 9 feet 11 inches and 6 feet 11 inches, and 8 feet 10 inches and 7 feet 7 inches. The tree is about 60 feet high, and has a beautifully-shaped top, with a spread of branches of almost 65 feet. It is in every respect a magnificent park tree, and, when in full leaf, has a rich, luxuriant appearance. The two largest Scots firs girth 11 feet 10 inches and 9 feet 10 inches respectively at 3 feet from the ground. There are four very fine Scots firs a little to the west of the house, the largest of them girthing 10 feet 3 inches at 3 feet from the ground. In 1860 this tree girthed 8 feet 7 inches at 3 feet, being a growth of 1 foot 8 inches since that date. The beech with the largest girth grows by the side of Wester Pond. The trunk, which is very short, girths 12 feet at its narrowest part. It has a spread of close, even branches of fully 80 feet.

Leaving the neighbourhood of the mansion, and proceeding along the Crieff approach, we meet with many valuable products of the forest. Here are the two best larch trees upon the property, growing close by the side of the approach. The larger of these is really a splendid tree, being fully 80 feet in height, and having a girth above the swell of the roots, being 3 feet from the ground, of 13 feet 8 inches, and 11 feet 10 inches at 5 feet. Availing ourselves of Mr Moray's former measurements of this tree, we find that in 1860 it girthed 11 feet 5 inches at 3 feet, and in 1876 it girthed 12 feet 8 inches at 3 feet. There has thus been an increase in the girth of this tree since the year 1860 of 2 feet 3 inches, and during the past seven years of about 12 inches. The smaller tree has a girth of 11 feet 10 inches at 1 foot, and 9 feet 11 inches at 5 feet, being a corresponding increase upon the former measurements. The largest elm upon the property also grows by the side of this approach. At 3 feet up it girths 13 feet 6 inches, and at the narrowest part of the bole, being 5 feet from the ground, the girth is 12 feet 9 inches. The roots are very irregular. In 1860 this tree girthed 11 feet 4 inches at 3 feet, being an increase of 2 feet 2 inches since that date. In another part of the policies there is an elm almost equally large, the girth at 1 foot being 13 feet 8 inches, and at 5 feet 12 feet 3 inches. Continuing along the Crieff approach, we next come to the largest lime tree upon the property. Above the swell of the roots, being 2 feet from the ground, this tree girths 10 feet 11 inches, and 8 feet 11 inches at 5 feet. In the same neighbourhood, and chiefly at the west side of the house, there are a number of very good silver firs. The largest of these girths 16 feet 4 inches above the swell of the roots, being 2 feet from the ground; and at 5 feet the girth is 13 feet 11 inches. In 1860 the girth at 3 feet was 12 feet 4 inches. The tree has a fine bole of 20 feet, after which it breaks into two heavy branches, and reaches an entire height of 80 feet, Close beside this fir, and growing by the edge of the approach, is the prettiest spruce upon the estate. It is beautifully feathered to the ground, and is very regularly tapered. At the narrowest part of the bole, being 3 feet from the ground, the girth is 10 feet 11 inches. It has a height of about 70 feet, and is altogether a most symmetrical tree. It is not, however, the largest spruce upon the property, as we met with another having a girth of 11 feet 8 inches at 3 feet from the ground. The largest Weymouth pine, also in this neighbourhood, girths 8 feet 4 inches at 3 feet up, with a fine bole of 20 feet. Beside it is the largest cedar of Lebanon, which girths 9 feet 5 inches at 1 foot, and 8 feet 3 inches at 5 feet,—the entire height being about 65 feet. In 1860, when this tree was measured at 3 feet from the ground, the girth was 6 feet $10\frac{1}{2}$ inches. There are also two very fine cedars of Lebanon growing in forced ground in front of the mansion, and making unusually good progress, owing to the richness of the soil.

As we draw nearer the lodge, the Crieff approach, which originally led to the old mansion-house, increases in sylvan beauty, the general effect being considerably enhanced by openings here and there amongst the trees revealing the "Muckle Burn," or the Cairny, rippling amidst moss-grown boulders, and gurgling over miniature waterfalls. Ever and anon we pause to listen to the enchanting music of the bubbling stream fresh from the mountains, or to admire the fair proportions and rich foliage of the trees and evergreens which line its banks. At almost every step we may rest in one of those fairy-like spots for which Hunt sighed so much—

> "Oh! for a seat in some poetic nook
> Just hid with trees and sparkling with a brook!"

Crossing the Crieff Road, we enter the North Glen of the Muckle Burn, one of the most charming retreats in a neighbourhood peculiarly rich in the most enticing forms of natural scenery. The course of the burn is in some places very rugged, the water having to leap over rocks of considerable altitude, and on both sides the banks are picturesquely adorned with forest vegetation. Neat walks have been cut along the greater part of each side of the burn. The walk up to the first waterfall has been in existence for a good many years, but about twelve years ago Mr Moray continued the road to the upper fall,—a distance of one mile and a half from the public road,—and erected a number of rustic seats at the most desirable points. The burn is crossed at various places by bridges, also of rustic design, from which the visitor can look with safety upon the violent passage of the torrent beneath, as it rushes over the precipitous rocks to the deep pools which it has scooped out for itself far down in the Glen. The lower part is enriched by some splendid trees of various kinds. There are several fine larches of about eighty years of age. One of the trees that struck us as being amongst the most remarkable in the Glen is a splendid ash rising to a height of fully 100 feet. It is situated immediately below the first waterfall, so that there is always plenty of sap about the roots, without the water ever becoming stagnant. There are many nice openings amongst the trees, from which delightful peeps are got of the surrounding country. The best view is met with almost half-way up, where the visitor can rest upon an oak seat, and leisurely admire a magnificent panoramic scene. The view embraces the romantic Castle of the Drummonds and Drummond Park, studded with numerous heavy trees, and surrounded by dark plantations of fir. South-east of Drummond Park we see the blue smoke of Muthill curling amidst embowering woods, the ground behind gradually rising until it assumes the proportions of a considerable hill. The Ochils are seen in the extreme south-east dipping into the plain below Auchterarder.

Having admired the beauties of the "Muckle Burn," we re-enter the Crieff approach, and proceed along the high walk which overlooks it, and again find much that is interesting. Not the least of the attractions is the magnificent view which is here to be had of Lady Fanny's Grass Walk,—three-quarters of a mile in length and about 45 feet wide,—the entire walk being lined with magnificent trees. The Cairny runs along the east side of the Walk, and on the east side of the burn there is a row of splendid limes. On the west side of the burn there is another row of equally good lime trees, the burn flowing quietly between the two rows of limes. On the west side of the Walk there is a fine row of beech and oak trees, mixed, so that there are altogether three rows of trees, one row being on either side of the Walk, forming a close avenue, and the other lining the outer side of the burn. This handsome walk is supposed to have been constructed about 1750, by Lady Frances Montgomery, sister of Lady Christian Moray, wife of James Moray, the thirteenth laird of Abercairny, and great-grandfather of the present proprietor. Lady Frances stayed at Abercairny with her brother-in-law for some time after the death of her sister, and took charge of the children. The avenue, as already stated, has a straight course of three-quarters of a mile; but this is not the entire length of the Walk, as it afterwards takes a sharp turn to the east, and continues for about a mile farther. Proceeding along the high walk above the Crieff approach, we come across the second largest of the silver firs upon the property. At 1 foot from the ground the girth is 16 feet, and at 5 feet the girth is 13 feet 4 inches. Beside it is a beech girthing 10 feet 9 inches at 1 foot, and 10 feet 6 inches at 5 feet. A little farther on we come to the shrubbery, which contains a number of rare and beautiful plants. The shrubbery covers a considerable space, and has been greatly enlarged by the present laird. It is intersected by two grass walks of 30 feet wide. Amongst the most notable of the trees here is a *Wellingtonia gigantea*, coming away nicely, and now about 35 feet high. In the middle of the shrubbery there is a clump of silver firs of large size. There are also fine specimens of most of the newer coniferous plants, including *Picea nobilis*, *Picea Nordmanniana*, *Abies Hookerii*, *Pinus excelsa*, *Pinus monticola*, *Pinus Cembra*, *Taxodium sempervirens*, *Abies Menziesii*, *Cupressus Lawsoniana*, &c. The largest *Taxodium sempervirens*, which is situated in the flower garden, is 35 feet in height, and has a girth of 5 feet 2 inches at 1 foot, and 5 feet at 3 feet from the ground. The largest *Abies Menziesii* is 42 feet in height, the girth at 1 foot being 4 feet 3 inches, and at 3 feet 3 feet 10 inches. The largest *Cupressus Lawsoniana* reaches a height of about 35 feet, and at 3 feet the girth is 3 feet 7 inches. On leaving the shrubbery we find ourselves at the back of the mansion-house, where the largest beech upon the property grows. At 1 foot the girth is 12 feet 6 inches, and at 5 feet 11 feet 9 inches. There is a good bole of 8 feet, at the top of which the tree breaks into three large limbs. The tree is most shapely in form, and is exceedingly ornamental.

The estate of Inchbrackie, immediately adjoining Abercairny, was purchased by Mr Drummond Moray in December 1882, the property having previously belonged, for eleven generations, to the Graemes, who are in direct male descent from William, first Earl of Montrose, by his third wife, Christian, daughter of Thomas Wavan of Stevenson, and relict of Patrick, sixth Lord Hallyburton. The first Earl of Montrose, as is well known, was killed at the Battle of Flodden. The remains of the ancient Castle of Inchbrakie, which are surrounded by a moat, are in the vicinity of the present house. The Castle is understood to have been destroyed by Cromwell, to punish the proprietor, Patrick Graeme, an active officer in the army of his cousin, James, fifth Earl of Montrose, who was created Marquis on 16th May 1644. A large yew tree within the moat is said to have afforded concealment by its thick foliage to this celebrated Marquis of Montrose during the troublous times of 1646. The girth of this tree at 1 foot from the ground is 11 feet 6 inches, and at 5 feet the girth is 10 feet 7 inches. The length of bole is 7 feet, the entire height of the tree about 35 feet, and the spread of branches 39 feet. The finest larch on this property grows on slightly rising ground. It girths 12 feet 5 inches above the swell of the roots and below any of its large curved branches, about 3 feet from the ground. There is a very fine old ash tree in "Front Park," in a group of old sycamore trees, the girth being 14 feet 9 inches at 3 feet from the ground. There are also a number of handsome oaks, limes, and beeches. One of the largest of the latter girths 11 feet 8 inches above the swell of the roots. Close to the mansion there is a fine old avenue about 80 yards in length, known as "The Beech Walk." The total amount of wood upon this property is about 76 acres.

Little more than a mile beyond the park wall of Abercairny lies the village of Fowlis-Wester, snugly notched into a natural cleft in the hillside. One or two graceful and picturesque sycamores stand here and there amongst its old-fashioned thatched houses and straggling kailyards. Fowlis at one time seems to have been filled with busy weavers, and to have been surrounded by a good many smaller villages or clachans, each of which held its looms, and had its acres or crofts. These neighbouring clachans are now entirely swept away, and the only trace of their former whereabouts is here and there a group of old ash trees. The last to disappear of such villages was Castleton, on the Fowlis-Wester estate of Sir Patrick Murray, in the midst of which stands a huge grass-grown mound, the ruins of the ancient castle of the Earls of Strathearn. Fowlis is in an interesting transition state, passing from a mouldering clachan into a pretty, tidy village. This is due to the fostering care of the Laird of Abercairny, under whose wise and thoughtful management, encouragement is given to thrifty labourers to occupy well-built slated houses with trim surroundings, while the old thatched cabins are patched and propped as well as may be for the sake of the poor old people, who not only sit within them rent free, but tax the ingenuity and generosity of their landlord to the utmost for means and devices by which the weather may be kept out of their dwellings. Among the modern structures of the village is a very well-built and commodious schoolhouse. Near it is a schoolmaster's house, which serves also as Post Office, Registrar's Office, and as the headquarters of parish business generally. A good inn, with neat porch, and trimly kept flower borders, and a bright sunny aspect, occupies a central position on one side of what may be called the Village Square. In the centre of this Square stands the most ancient relic certainly of former times. It is a stone cross, about 10 feet high and 3 feet broad, similar to those which are seen at Meigle, Glamis, and elsewhere. On one side are figures of men and animals in relief; on the other, the carved tracery which follows the outlines of the cross is quite distinct and very beautiful. This cross stands carefully surrounded by an iron railing, and is in excellent preservation, and well worthy of inspection. A picturesque gateway leads into the churchyard. The arch is semi-circular, like a Roman arch; and a triangular stone above the centre peeps out from among the ivy, with its boldly-relieved text from the old Scotch version—

"TAK HEID TO THY FOOT WHEN THOV ENTREST INTO THE HOWS OF GOD. 1644."

Above this inscription hang three leaves, points downward, and bound together at the top. Three brothers are said to have come from France,—their cognisance the three leaves (*feuilles*),—and to have settled, one here, another at Fowlis-Easter, near Dundee (which, like the estate of Fowlis-Wester, belongs to the Ochtertyre family), and the third at Fowlis, in Ross-shire. According to this legend Fowlis is a corruption of *Feuilles*. The church is a long, low-roofed building. The walls are old, and the slates rough and clumsy, but much has been done for the inside, which is neat and comfortable, and it held when we last had the pleasure to be there, a large and respectable congregation. The kirkyard is neatly kept.

In the middle of the older portion of the village (where the old wives and the old thatch-roofs prevail) stands one of the handsomest and most widely-spreading dark-leafed sycamores to be seen anywhere. Its site is the summit of a knoll, on which the converging ends of several kailyards belonging to the surrounding houses meet, and it can be seen from all points and from great distances. It is known as "The Tree of Fowlis." The story has come down, and lives in the clachan, that it was planted by a man of Fowlis on a Sabbath-day, and that it was planted " wi' his thoom." In 1860, when this tree was measured by Mr Drummond Moray, it girthed 12 feet 9 inches above the swell of the roots, being about 3 feet from the ground. The girth above the swell of the roots is now 17 feet, being a growth of 4 feet 3 inches since the year 1860. At 5 feet the girth is now 14 feet 3 inches. There is a fine bole of 15 feet, a spread of branches of 75 feet, and a total height of about 70 feet. It is in every respect a most magnificent tree, and is regarded by the people of Fowlis with becoming veneration. On the lawn in front of the manse stands a very fine specimen of *Araucaria imbricata*, which has produced a crop of cones. The tree is about 25 feet in height, and remarkably graceful in its ramification. It was planted by the Rev. Mr Hardy upwards of twenty years ago. The soil in which it is rooted is very stony. It stands about 350 feet above sea-level, and is accordingly comparatively safe from the influence of hoar-frost, from which other trees of the species in the neighbourhood suffer much. Being planted above a "ha ha" dyke, the moisture escapes readily from around it.

The Druidical remains of Fowlis hardly belong to our present line of inquiry, but they have exercised the ingenuity of many learned archæologists. The traces of the presence of the Druids in this parish are so ample that there can be no doubt the district must have been a favourite seat with this ancient people. It is doubtful, however, whether everything popularly attributed to the Druids really belonged to them, and anyone who has spent a little while in examining the Cross, and the church gateway, and the old schoolmaster's Latin epitaph,* and the Tree of Fowlis, and the old biggins and the modern cottages, and who has interviewed a few of the inhabitants, will do well to finish off with a sceptical investigation of a huge flat boulder, on the surface of which either art or nature has carved deep trenches. The more devout affirm that the Druids used this as a sacrificial altar. Those affected by modern rationalism declare that the weather has washed out the softer strata of the stone. A short climb to the top of the wooded hill places the visitor before a genuine and indisputable Druidical relic—a double circle of upright stones, one circle within the other.

Leaving the interesting village of Fowlis, we proceed to examine the extensive woods by which it is surrounded. The plantations on the estate of Abercairny extend to more than 1500 acres, and are well distributed over the property. About 400 acres have been planted by the present proprietor, about 300 acres of which are on land where no trees grew before,—the remainder being on land which had formerly produced a crop of forest trees. The largest tract planted of late years is on Pitlandy Hill, about a mile north of the village of Fowlis, and on the east side of the road leading to Buchanty, where about 170 acres have been planted. Before this was planted, the ground was thoroughly drained with surface drains 18 feet apart, 2 feet wide at mouth, and 18 inches deep, and fenced with six wires on wood posts, 6 feet apart, with iron straining-posts at convenient distances. The plants used were Scots fir, larch, and spruce, with a few silver firs. A great part of it was planted in the spring of 1877, which, it will be remembered, was a very trying spring on newly-planted larch, and, in conse-

* OLD SCHOOLMASTER'S LATIN EPITAPH.—The schoolmaster referred to died in 1746, and his sons seem to have composed the elegy, which says of the old dominie—

" DE QVO QVOD SUPEREST
MONUMENTA APOD VIVOS MAJORE
FIDE EXHIBEBVNT QVAM
EPITAPHII ELOGIA SVSPECTA."

The stone records that the surviving mourners are his widow, "ELIZ. REID," and his sons "PAT. JO. GVL. ET ALEXR.," and the closing line is

" A.M. OMNES,"

as to the precise meaning of which critics are said to be divided in opinion.

quence of the site being very much exposed, a good many of the plants died. Those planted more recently have stood out much better. The Scots fir and spruce are now coming forward, but the Scots firs have been very much injured in consequence of black game eating the buds. Near Connachan Shooting Lodge, on the farms of Connachan and Fendoch, upwards of 40 acres were planted about twelve years ago in differently-shaped clumps, varying in size from half an acre to 11 acres. Before being planted, the sites were all fenced with four-barred fences on stobs driven 3 feet apart. The chief objects the proprietor had in view were, we believe, to beautify that part of the estate, and to afford shelter to the tenants' flocks in inclement weather, and they already to a great extent answer the purposes for which they were intended. They were planted chiefly with Scots fir, larch, and spruce. The larch has suffered very much of late years from the disease, but the other kinds continue to progress satisfactorily.

The young plantations on the land, which has already produced a crop of forest trees, are in a fair thriving state. No trace of the *Hylobius abietis* has been found on the Scots firs. The absence of this pest is no doubt partly due to the land having lain merely as pasture land for several years before it was replanted. The larch, however, has not done so well as could be wished; but on filling up the ground after the first planting Scots firs were chiefly used, as they showed an inclination to succeed best. The other plantations on the property consist also for the most part of Scots fir, larch, and spruce, with, in many cases, a mixture of silver fir, sycamore, ash, and oak, particularly round the margin of the woods and on the sides of the public roads. The Scots firs predominate, and more especially on the south side of the property, where they are about 80 years of age, and are tall, clean-grown, and in a healthy state. The wood containing the greatest number of larch trees is the Low Moor Wood, which is about 50 years of age, and extends to about 230 acres. Numbers of the larch trees are yearly being cut down for estate purposes. The soil varies very much. Where it is thin, and on a gravelly and rocky subsoil, they are showing symptoms of decay; but where the soil is loam, on a clay subsoil, they are still in a thriving state. The Scots firs here, and in several of the other plantations, are suffering very much from the borings of an insect. The injury is chiefly on the stems of the trees, where the insects find their way into the alburnum or sap wood, and perforate it in every direction, thus interrupting the flow of the sap, and ultimately killing the tree. In the policy woods, hardwood trees predominate, and consist principally of oak, beech, elm, lime, sycamore, ash, &c. There are, however, several very fine Scots fir, larch, silver fir, and spruce trees, both by themselves and mixed up with the hardwoods. There are also a number of acres of oak coppice, which are still kept as such, more for the sake of variety than for profit. All the plantations are regularly thinned, and whatever is found to be dead, or in an unhealthy state, is taken out. The principle is to thin often, and to thin sparingly. The plantations are mostly all fenced. On the north side of the property they are fenced with stone dykes, wire on wooden posts, and stobs and rails; and on the south side of the estate they are principally fenced with stone dykes, wire on wood posts, and hedges. With reference to hedges there are few estates in Scotland, if any, where the hedges are kept in such admirable order as at Abercairny.

Mr Drummond Moray has all his life taken a deep personal interest in the woods and remarkable trees upon his extensive property, not only at Abercairny, but, as we shall afterwards show, at Blair-Drummond and elsewhere. This is not only seen in the valuable comparative measurements we have been enabled to give of the more notable trees, but also in the general management of the plantations. Throughout the whole property the utmost taste and skill are displayed in this important department of estate work, due regard being paid to the beautifying of the landscape and to the value of the timber.

XXI.—BLAIR-DRUMMOND.

History of the Property—The Moss of Kincardine—Discovery of Marine Remains—An Ancient Forest—Roman Relics—The Reclamation of the Moss—Lord Kames—The First Mansion-House—The New House—The Policies—Remarkable Oak and Beech Trees—A Memorial of Benjamin Franklin—Remarkable Spanish Chestnut, Sycamore, Lime, and Birch Trees—Remarkable Fir and Larch Trees—Doune Castle—The Silver Firs—Havoc caused by Great Storms—Progressive Growth of Common Trees—Burnbank—The Plantations.

THE estate now known as Blair-Drummond is composed principally of what was formerly the Barony of Kincardine, and of other smaller properties which have from time to time been acquired by the different proprietors since the estate came into the hands of the present family. The first proprietor of this family was George Drummond of Blair, in the Stormont, which property of Blair, and previously the property of Ledcrieffe, he and his ancestors had held in a direct line for eight generations, commencing in 1486 with Walter, the third son of Sir Walter Drummond of Cargill. The present proprietor, Charles Stirling Home Drummond Moray, is the ninth in direct succession from George Drummond, the first proprietor of Blair-Drummond.

Going back to an earlier period, the first proprietors we read of are the Montefixes or Muschets, a very old family, said to have sprung from the Earls of Montfort, who were the Dukes of Bretagne, and to have come from France to England with William the Conqueror, and from England to Scotland with the Princess Margaret, who became the Queen of Malcolm Canmore. The property passed to the Drummonds, in the beginning of the fourteenth century, by the marriage of Sir John Drummond, ancestor of the noble House of Perth, to Lady Mary Muschet, eldest daughter of Sir William Muschet, Justiciar of Scotland. In 1684, George Drummond of Blair acquired, by purchase from the Earl of Perth, the lands of Drip, Cambusdrennie, and others, in the parish of Kincardine, which were erected into a free barony, and called Blair-Drummond, from the name of the purchaser, and that of his paternal estate of Blair, in the Stormont. It was not, however, until the year 1714 that he also purchased from the Earl of Perth the other parts of the Barony of Kincardine, on which Blair-Drummond House was afterwards built, and the park and pleasure-grounds laid out and planted. At the time of the purchase there was no mansion-house on the property. The old Castle of Kincardine, the residence of the former owners, the Muschets, which is believed to have stood in the vicinity of the old church at the western extremity of the pleasure-grounds, had gone to decay, and the land about it had been in a great measure divested of trees, with the exception of the few which were usually raised, according to the custom in earlier times, on the turf dykes that surrounded the small corn-yards of the tenants. Many of these trees are still to be seen at various parts of the property standing in clumps, which, from their age, and the care taken by the successive proprietors to protect them, form a beautiful and effective feature in the landscape.

The property lies in the ancient Stewartry of Monteith, and on the banks of the Rivers Forth and Teith, and is composed partly of the rich and fertile carse land of the district, and partly of dry field soil, in all its different qualities; but a portion of the land, called the Moss of Kincardine, to be afterwards referred to, was covered with deep moss, and has been rendered famous from the extraordinary measures taken to reclaim it. It is generally supposed that the whole of the carse lands here were at a not very remote period, geologically speaking, a continuation of the Firth of Forth. Marine remains of various kinds have been discovered at different times. Not only do beds of oyster, mussel, cockle, and other marine shells appear at various depths between the surface of the clay and the channels of the rivers, but the remains of the more advanced denizens of the ocean have frequently been found. Only a few years ago a portion of the skeleton of a whale was found about a quarter of a mile from the manse, embedded in the clay, which had formerly been covered with moss, and lying on another stratum of moss below the clay. This interesting relic of the prehistoric ages is deposited in the Edinburgh College Museum. The writer of the New Statistical Account, in recording this discovery, mentions another very curious circumstance, of which we have never seen any explanation. "It is very remarkable," he says, "that a small piece of deer's horn, with a hole bored in it, was found along with this skeleton, of exactly the same description as a piece of horn which Mr Bald mentions as found with the skeleton of the Airthrey whale, now in the Edinburgh College Museum." In 1766, when Henry Home, the celebrated Lord Kames, became connected with the property, his wife having succeeded her nephew as heiress of entail, he immediately directed his attention to the improvement of the Moss. The Moss of Kincardine consisted of over 1800 Scotch acres, of which nearly 1500 were on the estate of Blair-Drummond alone. The Moss was of considerable depth, and rested on a subsoil of rich clay, consisting of strata of grey, reddish, and blue colour, of the same quality as the Carse along the Forth from Stirling to Falkirk. On the receding of the sea, and the subsiding of the waters of the Forth, the Teith, and the Goodie, by the formation of deeper channels, this rich flat district was soon covered with oak, birch, alder, and hazel, as well as a few firs. Some of the oaks, the remains of several of which have been found, were of enormous size. In a paper read before the Royal Society of Edinburgh in 1793, it is recorded that "forty large oak trees were lately found lying by their roots, and as close to one another as they can be supposed to have grown." One of these oaks measured 50 feet in length, and more than 3 feet in diameter, and 314 circles, or yearly growths, were counted in one of the roots. In another part of the Moss an oak was found that measured 4 feet in diameter. In the new Statistical Account it is recorded that in 1823 an oak of the following dimensions was found in Blair-Drummond Moss:—Length of bole, 41 feet; circumference at the surface of the ground 16 feet, and at the top 9 feet. One branch, 18 feet in length, girthed 5 feet 4 inches. The solid contents of this tree were over 360 feet. The height of the original stem of this tree is believed to have been about 60 feet. A large oak was also dug out in 1826, and was converted into some beautiful pieces of furniture, which now adorn Blair-Drummond House. From the strength and consistency of the roots, and as they were cut about $2\frac{1}{2}$ feet from the ground, the part where the tree is easiest felled, it is not supposed that these monsters came to a natural end, or were destroyed by gales. It is generally believed that they were cut by the Romans, in their efforts to dislodge the hardy natives, —a supposition which finds corroboration in the circumstance that marks of the size of a Roman axe have been discovered on some of them, where they had been haggled. It is well known that the Romans cleared large tracts of the ancient forests, and their presence in this district is testified by numerous remains, several brass camp kettles, spear-heads, &c., having been found in the Moss within the past century. It has been calculated that the Moss of Kincardine must have been about 1760 years old when it was removed. This computation has been made from the fact, that it was about the year 81 of our era that the Romans established a regular chain of posts in the direction of Kincardine Moss, the Pass of Leny, and the station of Ardoch, where the forests formerly stood; and about fifty years thereafter the Wall of Antoninus was built nearly in the same line. Lord Kames did not make very much progress with the clearing of the Moss, owing to the want of sufficient water-power; but he succeeded in demonstrating the practicability of the plan of floating which he had adopted. When his son, George Home Drummond, succeeded him in 1783, he was specially encouraged to carry on this work by the ingenuity and perseverance of Mr George Meikle, of Alloa, the son of the inventor of the thrashing-machine, who invented a water-wheel for raising a large supply of water for floating the moss. The original model of this wheel is amongst the most cherished relics at Blair-Drummond House, and gives one a capital idea of the ingenious method adopted. The wheel, which was lined with buckets round the whole inner circumference, made about four revolutions in a minute, and in that time raised from 40 to 60 hogsheads of water. This it discharged into a trough connected with it, 17 feet above the surface of the stream which supplied the water and turned the wheel. From this trough the water flowed partly in pipes and partly in an open aqueduct 1754 yards in length into the reservoirs in the Moss. " The water-power being thus procured, the floating by means of spade labour became comparatively easy, and the Forth served as a recipient for carrying the Moss into the Firth. By the end of October 1787, the wheel, pipes, and aqueduct, were all finished, and the result realised the most sanguine expectation of all parties. The total expense exceeded £1000. The tenants voluntarily engaged to pay the interest of the money so expended, but the proprietor generously relieved them from this engagement." The labours of the great wheel commenced in 1787 and ended in 1839, and the workmen necessary to carry out the operations were procured chiefly from the parishes of Callander, Balquhidder, and Killin. The advantage of this great undertaking was speedily apparent, many of the settlers on the Moss afterwards becoming farmers on a large scale. It is pleasant to learn from the New Statistical Account of the parish that the exertions of the laird of Blair-Drummond were heartily appreciated by his tenants.

While the reclamation was in progress, the operations attracted a great deal of attention all over the country, and exhaustive descriptions were published of the methods adopted for carrying out the work, and the results which were achieved. All these descriptions agree in attributing great credit to Lord Kames, and his son, George Home Drummond, for their public-spirited enterprise, and the ability they brought to bear upon an exceedingly difficult problem.

Lord Kames was not only distinguished as a great land improver. He also won for himself fame in the special sphere to which he devoted his life. He was the son of George Home of Kames, in Berwickshire, and was born in the year 1696. He studied for the Scottish Bar, to which he was called in 1724, and his remarkable talents and perseverance soon procured his advancement. He was raised to the Bench in 1752, his title being taken from the place of his birth. In 1763 he was made a Lord of Justiciary. He was a Commissioner of the Forfeited Estates after the Rebellion of 1745, and was a member of the

Board of Trustees for the Encouragement of the Fisheries, Arts, and Manufactures of Scotland. He was also a voluminous and able writer on many important subjects, legal, philosophical, and agricultural. Some idea of his character and disposition may be gathered from the following quaint inscription on an obelisk erected by himself on a prominent eminence to the south-west of the house:—

> "For his neighbours
> As well as for himself
> Was this obelisk erected by Henry Home."

> "Graft Benevolence on Self-Love,
> The fruit will be delicious."

This eminence bears evidence of having been used as a Roman sentinel station, as the remains of a fosse are plainly visible. The mound commands a complete view of the low country around Blair-Drummond, and it would be almost impossible for the Romans to have been taken unawares with such a splendid outlook. Notwithstanding the light, sandy soil, magnificent trees now surround this old military outpost.

The erection of the first mansion at Blair-Drummond was commenced in the year 1715, but, shortly after it was started, the operations had to be suspended for a year, owing to the disturbed state of the country. This house, which occupied a lower site than the present one, was a very plain structure, four storeys high, with seven windows in each row, and wings for the offices and servants' accommodation. After being occupied for upwards of a century and a half, it was taken down in 1870 by the late proprietor, George Stirling Home Drummond, who at the same time erected the new house, which is situated on a beautifully-wooded piece of rising ground, commanding a most extensive and varied view. The plans were designed by Mr J. C. Walker, Edinburgh, and the style is Scottish Baronial, without unnecessary ornamentation. In the east front, which extends to 172 feet, there is a magnificent square tower rising to a height of 85 feet. The principal entrance, which consists of a plain, circular-headed doorway, with rope cable, is in the basement of the tower. The doorway is surmounted by a panel with armorial bearings. In the third stage there is a large oriel window, supported on corbels, and decorated with grotesque heads, the whole being finished with a stone roof. The tower is crowned with angle turrets, and a bold corbelled parapet. Connected with the tower, from the basement floor, is a circular stair tower rising to a height of 123, and finished with an ornamental finial. The south front is nearly 140 feet in length and contains drawing-room, dining-room, and library. These are entered from a spacious corridor, extending to 100 feet, a handsome balustraded stair leading into the flower garden at the west end of the corridor. The south front is ornamented by two circular towers, the one being finished by a gable and the other with a pointed slate roof. These towers contain stairs leading to a gallery. The library measures 50 feet long by 22 feet wide, and is lighted principally by a large oriel window in the centre. Folding-doors lead to the drawing-room. The entrance hall is $46\frac{1}{2}$ feet long by 20 feet wide. All the walls, to the height of 10 feet, are elaborately panelled with wainscot. The exterior walls are built of freestone from the celebrated Plean Quarry, near Stirling, and are pointed with Portland cement, the entire mansion being one of the most finished and beautiful in Scotland. From the top of the square tower a view is obtained which for extent, variety, and historic interest, cannot be excelled. Towards the south-east the ancient Castle of Stirling, the Abbey Craig, the field of Bannockburn, and the Rock of Craigforth, all crowd in rich variety upon the vision, and recall the stirring events with which they are inseparably associated. On a clear day, Borrowstounness, Arthur's Seat, and the hills around Edinburgh, are also visible. Eastwards, we look upon the Ochil range, including Dumyat, and mark the bloody field of Sheriffmuir. To the west and north, the view embraces a large stretch of the Grampians, including such snow-capped peaks as Ben Lomond, Ben Venue, Ben Vorlich, Ben Ledi, Stuchachrone, Uamvar, Ben More, &c. On the south the view is bounded by the Lennox Hills, which run with little interruption from the Castle of Stirling to the Castle of Dumbarton,—the greater part of this extensive range being within sight. The tortuous Forth, too, glides beneath our feet, and may be followed as far as eye can reach till it widens into the Firth.

Having satiated our minds and eyes with the more general surroundings of Blair-Drummond, we now proceed through the policies, under the kind conductorship of Mr Ballingall, the factor, and take a nearer view of the beauties of the place. We no not proceed very far before we learn that the great trees in the parks do not belie their reputation. Everywhere we meet with venerable giants; and one wonders, indeed, how many of those trees could have attained such a size, within so short a period—

> "Those hoary trees, they look so old,
> In truth, you find it hard to say
> How they could ever have been young,
> They look so old and grey."

It is not only the size of the Blair-Drummond trees that renders them remarkable. Their shapeliness is quite as noteworthy as their size. This is partly accounted for by the fact, that no cattle have ever been permitted to graze upon the parks. Sheep alone are allowed to pasture, so that the trees have full scope to spread their branches to the ground, and develop all their natural beauty and characteristics. Splendid specimens of the ordinary varieties of trees are to be found all round the mansion. The first great tree to which we turn our attention is a magnificent oak situated about 150 yards east from the house. At 1 foot from the ground the girth is 18 feet 2 inches, and at 5 feet the girth is 15 feet 8 inches. The bole is about 30 feet, and it carries a grand head, the tree being altogether a fine specimen, without blemish. We are again indebted to Mr Drummond Moray's care over the trees upon his property for the comparative measurements we are able to give, showing the growth of the more notable trees during the last twenty-three years. In 1860, the oak we have just referred to girthed 14 feet 9 inches at 3 feet from the ground. The girth at the same height is now 16 feet 4 inches, being equal to a growth of 1 foot 7 inches during that period. This is the largest oak upon the property, but there are a large number closely approaching it in size. The girths of some of the others which we measured are as follows:— At 1 foot, 17 feet 7 inches, and at the narrowest part, being 3 feet from the ground, 14 feet 6 inches; one standing near the spot where the gate of the old stable court was situated girths 16 feet 6 inches at 1 foot, and 12 feet 4 inches at 5 feet; a very picturesque oak in the haugh near the Mill Lade girths 15 feet 9 inches at 1 foot, and 12 feet 10 inches at 5 feet; and another girths 15 feet 6 inches at 1 foot, and 12 feet 4 inches at 5 feet. These are merely samples of a large number of oaks distributed all over the policies.

Although oaks are very numerous, the beech—which some one has described as at once the Hercules and Adonis of our Sylva, vying with the oak itself in magnificence and beauty—is still more conspicuous. The finest of these are situated along the side of the Teith, where there is a grand row fully half a mile in length. These were planted between the years 1725 and 1730 by James Drummond, the second laird of Blair-Drummond, who died in 1739. All these trees are large and picturesque, and are a great ornament to the river side. At first the ground here was subject to the periodical floods of the Teith, but it has now, at considerable expense, been secured from the encroachments of the river by a substantial stone-faced embankment. This embankment, and others at various parts of the river, as well as many other valuable improvements and additions to the estate, were made by the late Henry Home Drummond, the father of the present proprietor, who was so long associated with the public business of the county of Perth, and which he represented in Parliament for upwards of twelve years previous to 1852. At a former period he represented Stirlingshire for ten years. The largest beech in the row by the river side girths 20 feet at 1 foot; and at 5 feet from the ground, being about the narrowest part of the bole, the girth is 15 feet 4 inches. Another girths 18 feet 9 inches at 1 foot, and 13 feet 11 inches at 5 feet. The girth of some of the other trees in this row is as follows:— 17 feet 5 inches at 1 foot, and 14 feet 2 inches at 5 feet; 16 feet 1 inch at 1 foot, and 12 feet 10 inches at 5 feet, being the narrowest part of the trunk; and 14 feet 9 inches at 1 foot, and 12 feet 8 inches at 5 feet. Amongst the beeches in the policies generally, notice may be taken of three very fine trees growing at the foot of a sloping bank a little to the east of the house. The largest of these girths 18 feet 6 inches above the swell of the roots and 15 feet 3 inches at 5 feet up. It has a beautiful clean stem of about 25 feet. In 1860 this tree girthed 14 feet 10 inches at 3 feet, the present girth at 3 feet from the ground being 16 feet—a growth of 14 inches since that date. The other two girth 16 feet 10 inches above the swell of the roots, and 14 feet 4 inches at 5 feet; and 15 feet 10 inches above

BLAIR-DRUMMOND BEECHES.

the swell of the roots, and 14 feet 3 inches at 5 feet. All these trees are about 80 feet in height. A few yards east from the largest of the oaks we have described, there is also a beech of huge proportions and noble bearing. Immediately above the swell of the roots the girth is 20 feet 6 inches, and at 5 feet the girth is 17 feet 3 inches. There is a good bole of 12 feet, a spread of branches of nearly 90 feet, and an entire height of about 70 feet. In 1860 this tree girthed 16 feet 1 inch at 3 feet, the girth at that height from the ground now being 17 feet 7 inches, being a growth of 1 foot 6 inches since 1860. There is also a grove of large-sized beech trees growing to the east of the house.

A little to the south of the mansion there is a very interesting memorial of one of the world's great men, being no less a personage than the celebrated Benjamin Franklin. This consists of a clump of trees,—two sycamores, an elm, and a laburnum,—planted by the hand of the distinguished American while on a visit to Lord Kames. All these trees are of large size, but are not specially remarkable in this respect. In the same neighbourhood there is a very curious lime tree. In the year 1846 it was blown down, and the foresters proceeded to cut it up; but instead of first cutting it off at the root, they cut it across at the top of the bole. Being thus freed of its great top weight, the stem still attached to the root at once sprung back into its original position where it was left standing, and it soon sent out vigorous young sprouts, and it has now a very fine top, with no indication of its former wreck. The largest of the lime trees girths 14 feet 2 inches at 5 feet from the ground. The largest Spanish chestnut on the property is situated at High Daira, the girth at 1 foot being 12 feet 6 inches and at 5 feet 9 feet 3 inches In 1860 this tree girthed 9 feet 7 inches at 3 feet up,—the present girth at that height being 10 feet 8 inches. A still larger tree was blown down by a gale on 26th November 1880. It stood in front of the garden, and girthed 13 feet 8 inches at 1 foot, and 11 feet 5 inches at 3 feet. The largest of the sycamores grows in the park by the side of the Doune approach, the girth at 1 foot being 21 feet 6 inches, and at 5 feet 17 feet 10 inches, with a bole of 8 feet, surmounted by two great limbs. There are a large number of very beautiful birch trees scattered over the grounds, the largest of those we measured girthing 9 feet 3 inches at the narrowest part of the bole.

Coming to the trees belonging to the fir tribe, we find many specimens of the commoner varieties which claim our attention. Scots firs, if not very numerous, are of respectable dimensions. A very fine specimen is growing in the park near the house, the tree being both large and picturesque. At 1 foot above the ground the girth is 14 feet 5 inches, and at 5 feet the girth is only 3 inches less. Another of the same species close beside it girths 11 feet 3 inches at the narrowest part of the bole. A little to the south of the house we have another very good Scots fir, the girth at 1 foot being 13 feet 3 inches, and at 5 feet 11 feet 5 inches, the bole being about 15 feet. There are many fine larches throughout the grounds. The first we measured at random girthed 13 feet 4 inches at 1 foot, and 9 feet 11 inches at 5 feet. The largest of this variety which we found is a very well-proportioned tree growing near the bank of the river, opposite the Castle of Doune. The girth clear of the swell of the roots is 13 feet 10 inches, and at 5 feet up the girth is 11 feet 10 inches.

While referring to the Castle of Doune, it may be mentioned, in passing, that the ancient stronghold is still surrounded by some of the fine trees which adorned it in the latter days of its glory. The imposing ruins of this ancient baronial edifice form one of the most interesting and picturesque objects in the landscape of the district for many miles around. At almost every turn charming peeps are obtained of the great roofless structure, embowered amidst richly-foliaged timber, unconsciously reminding us of the "good old days," when the Lords of the Regality reigned supreme. The Castle stands on an elevated peninsula at the point where the Ardoch flows into the Teith, and although the roof has long since succumbed to the ravages of time, the walls—40 feet high and 10 feet thick—are still entire. The front faces the north, and at its north-east corner there is a spacious tower, rising to a height of 80 feet, and from which a most extensive and gorgeous view is obtained. Another tower stands at the opposite corner, but it does not exceed the height of the walls. The principal entrance is beneath the main tower, and the visitor has still to pass through its "ponderous grated gate, within a heavy iron-studded folding door." The rampart and fosse by which the Castle was defended are still traceable. The age of this venerable stronghold is uncertain, but it is considered probable that its founder was Walter Comyn, the fourth Earl of Monteith, who flourished in the reigns of Alexanders II. and III., and died one of the

Regents of the kingdom in 1258. In 1425, after the execution of Duke Murdoch by James I., the Castle of Doune and the lands connected with it were annexed to the Crown, which retained them till 1502. The Castle now belongs to the Earl of Moray, and gives the title of Lord Doune to his eldest son. The present Earl has shown great interest in this ancient structure, and has been at the expense of restoring the judgment hall, and otherwise improving the appearance of the Castle and its surroundings.

Doune possesses another object of more than ordinary interest, viz., the Bridge of Teith, which spans the river just opposite, and within fifty yards of, the north entrance lodge to Blair-Drummond grounds. The bridge was erected in 1535 by Robert Spittal, tailor to Margaret, Queen of James IV., and has rather a curious history. It is traditionally stated that Spittal, who appears to have been a man of considerable wealth and energy, reached the ferry without the wherewithal to pay the pontage, and the ferryman, doubtless in ignorance as to who the traveller was, refused to row him over the river. Spittal, it is said was so indignant at this that he resolved to erect a bridge at his own expense, for the convenience of travellers in the future. The bridge is a substantial erection, and the donor's name is perpetuated upon the parapet. The armorial bearings of England and Scotland on separate shields, surmounted by crowns, are carved inside the eastern parapet, in testimony, no doubt, of the donor's connection with royalty, humble though it was. There is also the following inscription, in the centre of which is a shield with a device resembling a spread eagle, and in the base a pair or large scissors formed *en saltier*:—" IN . GOD . IS . ALL . MY . TRUST . QUOD . ——TTEL . THE . TENTH . DA . OF . SEPTEMBER . IN . THE . ZIER . OF . GOD . MVXXXV . ZIERS . FUNDIT . WAS . THIS . BRIG . BE . ROBERT . SPITTEL . TAILZER . TO . THE . MAIST . NOBLE . PRINCES . MARGARET . —— NG . JAMES . THE . FEIRD . —— OF . ALMIS." When the tricentenary of the bridge was celebrated in 1835 a facsimile of the above inscription was engraved for distribution, and the following was added to it as a translation:—" In God is all my trust, quoth Spittal, the 10th day of September, in the year of God 1535 years, founded was this Bridge by Robert Spittal, tailor to the most noble Princess Margaret, Queen to James the Fourth.—Of alms." Spittal has also left several other important benefactions, including the erection at his own charge of the bridges of Bannockburn and Tullybody, and the Spittal Hospital in Stirling for the relief of decayed tradesmen. The original structure at Bridge of Teith still remains, but it was widened in 1866 by the Road Trustees to meet the increased traffic.

Blair-Drummond is exceptionally rich in silver fir trees, many of them being of great size and fine form. The first of this variety which we tested with the tape girths 18 feet 7 inches at 1 foot, and 13 feet 11 inches at 5 feet. It has a splendid bole of 25 feet, and at the fork divides into two great limbs, the entire height being about 100 feet. In 1860 this tree girthed 12 feet at 3 feet from the ground; and at present the girth at the same height is 14 feet 2 inches, there being thus a growth of 2 feet 2 inches within that period. The tree is situated at High Daira, and grows in gravelly soil. It is believed to be about 100 years of age. At the cross roads there is a much larger specimen, the girth at 1 foot being 18 feet 8 inches, and at 5 feet 15 feet 3 inches. In 1860 the girth at 3 feet above the ground was 13 feet 2 inches, the present girth at that height being 15 feet 6 inches—a difference of 2 feet 4 inches.

There are a number of specimens of the Douglas fir scattered over the property, but most of them are small. The largest of these was planted in 1843, and has attained a height of about 60 feet,—the girth at 1 foot being 7 feet 1 inch, and at 5 feet 6 feet 5 inches. Another very good specimen, planted in 1855, has reached a height of about 30 feet, but it unfortunately lost its leader a few years ago. The girth of this tree at 3 feet from the ground is 4 feet 6 inches. There are several spruce trees of good proportions. Many of the spruce and silver fir trees, in the neighbourhood of an artificial piece of water, are the abode in spring, summer, and harvest of the sombre heron, whose ponderous nests, containing a noisy and voracious brood, weigh down the more lofty branches, leaving some parts at the top exceedingly bare.

The storm which wrought so much havoc in Scotland on the

28th December 1879, was exceptionally disastrous in its effects within the picturesque policies of Blair-Drummond. The damage was not confined to young trees, but many centenarians of great size succumbed to the force of the gale. The following note of the dimensions and probable age of some of those blown down, will illustrate the nature of the damage on that eventful night, the girth being taken 30 or 40 inches above the ground, free of the swell of the roots :—

	Girth ft. in.	Height in feet.	Probable age in years.
Silver fir on top of Sandhill,	11 6	104	100
Do. on sloping bank below do.,	10 0	108	100
Do. on Doune approach near well,	12 1	116	100
Do. do. cross roads,	11 3	81	100
Do. do.	10 0	85	100
Do. do. Merlius Ford,	12 0	104	110
Scots fir in park,	9 0	80	100
Do. on Doune approach cross roads,	7 11	80	90
Larch west of house,	12 0	90	100
Do. near front door,	9 2	118	146
Beech near entrance-gate of house,	12 0	104	160
Do. in park,	11 9	100	160
Poplar near Bridge at Haugh,	12 4	106	90

From the notes preserved by Mr Drummond Moray, we learn that several very large trees were blown down by a gale in the month of March 1856, amongst the largest of these being a beech which girthed 17 feet 3 inches at the narrowest part of the bole. In 1843 some very large trees were also blown down, one of the beeches girthing 15 feet 1 inch at the narrowest part of the bole.

Amongst the notes made by the former proprietors is the following very interesting table, showing the progressive growth of the principal varieties of trees growing at Blair-Drummond :—

	1816. ft. in.	1823. ft. in.	1834. ft. in.
Beech at narrowest part of bole,	12 1	13 11	15 1
Oak, 3 feet from the ground,	10 5	11 8	12 8
Larch, do.,	8 2	8 7	8 11
Chestnut, do.,	7 9	8 3	8 5
Birch, 5 feet from the ground,	6 10	7 10	8 3

The adjoining estate of Burnbank, which was added to the Blair-Drummond property in 1752, also contains some splendid old timber, principally ash and sycamore. Some of the ash trees recently cut down were found to be over 300 years of age. The finest of this variety now standing girths 14 feet 9 inches at 1 foot, and 13 feet 2 inches at 5 feet, with 30 feet of a stem, and an entire height of 80 feet. Another good specimen stands at the door of the old mansion-house. It has an extraordinary growth about the roots, but, free of the swell, the girth is 13 feet, and at the narrowest part of the bole, 12 feet 10 inches. Another girths 13 feet 6 inches at 3 feet from the ground. A fine walnut tree adorns the old orchard, where it had doubtless been planted for the sake of its fruit. The girth at one foot is 15 feet 7 inches, and at 5 feet 13 feet 3 inches. The height of the bole is 6 feet, above which it branches into two great limbs, the entire height being 75 feet, and the spread of branches about 50 feet. The soil is rich alluvial, the subsoil clay, and the exposure free. There are a good many fine old Scots fir trees, one of those in the orchard girthing 10 feet 2 inches above the swell of the root, or 2 feet from the ground, and 9 feet 6 inches at 5 feet. A sycamore, also in the orchard, girths 13 feet 7 inches at 1 foot, and 10 feet 5 inches at 5 feet. There is another beside it of almost exactly the same dimensions, while within a very short distance is another girthing 12 feet 5 inches at 3 feet up. A Scotch elm, girthing 14 feet 2 inches at 3 feet from the ground, may also be included in the list of the more remarkable trees, as it is in every respect a grand specimen. This elm stands at some distance from the others, but within the Barony of Burnbank. This estate, like the Barony of Kincardine, was formerly the property of the Muschets. There is an old tombstone of this family in the orchard near where the mansion-house stood. It bears the following inscription:—" Here lyes the corpes of Margaret Drummond, third daughter of the laird of Invermay, and spouse to Sir George Muschet of Burnbanke: Her age 26: Departed this life in the wisitation, with her three children at Burnbanke, the 10 of August 1647."

The quality of the land at Blair-Drummond and Burnbank, being a good loam or clay, is rather too valuable for extensive plantations, and consequently a comparatively small part of the property is laid out in this way. Still, there are several good strips and clumps of wood planted in the most exposed and least valuable parts of the estate for shelter and ornament. The ornamental woods extend to about 200 acres, the oldest portion being planted between 1715 and 1720, when the old house was in course of erection, and the rest about the beginning of the present century. At Lochhills there is a very fine wood of 70 acres, containing some good beeches and oaks, the oldest of them being about 120 years of age. At Burnbank there are about 40 acres of copse, mixed with splendid oak and fir; and at Thornhill there are about 80 acres, principally fir. These plantations do not amount to many acres in the aggregate, but they help materially to heighten the effect of the landscape, and "fill the air around with beauty."

XXII.—ARDOCH.

The Proprietors of Ardoch—The Policies—Ardoch House—Aged Sycamores—The Plantations and Remarkable Trees—The Roman Camp—The Walk along the side of the Knaick.

ALTHOUGH Ardoch is more associated in the popular mind with the remains of the great Roman Camp than with anything that is marvellous in forestry, yet, like the other properties of Charles Stirling Home Drummond Moray, it is not without interest to the arborist. Here he will not only find abundant evidence of recent improvements in the cultivation and embellishment of the land, but can feast his eyes upon not a few of those venerable monarchs which excite so much admiration. The Roman Camp, however, adds considerably to the attractions of the property, and draws visitors from all parts of the world to examine its wonderful construction, and moralise upon the hollowness of earthly greatness. The history of the property is also most interesting. We do not propose to go back to the period of the Roman occupation,

"When wild in woods the noble savage ran,"

as that would be rather unprofitable, and void of practical interest. It is sufficient for our purpose to show how it came into the possession of the family of the present proprietor. The estate was originally acquired by the Sinclairs from the Abbot of Lindores, and came into the possession of the Stirlings by William Stirling (who received from his brother, Sir James Stirling of Keir, on 10th May 1543, a charter of the lands of Dachlewne, &c.) marrying Marion Sinclair, only daughter and heiress of Henry Sinclair of Nether Ardoch, and his wife, Beatrix Chisholm, who received a charter from her cousin, William Chisholme, Bishop of Dunblane, of Chapel Land, &c., to her in liferent, and to her husband and their children in fee.

This branch of the Stirlings continued in direct male succession for nearly three centuries. The great-grandson of William was created a Baronet in the reign of Charles II., and that title remained in the family for nearly a century and a half, but became extinct through the failure of male heirs, although the first Baronet was the eldest of thirty-one children, and Robert, one of his brothers, lived to the patriarchal age of 112 years. Henry Stirling, the eldest son of William and Marion, acquired Over Ardoch in 1574, and died in 1628. He was succeeded by his eldest son, William, who, in 1592, married Margaret, daughter of James Murray, fiar of Strowan, and by whom he had the thirty-one children already alluded to. He died in 1651 or 1652, and was succeeded by his eldest son, Henry, who was created a Baronet on the 2nd May 1666, and died three years afterwards. His son, William, a minor, succeeded him. The uncle (Robert) of the youthful proprietor was appointed his guardian, and managed the estate from 1669 till 1683. Sir William died in 1702, but his uncle, Robert, lived till 1716, when he died at the great age of 112. He is still spoken of as "The Tutor of Ardoch." Sir William's son, Sir Henry, and his grandson, Sir William, succeeded each other. The latter having no male issue, the baronetcy went, on his death in 1799, to his brother, Thomas, who entered the army in 1747, became a General in 1780, and died unmarried in 1808, when the baronetcy became extinct. Sir William, the seventh proprietor, who died in 1799, was succeeded by his eldest daughter, Anne, born in 1761. In 1778 she married Captain, afterwards Colonel, Charles Moray of Abercairny, and at her death in 1820, she was succeeded by her second son, Major William Moray Stirling, who was severely wounded at the Battle of Waterloo. He married, in 1826, Frances, third daughter of Archibald, first Lord Douglas, but died without issue in 1850, having succeeded his elder brother, James Moray, in the estate of Abercairny in 1840. In both estates he was succeeded by his elder sister, Christian, who, in 1812, married Henry Home Drummond of Blair-Drummond. She died in 1864, and her eldest son, George Stirling Home Drummond, also succeeded to the estate of Ardoch. He died without issue in June 1876, and was succeeded by his younger brother, Charles Stirling Home Drummond Moray, the present proprietor.

Each of these successive proprietors has contributed something to the improvement and embellishment of the property, which, from an arboricultural point of view, at all events, contrasts most favourably with a good deal of the surrounding country. The property is bounded on the south by the River Allan, and on the west by the River Knaick until it reaches the village of Braco, where it enters the park of Ardoch, and flows through the policy grounds until it joins the Allan at a substantial stone bridge which crosses the road near Greenloaning Railway Station. Entering the policies by the south approach, about one mile from Greenloaning Station, we cross the Knaick by an iron bridge, erected a few years ago. Until about twenty-five years ago, there was no bridge here whatever, and the river had either to be forded, or a long detour made round by the bridge about a mile farther up the river. The fording of the river was often impossible, and at times dangerous. The Knaick frequently rises with alarming rapidity, and carries everything before it with irresistible force. It has sometimes been seen coming down its bed in the form of a wall of water three feet high, and even more, and with such speed that one could scarcely follow it. In such circumstances, narrow escapes are not uncommon even yet; and before a bridge was erected the water could only be crossed at considerable risk, especially after heavy rains. The first bridge was built with a pier in the centre, but it was the cause of much obstruction to the progress of the river by intercepting trees and bushes washed down by the floods. The present bridge consists of a single span, the iron girders, 64 feet wide, resting upon strong concrete piers, built upon the banks. After crossing the bridge, we pass along a fine avenue, principally of beech, ash, and sycamore,—one of the beeches which we measured as a sample girthing 14 feet 7 inches at 1 foot, and 12 feet at 5 feet, with a bole of 10 feet supporting a luxuriant crop of heavy branches. There are also a good many oak trees, but they are not so old as the beeches.

This avenue leads directly to Ardoch House, a substantial

mansion, of a plain design, and occupying a site in the immediate vicinity of where the old house stood. The present mansion was built by Sir William, the seventh proprietor, and one of the stone corners over the door bears the date 1793. A large wing was built at the east end of the house by Major William Moray Stirling about 1828, by which the accommodation was considerably increased. One of the apartments of the mansion contains some valuable relics in the shape of portraits of Peter the Great and Catherine I. of Russia, presented by Peter the Great himself to Admiral Thomas Gordon, Governor of Cronstadt, whose daughter, Anna, Sir Henry Stirling, third Bart., married at St Petersburg, 21st December 1726. The Admiral occupied a very distinguished position in the Russian navy. One of the most interesting features in the policies is to be found immediately around the house. This consists in a large number of very old trees, — probably about 500 years of age,—which appear to have been planted on turf dykes, and look as if they had been so placed to afford shelter to the original mansion-house. The age of these trees can only be matter of conjecture, but that they are exceedingly old can be learned from more than their looks. The late John Ramsay of Ochtertyre, Stirlingshire, mentions in his *Dissertations on Trees*, 1801, that " there was at that time at Ardoch a sycamore in girth 11 feet 11 inches. Whether its want of magnitude be owing to the thinness of the soil, or to the indiscreet lopping of its branches, is uncertain, but it is venerable on account of its great antiquity. The Tutor of Ardoch told the Lord Justice-Clerk Tinwald that he knew no change upon the old trees since he could remember." It is those same trees that are still standing round the mansion-house. Although very old they are still in vigorous health, and some of the sounder ones are making fair progress. Most of them, however, are gnarled and knotty in the trunks, and bear evidence of having passed maturity. The largest of those immediately around the house girths 18 feet 4 inches at 1 foot from the ground, and 13 feet 1 inch at 5 feet. The following are the measurements of a few of the others :—17 feet at 1 foot and 13 feet 1 inch at 5 feet, a very good tree, with a fine bole of 15 feet, and splendid limbs ;

17 feet at 1 foot and 13 feet at 5 feet; 15 feet at 1 foot, and 12 feet 6 inches at 5 feet; 12 feet 6 inches at 1 foot, and 12 feet 4 inches at 5 feet; 16 feet at 1 foot, and 10 feet 6 inches at 5 feet; 13 feet 6 inches at 1 foot, and 10 feet at 5 feet—a valuable tree, without knots, and a beautifully clean bole. There is another rather remarkable tree in the lot, with a great swell round the roots, the girth being no less than 24 feet—the girth above the swell being 11 feet. Another valuable sycamore close beside the above has a magnificent bole of about 25 feet, with a girth of 15 feet 1 inch at 1 foot, and 13 feet 2 inches at 5 feet.

Leaving the immediate neighbourhood of the mansion-house, but before losing sight of it, we meet with many noble specimens of different varieties of trees that are equally as interesting as the sycamores of which we have already spoken. The first of these to which we may refer is a very good ash, but singularly coarse in the bark. At 1 foot up the girth is 14 feet 8 inches. and at 5 feet the girth is 12 feet 6 inches, there being a splendid bole of close upon 30 feet. There are a great many Scots firs scattered throughout the policies and in the adjoining woods, but none of them are very aged. One of those in the policies which may be taken as a fair sample, girths 9 feet 4 inches at 1 foot from the ground, and 8 feet 1 inch at 5 feet from the ground. A few larches also grow around the house, those which we measured girthing 10 feet 3 inches at 1 foot, and 8 feet 1 inch at 5 feet; 9 feet 8 inches at 1 foot, and 7 feet 7 inches at 5 feet.

There is a considerable extent of woodland over the property, the most of which appears to have been planted about 80 or 100 years ago. The soil, generally, is a lightish loam, with patches of moss—the subsoil being almost all gravel. The plantations are mixed with all the common varieties of timber, and some fine specimens are occasionally to be met with. In the Moss Wood there is a particularly good birch, girthing 9 feet 9 inches at 1 foot and 7 feet 4 inches at 5 feet. There is an extraordinary growth at the bottom of the stem, evidently caused by the character of the soil, which is here very damp. A larch in the same wood girths 10 feet 10 inches at 1 foot, and 7 feet 10 inches at 5 feet. There is also some very fine timber growing on the margin of the lake, a beautiful sheet of water of 15 or 20 acres on the lower part of the grounds, and about half a mile south of the house. Amongst the more notable trees here is a splendid silver fir girthing 10 feet 8 inches at 1 foot, and 9 feet 6 inches at 5 feet. A great many large silver firs were blown down in this wood by the gale of 28th December 1879, which seems to have been experienced here in all its fury. The stool of one of these girths 9 feet 3 inches, while another measured $5\frac{1}{2}$ feet in diameter,—the rings on both of these trees indicating that they were over 100 years of age. A clump of large ash trees grow in the Black Dyke Park, and appear to mark the site of a quondam farm steading. Behind the gardener's house there is a very good sound sycamore girthing 15 feet 4 inches at 1 foot, and 13 feet at 5 feet; and a beech girthing 12 feet 7 inches at 1 foot, and 9 feet 6 inches at 5 feet. The garden is situated a short distance to the north of the house, and drains into an ornamental pond of about 5 acres in extent, studded with picturesque islands, and spanned by a rustic bridge connecting the grounds around the house with the garden.

From the sycamore and beech trees at the back of the garden to the Roman Camp is but a few steps, and as there is certainly no more interesting object in the district than this relic of the Roman occupation, we cannot neglect to take special notice of it. It is believed to be the most perfect of all the remains of Roman Camps in Britain, and it is now so thoroughly protected that it is likely to remain so. A portion of it was at one period destroyed by the plough, but Sir William Stirling, the seventh proprietor of Ardoch, had the whole of the Middle Camp enclosed with a stone wall, so as to prevent such a misfortune in the future. The manner in which the camp has been constructed has contributed very much to its preservation. Instead of the ramparts being built up with the earth dug out of the trenches, all the soil excavated was carried away to a distance, so that the ramparts were formed out of the natural ground. But for this circumstance we probably would never have been able to trace those

"Mouldering lines
Where Rome, the Empress of the World,
Of yore, her eagle wings unfurled."

The outlines of four places of encampment are still discernible, although in different degrees of completeness. The greatest of these is known as the North Camp, and it measures about 2800 feet in length and 1950 feet in breadth, or about half a mile long and one-third of a mile broad. It has consisted of a single rampart and ditch, and the outline of the entire work can be traced without much difficulty. The whole of the eastern side is well defined by the Roman Causeway, and at one part, in a plantation on elevated ground, the entrenchments are quite entire. "In the same spot may still be seen the remains of a square redoubt within the lines, raised probably for the purpose of enabling the soldiers by military engines to discharge stones or darts, clear of the top of the rampart, in defence of the gate beside which it has been placed, and for the further protection of the gate is to be found a traverse without the lines. Of the western side about a third of the length is in excellent preservation. And of the south side the half remains nearly in as perfect condition as when the soldiers raised it, together with a portion of the Decuman Gate." It is said by the Rev. John Macintyre, Greenloaning,—from whose exhaustive lecture delivered to the Stirling Field Club on 5th October 1880, we have gathered the most of our particulars as to this camp,—that there is very great reason to conclude that "Agricola occupied this camp previous to the division of his army. It consisted of the 2nd, 9th, and 20th Legions, and a *tertiata castra*, or camp for three legions, was not square but oblong, being one-third longer than it was broad, and such is the form of this camp. Its dimensions, also, perfectly agree with the probable magnitude of that army, for the legion, according to Polybius, consisted of 8400 infantry and 900 cavalry, making the combined force under Agricola to amount to nearly 28,000. Detachments of the army no doubt were stationed in different parts of the country as garrison troops, but the Roman general had along with him a body of British auxiliaries, who perhaps did more than supply the parties engaged in such duty ; since we find that instead of 2700 horsemen, the '*justus equitatus,*'— the proper proportion of cavalry for three legions,—no less than 3000 horsemen were present at the Battle of the Grampians. To such an army this camp, allowing the moderate computation of 100 feet of intervallum within the lines, would afford 160 superficial feet to each soldier—the precise space which Polybius speaks of as absolutely necessary, taking infantry and cavalry in their relative proportions." The West or Lesser Camp is situated close beside the former, lying partly within and partly without its western side. Only two small portions now remain entire, although the line of the fortifications may be beautifully traced for a considerable distance by a difference in the soil and in the shade of the grass. The length of this camp, taken probably when more complete, is 1900 feet, and its breadth 1340, and it is estimated that it would afford accommodation for 12,800 troops. The Third or Middle Camp lies directly south of the great one, and, from its being enclosed with a wall, it is in better preservation than the others. The fourth of these ancient works is commonly described as " The Camp," because it is the one best known, on account of the great number and completeness of its entrenchments. There are six rows of ditches, and seven ramparts. Its length from north to south, within the present defences, is 510 feet, and its breadth from east to west is 435 feet, including the raised platform of the prætorium, which by itself is 84 feet long and 87 feet broad. Near one of the sides of the prætorium, a shaft is popularly supposed to have proceeded in a slanting direction to a great depth. It cannot, however, now be discovered, though some ingenious attempts have been made for the purpose, — a gentleman who rented Ardoch House and grounds about the year 1720, from the proprietor, then in Russia, having caused the opening to be covered with a millstone at some distance beneath the surface of the soil, to prevent the escape of foxes and rabbits, and the loss of his dogs. The popular tradition is that the shaft had a subterranean communication with the Grinnan Hill, distant a quarter of a mile. Hence the old rhyme—

> " From the Roman Camp at Ardoch,
> To the Grinnan Hill of Keir,
> Are nine king's rents,
> For seven hundred year."

It is more reasonable to suppose, however, that the shaft would be sunk for the purpose of drawing filtered water from the Knaick, which flows close by at a depth of about 50 feet from the Camp, and from which it might be lifted up by the old Roman method of a horizontal wheel and chain. Few articles of value are known to have been discovered about the Camp, although a tradition exists that there is no lack of buried treasure. It is stated in the Statistical Account of the Parish of Muthill, in which the Camp is situated, that many years ago, in order to ascertain whether there were really any treasures or objects of antiquarian interest in the shaft, a man who had been condemned to suffer death by a neighbouring lord, agreed, on condition of being pardoned, to be let down by a rope and make search. He was so far successful at first as to bring up from a great depth some Roman spears, helmets, fragments of bridles, and other articles, but upon being let down a second time he perished from foul air. It is added these things lay in Ardoch House for many years, but were all carried off by soldiers in the Duke of Argyle's army after the Battle of Sheriffmuir in 1715, and could never be recovered. Certainly there are no remains now known to be in preservation with the exception of one most interesting relic, a monumental stone, dug up near the prætorium, and presented by the late Sir William Stirling of Ardoch, many years ago, to the University of Glasgow, and is to be seen in the museum there. The inscription tells its own tale—

>   DIS MANIBVS
>   AMMONIVS DA
>   MIONIS. COH
>   I HISPANORVM
>   STIPENDIORVM
>   XXVII HEREDE
>   F. C.

Dis Manibus Ammonius Damionis cohortis primae Hispannorum Stipendiorum Viginti Septem Heredes Faciendum (or Fieri) Curaverunt.

Sacred to the gods below, Ammonius, the son (or servant) of Damio, of the first Cohort of Spanish Auxiliaries. Aged 27. His Heirs caused this monument to be erected.

Mr Drummond Moray has repeatedly had applications made to him to permit excavations in search of the supposed buried treasure or antiquities, but all of these are firmly resisted. The wisdom of such a resolution is so apparent that it is not a little surprising that any applications calculated to destroy so interesting and important a relic of the first great events in Scottish history should ever be made. On the occasion of the Queen's first visit to Scotland in 1842, both Her Majesty and Prince Albert manifested much interest in the Roman Camp at Ardoch, and paid a visit to the place on their way south. When it became known that the royal visitors were to come along the Muthill road from Drummond Castle to Ardoch, Major Moray, the then proprietor, had a small stone gateway hastily constructed in the wall which separates the public road from the Camp, so as to facilitate the entrance of the royal party. A large party of the tenants and others from the surrounding districts assembled, and gave the Queen and the Prince Consort a most hearty welcome when they alighted from their carriage and perambulated the grounds that were once so thronged with the warriors of a power of which little now remains except a name, but a power, nevertheless, which has exercised a vast influence upon the history of Europe.

After examining all that remains of the great Roman Camp, we proceed by a beautiful walk along the side of Knaick from the bridge at the North Lodge to the Mill of Ardoch. The present bridge over the Knaick at this point is comparatively new, and replaced one of General Wade's bridges, still standing beside it. The walk along the river side is one of the most picturesque in Perthshire. The water rushes rapidly from rock to rock, and here and there makes a bound over a considerable precipice. At one point, a little above Ardoch Mill, a fine view of three pretty large waterfalls are seen at a glance. The whole of the three waterfalls can be embraced in one view on looking between two trees growing side by side upon the banks, the trees themselves adding to the beauty of the picture. One of the most rugged parts of the walk is known as "St Ringing's Loup," although it is hardly entitled to that designation now, as the water has cut through the soft stone to such an extent as would require unheard of agility to enable any one to leap from the one side to the other. The banks of the river, on the

Ardoch side, are planted with larch, oak, beech, and other trees of the commoner varieties, adding very much to the natural grandeur of the scene.

With Ardoch we complete our survey of the properties of C. S. H. D. Moray. With regard to all of them, it may be said that they afford abundant material for the study of arboriculture, and present, as we have endeavoured to show, numerous features of general interest. The many noble trees which have survived the withering blasts of centuries at Abercairny, Blair-Drummond, Ardoch, &c., speak eloquently on behalf of the exertions of the former proprietors to ornament and enhance the value of their estates; while the many acres of thriving young plantations that clothe the rugged hillsides, utilise the more unprofitable ground, or adorn the policies, testify to the labours of the present laird, and to his good taste. It is not within the scope of this work to refer specially to the agricultural lands, but it may be said that wherever there is a disposition to improve the woodlands, there is always an equally strong desire to render the agricultural lands all that a landlord can make them. No one can travel over these properties and fail to observe that this is specially the case here,—fields being well drained and fenced, and roads and paths kept in first-rate repair, while farmers' houses and workmen's cottages are commodious, neat, and comfortable.

XXIII.—KEIR.

Situation—Origin of Name—Stirlings of Keir—Peculiar Ways of Spelling the Family Name—Acquisition of the Property of Keir—Additions to the Family Estate—History of the Property—Burning of the Old Tower of Keir—Building of the present Mansion—The Woods in the 17th Century—Early Improvements—Improvements by Sir William Stirling Maxwell—The Policies—Memorial Trees—Extensive Planting of the Newer Conifers—The Keir Araucaria—The Pinetum—Extent of the Property.

KEIR, around which cluster so many historic and literary associations, is beautifully situated about a mile to the west of the fashionable mineral spa of Bridge of Allan, and between it and the ancient City of Dunblane. Its natural beauties are many and varied, and the landscape of which it forms a prominent feature is one of the finest in this part of Scotland. From the terraces of Keir House the prospect is indeed an extensive and splendid one. In front, at a distance of some three miles, rises, abrupt and bold, the rock which is crowned with the Castle of Stirling, so rich in historic associations, and occupying a site both romantic and beautiful. Sir Walter Scott has rendered classic

"Grey Stirling, with its towers and town."

But Stirling is not the only point of the landscape which is touched by the great magician's wand.

"The lofty brow of ancient Keir"

is lit up by the same touch of genius, and so are other spots in the Vale of Monteith, over the glorious expanse of which the eye wanders with delight, till it rests on the screen of mountain-peaks which bound the western horizon—Ben Lomond, Ben Voirlich, Ben Ledi, and the other mighty giants which "sentinel enchanted lands." To the left, the Abbey Craig, rising sheer from the plain, looks down on the old walls of Cambuskenneth Abbey and the scene of the Battle of Stirling Bridge;

while far away to the east the winding Forth pursues its tortuous but silvery course to the ocean. A lovely prospect truly both east and west, and one over which the eye roams with ever-increasing pleasure.

The origin of the name of Keir cannot be said to be altogether shrouded in mystery. Occasionally in the fifteenth century, and generally during the succeeding century, the place was designated "The Keir," but subsequent to that date the present name was commonly adopted. It has been variously spelt "Keyr," "Keyre," "Kere," "Keer," "Keire," and "Keir." About a mile to the south-west of Keir House there is the farm of "Auld Keir," which it is conjectured may have been the site of the original Castle of Keir. Keir appears to have been originally the name applied to the rude forts, which were at one time common in the district, and the presence of which are said to be still decernible by knolls of green surface covering a a great heap of loose stones. In the neighbouring parish of Kippen several places bear the name of Keir as a prefix, such as Keir-brae, Keir-hill, &c.; and in Dumfriesshire there is a parish called Keir, which the learned author of *Caledonia* says is derived from the British "Caer," signifying a fort. There seems no doubt, therefore, that the place owes its name to its ancient military importance, which is confirmed by the fact that a tower once stood close by the farmhouse now called Nether Keir.

Keir has for many generations been the home of that branch of the Stirlings which has the late Sir William Stirling-Maxwell for its most honoured representative. The earlier family muniments were, unfortunately, destroyed by fire during the war between James III. and his son, the Prince of Scotland, in the year 1488, but sufficient is known of the family history to prove its great antiquity, the descent being traced from the time of David I., who died in 1153. The first to appear in authentic history is Walter de Striuelyng, who is referred to in a charter by the monarch just mentioned, and supposed to have been granted about the year 1150. The peculiarity of the spelling of the name at this period will attract the attention of most readers, and curiosity will naturally be aroused as to its derivation. This curiosity will not be diminished when we mention that Mr William Fraser, who made a collection of the family papers for the late Sir William Stirling-Maxwell, has tabulated no fewer than 64 different ways of spelling the family name, with the dates of the papers in which each occurs. One of the most remarkable features in this table is the position which is occupied by what is now recognised as the orthodox method of spelling the name. In 1160 the name is spelt "Strevelyn"; and between this period and 1412 it is spelt in fifteen different ways, the orthography for that year being even worse than at the earlier date, viz., "Strewynlyng." This peculiar style continued till 1433, when it is spelt "Stirling." In the following year, however, the orthography again resumes its erratic course, and continues to present many grotesque combinations of letters till 1677, when the family name is spelt "Stirlling," after which, we presume, it returns to the more phonetic way which found favour in 1433. The etymology of the name is uncertain, but it is highly probable that the family name is derived from the neighbouring town of Stirling; and from *Nimmo's History of Stirlingshire* we learn that the county town having been from an early period a frontier stronghold, and the scene of many a struggle between contending armies, is said to have derived its ancient name of "Stryveling," which is supposed to signify strife, from this circumstance. Dr Rogers, in his work on the Chapel Royal of Scotland, says that Stirling derives its name from the fact of its being a high rock rising from a morass. Although the name may have originally been applied to the neighbouring town, as the scene of bloody conflicts, it is equally applicable to the early history of the family, which, doubtless, from its proximity to the scenes of the great struggles for Scottish independence, suffered very heavily from the unsettled state of the country. Although the descent of the family, as already stated, is traceable to the year 1153, Keir was not acquired by the Stirling family till 1448, when Lukas Stirling, who had previously possessed lands in Fife and Strathearn, purchased Keir from George Leslie of that ilk, ancestor of the Earls of Rothes. Sir William, the grandson of Lukas, got Keir erected into a barony by King James III. Sir William has been accused of being a party to the assassination of that monarch at the Battle of Sauchieburn, but the evidence is not sufficient to establish the charge. Sir John, the fourth Laird of Keir, added greatly to the family estates. Between 1517 and 1535 he acquired various lands in Perthshire, Dumbartonshire, Stirlingshire, Fifeshire, and Elginshire,—the lands in Perthshire including Kippendavie, Blackford, Brackland, &c. He took a prominent part in the public events of his time, and held the office of Sheriff of Perth in 1516. After the death of King James IV. at Flodden, he was entrusted with the custody of the young King's person. His property was forfeited in 1526 for his having appeared at the Battle of Linlithgow against the King's authority, but it was restored to him in the following year. He founded a chaplaincy to the Cathedral Church of Dunblane in 1509. Although evidently of a somewhat pious turn, his character appears to have been thoroughly in keeping with the lawless times in which he lived, taking part in several feuds and hazardous adventures, and was ultimately slain at Stirling Bridge. His son, Sir James, was the husband of the heiress of Cawder. He divorced his wife, but retained her estate, and thus added considerably to the wealth of the family. He was appointed by King James VI. one of the judges who tried Morton for the murder of Darnley, and pronounced sentence of death on the Regent. Sir George Stirling, the great-grandson of Sir James, was intimately connected with his kinsman, the first Marquis of Montrose, and was prosecuted in 1641, by the Committee of Estates, as one of the "Plotters." After the death of Sir George without surviving issue, the estates of Keir and Cawder were inherited by his cousin, Archibald Stirling, Lord Garden, a Lord of Session of some distinction in the reign of Charles II. The grandson of Lord Garden, James Stirling, was a keen Jacobite, and was tried for an alleged conspiracy in favour of the Stuart family in 1708, but was acquitted. Sir Walter Scott, in his *Tales of a Grandfather*, relates the following amusing anecdote in regard to the trial:—

"The laird of Keir was riding joyfully home, with his butler in attendance, who had been one of the evidence produced against him on the trial, but who had, upon examination, forgot every word concerning the matter which could possibly prejudice his master. Keir could not help expressing some surprise to the man at the extraordinary shortness of memory which he had shown on particular questions being put to him. 'I understand what your honour means, very well,' said the domestic coolly, 'but my mind was made up, rather to trust my own soul to the mercy of Heaven than your honour's body to the tender compassion of the Whigs!'"

James Stirling was forfeited in 1715, and deprived of his estates, which were afterwards acquired by his friends, and restored to his son, from whom they have descended to the present representative of the family, Sir John Maxwell Stirling-Maxwell, who was born in London on 6th June 1866.

The mansion-house of Keir is partly old and partly new. The older parts were probably erected about 1488, and took the place of the tower which was destroyed by fire in the course of a conflict with the Royal forces. The then laird of Keir espoused the cause of the nobles, headed by Prince James, against King James III. Shortly after the Battle of Sauchieburn, or, as it is sometimes called, the Field of Stirling, and after a skirmish with the Royal forces, in which the Prince's party were defeated, the Prince took refuge in the Tower of Keir, but he was driven out, and the place burned to the ground by his enemies. On 7th January 1488, soon after the accession of James IV., this laird received from the King's hands, the lands of Keir, the Tower and Place of Keir, and other lands; and on the same day a charter passed the Great Seal, by which the King, after narrating that " the Tower and Place of Keir had been burned by order of James III. when last at Striveling (Stirling), by the instigation of his evil councillors, by which all the old writs and evidents relating to the said lands had been destroyed," erected all these lands, and others, into a barony, to be called the Barony of Keir, and " to be held blanch for payment of a pair of gilt spurs at the Tower of Keire, on the Feast of St John the Baptist." When James IV. accepted the resignation of the Barony of Keir for a new erection, he had knighted the laird, for in the instrument he is styled " William Striveling," and in the charter " Sir William." On the 28th October 1488, he further granted £100 to " Schir Wilzeam of Stirling to the bigging of his place," as appears from the Treasurer's Accounts of the period. This grant of money, the erection of the Barony of Keir, and the knighthood, were given as a recompense for the support which the Laird of Keir had given to the cause of James IV., and for his losses at the hands of James III.

Whatever appearance Keir House may have had upon its "bigging" in 1488, we cannot now tell, but it was doubtless a very plain structure, and the grounds and woods by which it was surrounded could not fail to have been greatly destroyed by the operations of contending armies. By the 17th century, however, the property would appear to have been carrying valuable timber, because in the contract of marriage of Sir James Stirling, dated 1606, we find his father, Sir Archibald, in settling upon his son the estates of Keir and Cawder, reserving, with his own liferent of the property, the "power of cutting down the woods of Keir, Cader, and Brokland." Again, in a rental of Keir and Cawder, made up in 1632, it is stated that "the laird hes thrie woods, viz., the Wood of Keir, the Wood of Cader, and the Wood of Brockland, quhilk within schort space will be worth the sum of ten thousand punds money, quhilk woods were cuttit twenty years be umquhile Sir Archibald Sterling, quha got at his last selling of them, aucht thousand by (besides) the Ladies' part." It was somewhere between the years 1750 and 1760 that the series of improvements was commenced which culminated in the splendid work of the late Sir William Stirling-Maxwell. The credit of inaugurating this work is due to the lady of Archibald Stirling, Margaret Erskine, daughter of Colonel Erskine of Torrie, whom he married in 1751, shortly after his return from Jamaica, where he had acquired a moderate fortune as a merchant. Soon after her marriage, this lady planned, and partly executed, the Green Terrace on the south side of the park at Keir, on the top of the slope known as Camie Bank. The next laird was the one who succeeded through the failure of his eleven elder brothers. His son, James Stirling, was also distinguished for the great improvements he effected upon the property. In early life he was a Lieutenant in the 11th Dragoons, but left the service in 1793, when he succeeded his father in Keir. His principal occupation, like that of the late Sir William Stirling-Maxwell, was agriculture, and the breeding and rearing of cattle, in which he was eminently successful. During his tenure of the property, he greatly embellished the grounds, both by forming considerable plantations, and by making the north and south approaches, —the latter of which he carried on a bridge over Lecropt Burn, —and by building the two park lodges on the Crieff and Stirling Road. He also made two considerable additions to the mansion-house on the western side, including the drawing-room and gallery. Dying unmarried in 1831, he was succeeded by his brother Archibald, who resided at Keir with little interruption till his death, which took place somewhat suddenly on the 9th April 1847. During the sixteen years he held the property, he also devoted himself with great zeal to agricultural pursuits, and drained and improved large portions of his property both at Keir and Cawder. The breeding and rearing of stock, especially shorthorn cattle, had his special attention. He completed the drawing-room and gallery which had been added to Keir House by his brother, but left unfinished. At Cawder and Kenmure he added considerably to the value of his lands by the discovery of iron, coal, and freestone, which are still extensively worked.

The late laird, Sir William Stirling-Maxwell, did more, perhaps, than any of his predecessors to make Keir the magnificent place it now is, and to maintain the reputation of the family name. He was born at Kenmure on 8th March 1818. He graduated B.A. at Trinity College, Cambridge, in 1839, and M.A. in 1843. In 1849 he disentailed the family estates, and in 1852 he sold the estate of Hampden, in Jamaica, the remaining portion of a property which had for some years ceased to meet its expenses. His scholarly attainments were of a very high order, and he was the author of several important works displaying deep learning, including *Songs of the Holy Land*, *Annals of the Artists of Spain*, *The Cloister Life of Charles V.*, *Velazquez and his Works*, and *Notices of the Emperor Charles V. in* 1555 *and* 1556. He represented Perthshire in Parliament for many years—an honour which he possessed till his death, which took place at Venice on the 15th January 1878. Between the years 1849 and 1851, he effected several important improvements—a work which he continued to carry on, more or less, throughout his remaining lifetime. He completed the Green Terrace, which, we have said, was commenced by the lady of a former laird between 1750 and 1760, and made considerable other alterations upon the house and grounds. He removed the entrance from the east to the north, built a new set of offices, turned the old entrance-hall into a magnificent library, and added an imposing bay of five windows to the centre of the eastern front. The porch, gateway, and connecting arcade, and the fine terraces which surround three sides of the house, were likewise constructed by him, while he also added considerably to the pleasure grounds. These improvements were effected at enormous cost, and everywhere there is evidence of that fine artistic taste for which the late Sir William was so distinguished. Nowhere is this artistic taste more apparent than in the steading, the design of which is singularly beautiful and appropriate. The buildings, which were erected at a cost of about £8000, were commenced in 1858, but were not completed till 1861, the various portions bearing the dates at which they were built. In every department convenience has been primarily considered, and the ornamentation is neat and suggestive. At the entrance to the different sections, the heads of famous animals reared at Keir are carved in stone, and appropriate mottoes are conspicuous in the different parts of the building. In the dairy, for instance, "Cleanliness is next to godliness," is displayed in bold relief. On the clock tower, the visitor is reminded of the value of time by such wise saws as "Tak' tent in time ere time be tint," "It is later with the wise than he's aware," and "Hours are time's darts, and one comes winged with death." The family motto, "Gang Forward," is also conspicuously displayed. An excellent model of the steading, showing the minutest details, was prepared, at considerable expense, for the Kensington Museum, where it still is.

The most of those beauties at Keir which derive their charms chiefly from art, owe, as we have said, their existence almost entirely to the late Sir William Stirling-Maxwell. Nor are these beauties all at once apparent. It was part of his design that the attractions of the policies should be very gradually revealed, and that at every turn there should be something specially fitted to attract the eye—a noble or graceful tree, a peculiarly-constructed vase, or a far-reaching vista. Starting from the mansion, we cannot pass without admiring those fine upright cypress trees which cling to the walls up to the third storey, and impart to the building a look of freshness and novelty. Round the mansion there are several trees planted by distinguished hands, including a *Cupressus Lawsoniana* planted by the Queen of the Netherlands, and another of the same variety planted by Lord Beaconsfield. Many of the newer coniferous plants were being introduced to this country about the time Sir William succeeded to the property, and he early conceived it to be his duty to give them a fair trial, and accordingly planted them very extensively. Many of these are to be found in front of the vineries, a splendid and very substantial range of buildings about 300 feet in length, and built about 80 years ago. The first to attract our attention is a fine *Juniperus recurva*, or the weeping Indian juniper, a very graceful and beautiful drooping plant. It is about 50 feet in circumference, and reaches a height of about 30 feet. It is very seldom that such a fine plant is met with. Beside it is one of the original larches, an excellent specimen, girthing 16 feet 10 inches at 1 foot from the ground, and 10 feet 10 inches at 5 feet, and fully 90 feet high, most of it being bole. There is another very good larch, also one of the original introductions, girthing 13 feet 10 inches at 1 foot from the ground, and 12 feet 10 inches at 5 feet. We have also here the original plant of *Thujopsis dolobrata* (the hatchet-leaved *Thujopsis*), introduced into this country—Captain Fortescue having brought it from Japan as a present to Sir William. It is an exceedingly lovely plant, with flattened leaves of a bright green above, and silvery-white underneath, clasping the stem quite closely, and giving it a peculiar effect. It is described by Professor Thunberg as the most beautiful of all evergreen trees. This specimen was reared in a pot under the impression that is was not hardy, but was planted out in 1863. It is now 11 feet high, although only beginning to grow. It is very thriving, appears to be perfectly hardy, and is making a growth of 18 inches annually. There is only one specimen of *Abies Albertiana*, and it has reached a height of 30 feet. There is also a thriving specimen of the white cedar (*Chamæcyparis*), which has attained a height of 35 feet. We next come to the splendid *Araucaria imbricata*, for which Keir has been so long famed. This plant was sent to Keir as a seedling, about fifty-six years ago, from Glassniven Gardens, Dublin, by Mr Niven, the Curator, to his father, who was the gardener at Keir. It was reared in a pot and carefully protected during winter. It was planted out in May 1838, by Mr William Campbell, at that time foreman at Keir Gardens. It was then a strong, handsome, well-formed plant, about ten years old. The tree was, unfortunately, blown over by the disastrous gale of 28th December 1879. Two of the branches which were cut off at that time showed a girth of a couple of inches. The tree has again been set up, and seems likely to maintain its position as one of the finest specimens of its kind in Scotland. In height it measures exactly 44 feet 3 inches; at 1 foot from the ground the girth is 5 feet 4½ inches, at 3 feet the girth is 4 feet 8 inches, and at 6 feet the girth is 4 feet 4 inches. Amongst the other notable trees in the policies is a grand silver fir, 80 feet high, and girthing 11 feet 8 inches at 1 foot from the ground, and 9 feet at 5 feet. A *Cryptomeria Japonica*, a perfect specimen, feathered to the ground, has a girth of 6 feet 5 inches at 1 foot from the ground, and 5 feet 2 inches at 5 feet, and 75 feet in circumference. There are two exceptionally good specimens of *Abies Canadensis*, the largest of which girths 6 feet 10 inches at 1 foot, and 5 feet 7 inches at 5 feet. This tree crushes rather closely upon the large araucaria, for the benefit of which it had to be greatly pruned, very much to the regret of the late Sir William, who watched over these fine trees with a truly parental care. *Cedrus Deodara* has been extensively planted, and the plants are thriving very well. *Cupressus Lambertiana* is to be found with a height of about 80 feet, and there are several excellent specimens of *Cupressus torulosa*. There are a few specimens of *Wellingtonia gigantea* planted about 30 years ago. The first of these—planted by Sir William, on fine till—has a height of about 30 feet, and girths 8 feet 10 inches at 1 foot from the ground, and 5 feet 8 inches at 5 feet. Some of those growing on lighter soils are considerably taller, but they have not the same girth There are several specimens of *Thujopsis borealis*, one of the most valuable introductions of recent years, and all of them are healthy and graceful plants. There are very few Scots firs. The most noteworthy is one directly opposite the monument erected by Sir William Stirling-Maxwell in memory of his grandfather, who died suddenly at this spot. At 1 foot from the ground the girth is 10 feet 3 inches, and at 5 feet the girth is 9 feet 3 inches. There are a few hardwood trees of exceptional size. The most remarkable of these is the fine old Spanish chestnut opposite the mansion. It girths 22 feet both at 1 foot and 5 feet from the ground, has a spread of branches of 80 feet, and an entire height of 60 feet. An oak girths 15 feet 6 inches at 1 foot, and 12 feet at 5 feet from the ground.

The pinetum stands at the west side of the kitchen gardens, and was laid out by Sir William about 30 years ago, for the purpose of testing the value of the newer introductions. The first thing to attract notice is that each variety has been planted in pairs, so that if one failed the other might survive to carry on the experiment. Commencing at the bottom, the first pair, we find, are specimens of *Pinus excelsa*, then *Pinus Austriaca*, and *Picea nobilis*. The latter were seedlings, and are now 62 feet high, with a girth at 1 foot from the ground of 6 feet 10 inches, and at 5 feet 5 feet 7 inches. One of these was blown over by the gale of 28th December 1879, but was raised again, and is getting on very well. We next have *Abies Douglasii*, reaching a height of 70 feet, with a girth of 7 feet 4 inches, and 6 feet 8 inches at 1 foot and 5 feet respectively; *Pinus Cembra*, 40 feet high; *Pinus Webbiana*, 35 feet high; *Pinus monticola*, over 50 feet high, girthing 6 feet 11 inches and 4 feet 6 inches at 1 foot and 5 feet respectively. *Pinus Insignis*, a remarkably fine tree for the district, is over 50 feet high, and girths 8 feet 8 inches and 7 feet 10 inches at 1 foot and 5 feet from the ground respectively; *Abies Morinda*, 45 feet high, and richly feathered to the ground; and *Abies Menziesii*, girthing 7 feet 9 inches, and 5 feet 9 inches at 1 foot and 5 feet from the ground respectively.

Keir, as we have already indicated, is not the only property which belongs to the family whose history we traced at the beginning of this chapter. Keir itself is partly in the county of Perth and partly in the county of Stirling, but the family has to pay rates in two other counties, viz., Lanarkshire and Renfrewshire, their property being also situated in no less than 15 parishes. Besides Keir, there are also in Perthshire the estate of Quoigs, and the beautiful lands of Ardchullary, Annie, and Tombae, on the shores of Loch Lubnaig, picturesquely situated among the Braes of Balquhidder. There are here about 3390 acres, mostly moorland, copse, and water. In the western counties there is the valuable estate of Pollock, lying on the west side of Glasgow, and which is rapidly being feued. On the east side of Glasgow, there is the no less valuable estate of Kenmure, where the late Sir William was born; and adjoining it are the extensive properties of Cawder and Craighead. Cawder, Craighead, and Kenmure extend to about 5770 acres, and Keir and Quoigs to about 6242 acres. Of the latter about 650 acres are laid out in wood. The plantations are generally mixed, and, being carefully attended to, contain much valuable timber.

XXIV.—KIPPENDAVIE.

Attractions and Extent of the Property—Stirlings of Kippendavie—The Great Sycamore of Kippenross—Other Remarkable Trees—"On the Banks of Allan Water"—Pinetum—Plantations—Sheriffmuir—Improvements.

BOTH from an historical and an arboricultural point of view, Kippendavie, the property of Mr Patrick Stirling, ranks as one of the most notable properties in Perthshire. Within its boundaries is the site of the only battle fought in the Rebellion of 1715, and within its policies there has grown the largest sycamore tree in the country. Circumstances like these strike the imagination with some power, and the result is that when summer clothes the fields with beauty, and the balmy air calls forth the songsters of the grove to warble their sweetest melodies, there is no more favourite resort than Kippendavie. Attractive as the field of Sheriffmuir and the great sycamore of Kippenross may be, they do not stand alone. Throughout the whole property the visitor meets with new objects of interest and fresh charms at almost every turn. From the higher lands we look down upon the gorgeous plains watered by the Allan, the Forth, and the Earn. The precipitous rock from the top of which Stirling Castle frowns upon the peaceful valley below is a conspicuous object towards the west; while, in a still more westerly direction, "The lofty Ben Lomond" is seen presiding over a host of mountains, which stretch in an unbroken chain across the sky until they are lost sight of in the northern horizon. The intervals are filled in with a multitude of charming "bits" in woodland, river, and rural scenery, stretching over a wide range of country. The estate consists of four different properties,—Kippendavie, Kippenross, Cambusheenie, and Buttergask. Kippenross and Kippendavie are bounded on the east by the property of Lord Abercromby, on the north and west by

Keir, and on the south by Cromlix. Cambusheenie marches with Cromlix, and Buttergask is a small property by itself near Blackford. A considerable portion of the ancient cathedral city of Dunblane is feued from Kippendavie, and upon one of the most prominent eminences of this part of the estate the Dunblane Hydropathic Establishment has been built. The Stirlings of Kippendavie are a branch of the family of Stirling of Keir. Archibald Stirling, the third son of Sir Archibald Stirling of Keir, received from his father, by a charter, dated 5th August 1594, the estate of Kippendavie and other lands, and became the founder of this and other younger branches of the Keir family. John Stirling of Kippendavie, the sixth laird, who was born in 1742, added very considerably to the family possessions, Kippenross being amongst the additions. This estate—which is supposed to derive its name from Kippen Cross, one of the four ancient crosses of Dunblane—was for several hundred years the seat of a family of the name of Kinross, from whose creditors it was, in 1630, purchased by Mr Pearson, Dean of Dunblane. The Pearsons held it for nearly 150 years, the sixth laird of Kippendavie having purchased it from Mr William Pearson in 1778. The present mansion-house—which is situated on Kippenross, and is a fine substantial structure—was commenced by the last of the Pearsons who held the property, but was finished by Mr Stirling. In 1881 an additional wing was added to the house, and other important improvements carried out. In 1813 the sixth laird also acquired the superiority of Lanrick, Auchinbie, Shanraw, Woodland, &c., from James Stirling of Keir. The sixth laird conveyed the estate of Kippenross to his eldest son, Patrick, on the occasion of his marriage in 1810. Patrick predeceased his father, who was succeeded by his grandson, John Stirling. The late laird, who acquired a cosmopolitan reputation from his connection with great railway enterprises, also succeeded upon the death of his grandfather. The late laird died on the 27th July 1882, when he was succeeded by his eldest son, the present proprietor.

As already indicated, the chief arboreal feature on the property is the great sycamore of Kippenross. Unfortunately, it is now in ruins, but sufficient remains to attest its former greatness. The first damage it received was in 1827, when it was struck by lightning, and it never got the better of this injury. The late laird has affixed a brass tablet to the shattered trunk, bearing the following description of the tree before it received its fatal blow :—

> Cubic contents in 1821, 875 feet.
> 1841.
>
> | | Ft. | In. |
> |---|---|---|
> | Girth of the smallest part of the trunk, | 19 | 6 |
> | Do. where the branches spread, | 27 | 4 |
> | Do. close to the ground, | 42 | 7 |
> | Height, | 100 | 0 |
> | Extreme width of branches, | 114 | 0 |
>
> Kippenross, 10th June 1842.
> Aged 440 years.

The age of the tree was discovered from some old estate papers. It is known that this tree went by the name of the "Big Tree of Kippenross" as far back as the time of Charles II. In 1800 the tree was thus described by Mr Ramsay of Ochtertyre, in Stirlingshire :—" Regarding Kippenross, the sycamore at that place is now much the greatest in this country. The girth 2 feet from the ground on the north side is 22 feet; close to the ground, 30 feet; where the branches set off, 28 feet; at the smallest part, 18 feet. It has 18 feet of a bole. It has four vast branches, which spread to 70 feet. It is not an easy matter to measure this tree very accurately, owing partly to excrescences from its root, and partly to its standing on a slope, hence the various measurements which have appeared. It is the most ancient tree upon the place, and has long been stationary in its growth. In 1740 the late John Stirling of Keir was told by Widow Gillespie, who is past four score, who lived all her days about the place, that although all the other trees had grown a great deal within her remembrance, she knew no change upon the great tree which many people came to see as a curiosity." Mr Ramsay further says :—" Before quitting the sycamores of Kippenross, it is proper to observe that the soil on which the great one grows is a sharp and sweet loam, with about a foot thick on the top of gravel. It is so thin that a pear or apple tree would kancre in a couple of years." Many useful and ornamental articles have been made out of the wreck of this giant tree, the principal of these being two handsome chairs which adorn the hall of the mansion. There is still a very large sycamore growing near the house. The girth at 1 foot from the ground is 18 feet 5 inches; at 3 feet from the ground, 14 feet 8 inches; where the branches spread, 17 feet 1 inch; and the height of the bole is 13 feet 8 inches. In 1859 the measurement of this tree was as follows:—At 1 foot from the ground, 17 feet; and at 3 feet, 14 feet 1 inch. There are many other notable trees. A grand beech in the park girths 17 feet 4 inches at 1 foot, and 14 feet 4 inches at 5 feet, with a magnificent round bole of fully 25 feet, and an entire height of about 100 feet. A horse chestnut — which in 1859 girthed 12 feet at 1 foot, and 12 feet 6 inches at the spread of the branches—now girths 14 feet 10 inches at 1 foot, and 12 feet 10 inches at 5 feet, with a height of 80 feet. A black Italian poplar, with a height of 120 feet, has a girth of 17 feet 5 inches at 1 foot, and 14 feet 10 inches at 5 feet. One of the orginal larches, planted in 1738, has a girth of 18 feet at 1 foot, and 11 feet 6 inches at 5 feet, with a height of 80 feet. Another of the original larches has a girth of 16 feet 3 inches at 1 foot, and 10 feet 7 inches at 5 feet, and a height of 115 feet. This tree is chiefly remarkable for its fine proportions, as may be seen from the following measurements taken in 1859:—12 feet 6 inches in girth at 3 feet, 8 feet 6 inches at 27 feet, 7 feet at 54 feet, 1 foot 6 inches at 100 feet; entire height, 110 feet; branches spring at 20 feet from the ground; spread of branches, 55 feet; cubic contents, 515 feet. There is a very fine Scotch yew, girthing 7 feet 3 inches at 1 foot, and 5 feet 9 inches at 5 feet, with a bole of 13 feet. Amongst the other notable trees are an ash girthing 13 feet 7 inches at 5 feet from the ground; oak, 11 feet; silver fir, 9 feet; cedar of Lebanon, 6 feet 9 inches; elm, 11 feet 9 inches; several lime trees, 9 feet 6 inches, &c.,—all at 5 feet from the ground. There is a very good tulip tree in the shrubbery, the girth being 4 feet 8 inches at 1 foot, and 3 feet 11 inches at 5 feet. There is also a nice specimen of the entire-leaf ash and two good specimens of *Quercus concordia*.

One of the prettiest parts of the property is the walk along the banks of the Allan. This walk was made by Mr Hugh Pearson in 1742, and is reputed to be the first walk of the kind artificially made in Scotland. Mr Pearson was an ardent admirer and conserver of the trees upon the property, but after his death in 1749 great havoc was made with the trees by his eldest son, although the family was in no want of money. Some of the beech trees in this walk are of great size, but none of them reach the measurement of the one we have described as growing in the park. Proceeding along the river side, we come across the ruins of " My Lad's Mill," the scene of the tragic ballad, " On the Banks of Allan Water." The mill derives its name from a corruption of the old name, " Mill Lands," being the lands belonging to the mill. In the spring of 1863 several acres were laid out as a pinetum on the west side of the Darn Road leading from Dunblane to Bridge of Allan. The pinetum was well stocked with the most promising of the newer coniferous trees, the plants averaging $2\frac{1}{2}$ feet high when placed in the ground. Almost all the varieties have thriven very well. *Wellingtonia gigantea* has reached a height of 35 feet, with a girth of 9 feet at the ground. The best specimen of *Abies Douglasii* has a height of about 60 feet, with a girth of 4 feet 8 inches at 1 foot, and 4 feet at 5 feet. *Picea Nordmanniana*, *Abies Albertiana*, and *Pinus Cembra* have also prospered remarkably well. *Abies Menziesii* is not so healthy, the site not being suitable. *Pinus Austriaca*, *Pinus Laricio*, and *Picea Pinsapo* are thriving remarkably well. In referring to the newer coniferous trees, mention ought to be made of a splendidly-balanced specimen of *Araucaria imbricata*, 20 feet high, growing in the shrubbery.

The plantations, consisting chiefly of larch and spruce, are about 700 acres in extent. A large proportion of the plantations cover the battlefield of Sheriffmuir, which now presents a very different aspect from what it did when the two contending armies met with discomfiture. The old road from Dunblane to Sheriffmuir, along which the Scottish troops passed, has been closed up for a considerable time, but, at the personal expense of the late Mr Stirling, a new road was made by way of Stonehill. The great object of interest at Sheriffmuir is the gathering stones of the Scottish clans. These stones consist of three grey boulders, and the standard is said have been placed in a cavity at their junction. The stones bear marks of having been chipped by relic hunters; but, in order to preserve them, the late Mr Stirling enclosed them in an iron cage, consisting of 23 one-inch iron bars, which are so firmly fastened that it is now impossible for any one to carry away pieces of the stones. The following inscription has been engraved upon a brass plate affixed to the cage :—" The gathering-stone of the Highland army on the day of the memorable battle of Sheriffmuir, fought in November 1715. This grating has been erected to preserve the stone, by John Stirling, Esq. of Kippendavie, principal heritor of the parish of Dunblane. November 1840." We were sorry to notice that this plate has not escaped the hand of the Vandal—the date, "1840," being almost wholly obliterated by the scratchings of some mischievous wretch. The great storm of December 1879, and subsequent storms, have played sad havoc within the woods of Sheriffmuir and the adjoining plantation of Walter's Muir, about 80,000 trees having been blown down. A nursery, sufficiently large to provide all the young trees for the property, has been formed near Kippenross Sawmill, with the view of repairing the damage.

During the lifetime of the late proprietor a vast sum of money was spent upon improvements all over the property. The late Mr Stirling and his guardians had a number of small crofts broken up, trenched, and cultivated, and laid out into policy parks. A large amount was also expended upon farm buildings, drainage, and fencing, and a handsome new steading for the home-farm has just been completed. The late Mr Stirling also introduced a water-supply to Dunblane at a cost of over £3000. The present proprietor is forming a Clydesdale stud, based upon valuable Keir blood; as well as a Polled Angus herd, based upon the stock of the most noted breeders. He has entered into possession with the evident intention of acting up to the maxim that property has its duties as well as its privileges, and of doing what he can to improve the amenities and enhance the value of this beautiful and extensive estate.

XXV.—MONTEITH.

General Attractions of Monteith—The Earls of Monteith—The Forests of Monteith a favourite Resort of the Ancient Scottish Court—The "Law Tree"—Claimants for the Earldom—The "Beggar Earl"—Cardross — Cardross Mansion — Antiquities — Remarkable Trees—The Garden Policies—The Pinetum—The Glen—Cardross Noted for its Timber—Deforesting and Replanting—Big Wood of Cardross—*Abies Douglasii* the most Flourishing Variety—The Scenery—General Improvements—Efforts at Reclamation of Moorland—The Erskines of Cardross—Lochend—Lake of Monteith—Priory of Inchmahome—Queen Mary's Garden—Warning to Relic Hunters—The Old Spanish Chestnuts at Inchmahome—Castle of Talla—Rednoch—Blairhoyle—Glenny—Mondowie—Garden—Gartmore—Royal Visitors at Monteith.

IN entering upon the district of Monteith we are conscious of treading upon enchanted ground. In the days far off and dim,—almost before the rays of civilisation had begun to dawn upon a savage and warlike people,—the beauty and fertility of this part of Perthshire attracted the attention of a body of the monastic brotherhood,—the fraternity never being slow in detecting the most choice spots of country,—and in the most secure and picturesque spot of all the " varied realms of fair Monteith " they reared an " oasis in the natural and moral wilderness," from which for centuries they faithfully sought to brighten the minds of the untutored natives, both by precept and example. When, from their exceptional privileges, the monks attained to something like regal grandeur, their home became the occasional residence of the early Scottish Kings, whose chequered reigns here gave rise to some of the most notable events in our national history. In more recent times " Clan Alpine's warriors true " rendered the district for many years all but inaccessible to the stranger, and the stirring events which were enacted even during this period impart to almost every footstep its romantic tale. The district, in short, has in all ages been the cradle of an heroic race—a circumstance which is doubtless largely due to the unsurpassed magnificence of its natural attractions. The mountains which encircle it on every side rendered it hazardous in ancient times for hostile visitors to approach, and, at the same time, concealed the rich valleys which lie between, and which would afford an irresistible temptation to a nomadic race. The beautiful sheet of water which takes its name from the district provided a pleasant variety of diet to the natives and the monks of old, while its islands furnished a safe retreat in times of danger. In the present day, this charming lake, with its picturesque islands and sylvan shores, incalculably heightens the romantic grandeur of the district, and renders the landscape one of the most complete, in scenic effect, to be found in Scotland.

The district known as Monteith is situated chiefly in the south-west of the county of Perth, but is partly in the county of Stirling, and includes the parishes of the Port of Monteith, Aberfoyle, Callander and Leny, Kincardine, Kilmadock, Lecropt, Dunblane, and parts of Kippen—the whole of the valley watered by the Teith. In a narrower sense, Monteith may be said to embrace the tract of country immediately surrounding the Lake of Monteith, and it is to this locality that the name is more generally applied. In the twelfth century this attractive country supplied the title of an Earldom which was probably erected as early as any of the other ancient Earldoms into which Scotland was then divided. The first Earls of Monteith took their surname as well as their title from the district. The direct male line of the original Earls of Monteith failed at an early date, only three being known to have inherited the Earldom, viz., Gilchrist, Murdoch, and Maurice. The charter of the creation of the Earldom of Gilchrist is not known to exist; and, in the absence of any other authentic evidence, it can only be conjectured that the ancient Earldom embraced the larger portion of the district now known as Monteith. The latter Earldom, created by James I., in the fifteenth century, in favour of Malise Graham, formerly Earl of Strathearn, did not include all the lands of the original Earldom, which had been forfeited by Murdoch, Duke of Albany, as Earl of Monteith. On the contrary, the charter of creation of the new territorial Earldom of Monteith reserved to the King certain portions of the territory; among the places thus reserved being the Castle of Doune, which was the principal messuage of the ancient Earldom at the time of the forfeiture. The two daughters of the third of the original Earls, Maurice, were his co-heiresses, and by their marriages the territorial Earldom was carried successively into the great families of Comyn and Stewart, while the respective lords of these ladies obtained the personal dignity of Earls of Monteith, either in right of their wives or by special creation. The history of the original Earls of Monteith—like that of their neighbours, the Earls of Lennox and Strathearn—is an unfortunate one. The Earldom was almost from the first the cause of unnatural strife and keen legal contentions. As we learn from *The Red Book of Monteith*,—from which we derive many interesting particulars regarding the history of the district,—brother disputed with brother, and sister with sister, in successive generations, regarding their rights to it. Although the "Isle of Rest" was situated in the domains of the ancient Earls of Monteith, it did not prove symbolical of quiet enjoyment of their possessions. The Earldom was ultimately forfeited to the Crown, in the time of James I., upon the execution of Murdoch, Earl of Monteith, second Duke of Albany, and Regent of Scotland. No record has been preserved of the crime for which Murdoch, and others with him, including his son, were beheaded at Stirling. The pretext for his arrest is supposed to be the inattention of Duke Murdoch and his fellow-sufferers to a law passed at a previous Parliament, which may have afforded a colourable pretext for convicting them of treason. There is nothing to show, however, that they were moving sedition,—having attended Parliament, and performed their accustomed duties, until they were suddenly and unexpectedly arrested, tried, and executed upon the same day. It is conjectured, with some reason, that this cruel proceeding was prompted by the jealousy of the King, for dynastic reasons. For the same reason, the limitation of the power of the nobles, James I. seized several other Earldoms, but the confiscation of that of Strathearn was so marked an injustice that the nobles became greatly alarmed. To mitigate the severity of this seizure, James divided the forfeited Earldom of Monteith into two parts,—one of which he erected into a new Earldom of Monteith, in favour of Malise Graham, who was Earl Palatine of Strathearn, before his possessions attracted the cupidity of James. The eastern portion was reserved to the Crown, and was afterwards known as the Stewartry of Monteith. There is evidence that in the days of the First Graham, Earl of Monteith, the forests in and around Monteith were the favourite resorts of the Scottish Court, when at Stirling, for the sport of the chase, and, in order to make provision for himself and the lieges during the hunting season and at other times in Monteith, James III., on 8th February 1466, erected the town of Port into a burgh of barony, granting to Earl Malise a charter to this effect. The Port, however, does not appear ever to have been a burgh of importance. The cross of the burgh was the trunk of a very old hawthorn tree, which stood opposite the Manse of Port, and was called the "Law Tree," as it was there that legal business was transacted, as well as the ordinary sales of cattle and farm produce. The Grahams enjoyed the Earldom for nine generations—upwards of two centuries and a half. The most illustrious representative of this line was William Graham, the seventh Earl, who was a distinguished statesman in the time of Charles the First. As the lineal heir of Prince David, son of King Robert II., he secured the Earldom of Strathearn; but this claim, and an alleged rash boast that he had the "reddest blood in Scotland," and a better right to the crown than the King himself, so alarmed Charles that he revoked the grant of Strathearn, even sought to suppress in part his title of Monteith by a new title of Earl of Airth, and deprived him of all his high judicial offices. His eldest son, John, Lord Kilpont, was killed by James Stewart of Ardvorlich, while they were fellow-officers in the army of Montrose, at the Kirk of Collace shortly after the Battle of Tibbermuir. The only son of Lord Kilpont succeeded his grandfather as eighth Earl of Monteith and Airth; and on his death, without issue, in 1694, the title became dormant—a condition in which it still continues. A dignity so great was not likely to remain long without claimants, and the history of one of these has about it a melancholy touch of romance. This was William Graham, who became known as the "Beggar Earl." At an election of Peers on 12th October 1744, being then a student of medicine, he answered to the title of the Earl of Monteith, in respect of his being executor confirmed to the last Earl, who died in 1694, and he voted at several elections. His claim had no practical result, and he had a deplorable end, dying through penury and exhaustion, in 1783, on the roadside, near Bonhill, in the Lennox, while begging among the neighbouring farmers. The celebrated Captain Barclay Allardice of Ury and Allardice was also a claimant. Upon his death, the dignity was claimed by his daughter, but was opposed by Mr Graham of Gartmore, as the heir-male of the Graham Earls of Monteith, but no final decision has yet been pronounced.

Leaving the Forth and Clyde Railway at the Port of Monteith Station, we naturally bend our steps, first of all, to the ancient and historical estate of Cardross, the property of C. H. D. Erskine, well known as the Deputy Sergeant-at-Arms of the House of Commons. The name of the place is of Gaelic origin, and signifies "The fort on the promontory." There is upon the estate a Roman castellum, very entire, about 50 paces in diameter, and irregularly square, with inner and outer rampart, situated upon the edge of the Flanders Moss, then forming part of the ancient Forest of Caledon, which the Romans forced the inhabitants to cut down, the mark of the axe being found on trees lying at the bottom of the moss on the very spot where they were felled. The remains of a "corduroy" Roman road was laid bare some years ago in clearing moss on the Parks Farm, pointing in a straight line in the direction of the camp. The property is beautifully situated in the centre of this most charming district, and embraces within itself almost all that is most lovely in nature and most notable in history to be found in this interesting locality. The Grampians cast their shadows across its luxuriant plains, the tortuous Forth winds for miles along its southern marches, and the silvery waters of the Lake of Monteith wash its northern boundaries. In ancient times it was the scene of some of those great events which have given to Scottish history its undying fame, and in these days it is the silent witness of what Scotchmen can do to mitigate a severe climate and enrich an ungenial soil. The mansion-house is a very old building in the castellated style of architecture, and has been repeatedly added to and renovated. The house was garrisoned in Cromwell's time for ten years, and again in 1675, this time apparently as much for the purpose of annoying its proprietor as for securing the peace of the country. It is situated upon a plateau, about 30 feet above the level of the Forth, quite close to the river, and is surrounded by about 300 acres of policy park of the richest pasturage. The park was formed by the late Mr Erskine of Cardross towards the beginning of this century,—the ground having been before that time subdivided in small enclosures by deep sunk fences and high hedges, which, when taken away, made the present park, and set off the place and fine trees to advantage. The oldest portion of the house— the age of which is not known—consists of a square tower, and was in early times the residence of the Commendators of Inchmahome. The principal suite of apartments was built in the time of James VI. by the Earl of Mar, that he might give a suitable reception to that monarch, who had given him a grant of the estate on the suppression of the monasteries. There was a life-long attachment between the King and the Earl, both of them having been educated together under the widow of the Regent Mar, Annabella Murray. Standing as it does upon a plateau of some elevation, the house commands a splendid view of the district for many miles. Looking northwards through a beautiful vista of noble trees, the eye rests upon the ponderous form of Ben Ledi, and some of its equally ponderous comrades. The south front looks upon the highly-cultivated Carse of Stirling, and the rich pastoral hills of Kippen and Campsie. Due east, Stirling Rock and Castle rise abruptly upon the scene; while to the west we find, in Ben Lomond, that

> " A giant mountain stands
> To sentinel enchanted lands."

The interior of Cardross House is rich in artistic beauty and in relics of antiquity. In the drawing-room, begun in 1598 by the laird known as the last Commendator of the Priory of

Inchmahome, there is a very fine plastered roof, representing a model of a Dutch garden, and made by the same party of Italian artists who had been brought to this country by James VI. for the redecoration of Holyrood Palace. The dining-room had a roof of exactly the same pattern, but it, unfortunately, fell in while the bedrooms above were being restored by the grandfather of the present laird. Both rooms were originally hung with valuable tapestry. The family portraits are remarkable as being a complete collection of the ancestry of the present representative from the time of the Regent Mar to the present day. Almost all the portraits are originals, executed by the most eminent artists of their time, and include many notable characters in Scottish history. The unfortunate Queen Mary is represented by the portrait popularly known as the "Perthshire Red," from the colour of her dress. Amongst the relics which are most sacredly preserved is the sword which King Robert the Bruce is said to have left at Cardross during one of his visits. The heroic Scottish King paid several visits to this district, and spent some time at Cardross between his Coronation and the Battle of Bannockburn,—one account stating that he actually slept in Cardross House on the night previous to his greatest victory. Some historians mention that the renowned King also died here, but others are inclined to think that this event happened at Cardross in Dumbartonshire. The sword is a most extraordinary and formidable-looking weapon, measuring all over 6 feet $2\frac{1}{2}$ inches, with a blade of 4 feet $7\frac{1}{2}$ inches. The breadth of the hilt is $2\frac{1}{2}$ inches, and the weight is no less than 10 lbs. Another very interesting relic is a large camp kettle, discovered in a Roman camp, about one mile from the house. Not the least valuable of the relics is a pair of large wine-glasses, belonging to the Jacobite period, and which illustrate in a novel way the strong feeling which prevailed in Scotland in favour of the cause of the Stuarts. Both of the glasses are most beautifully and artistically chased, and bear the initials " J. R.," along with the following lines :—

> " God save the King, I pray,
> God bless the King, I pray,
> God save the King.

Send him victorious,
Happy and glorious,
Soon to reign over us,
   God save the King.

God bless the Prince of Wales,
The true-born Prince of Wales,
   Sent us by Thee.
Grant us one favour more,
The King for to restore,
As Thou hast done before—
   The family."

Leaving the house for the policies, we find ourselves surrounded on every side with objects which claim special attention. At the north front of the mansion there is one of the finest sycamores to be seen anywhere. Six inches from the ground the circumference is 20 feet 8 inches. Above the swell of the roots the girth is 16 feet 10 inches, and at 5 feet above the ground the girth is 14 feet 1 inch, there being altogether a magnificent clean bole of 19 feet 6 inches. At the top of the bole, the tree assumes a somewhat peculiar shape, being considerably flattened, and having three large branches springing out in such a way as to bear a strong resemblance to a three-branch candlestick. The tree is fully 80 feet in height, with a grand compact top. The lowest branches are 6 or 7 feet from the ground. This magnificent tree measured in 1845 as follows:—girth 6 inches from the ground, 19 feet; 3 feet from do., 13 feet 6 inches; $5\frac{1}{2}$ feet from do., 12 feet 10 inches; 19 feet from do., where it divides into three, 18 feet. Circumference of No. 1 limb, 9 feet 4 inches; No. 2 do., 7 feet 3 inches; No. 3 do., 7 feet 4 inches. Bole, 19 feet long. In 1863 the measurements taken at the same distances from the ground were 20 feet, 14 feet, 13 feet 5 inches, and 19 feet; limbs, 10 feet, 8 feet 2 inches, and 8 feet 3 inches; length of trunk, 19 feet 5 inches. There are many other scyamores of exceptional beauty and size. Most of the finer specimens are to be found in the "Plane-Tree Walk," leading from the south front to the banks of the Forth. The width of this path is about 15 feet, and the trees are about 20 feet apart. One of these trees has a girth of 10 feet 10 inches at the narrowest part of the bole, the average girth of the whole at the narrowest part of the bole being about 9 or 10 feet. In the same neighbourhood there is a favourite sycamore girthing 15 feet 6 inches at the narrowest part of the bole. This is a very old tree, and the veteran Admiral Elphinstone Erskine told us that he knew no difference upon the tree during the last 65 years. It is still in good health, and clothed with splendid foliage. Another sycamore has a girth of 12 feet 4 inches at the narrowest part of the bole. A very fine specimen of the variegated species girths 13 feet 4 inches at the narrowest part of the bole. There are several other specimens of hardwood trees remarkable for their great size. Amongst these, the elms take a prominent place. In the large park surrounding the house, and close to the banks of the Forth, there is a very picturesque elm, with a girth of 21 feet at 1 foot from the ground, above which it swells out considerably. Opposite it there is another elm which, if not so picturesque, is of greater commercial value, having a fine clean stem, with a girth of 21 feet 6 inches at 1 foot and 15 feet at 5 feet from the ground. Another girths 13 feet at the narrowest part of the bole. There are several good beeches. One with a grand stem girths 17 feet 8 inches immediately above the swell of the roots, and 13 feet 6 inches at the narrowest part of the bole. Oak trees are numerous and of large size. The first which we specially observed has a girth of 16 feet 3 inches at 1 foot from the ground, and 11 feet 4 inches at 5 feet, with a bole of 20 feet; and the second a girth of 16 feet at 1 foot and 13 feet at 5 feet. The largest of the ash trees girths 22 feet at 1 foot, and 14 feet at 5 feet—a capital tree, with a bole of 20 feet. The following measurements are also worthy of being recorded:—No. 1, 21 feet 6 inches above the roots, and 16 feet 6 inches at 5 feet, with a splendid bole; No. 2, an old tree, beginning to decay, 20 feet 6 inches at 1 foot, and 15 feet 6 inches at 5 feet; and No. 3, 19 feet above the roots, and 13 feet 9 inches at 5 feet from the ground. There are two beautiful sister ash trees, directly opposite each other, and girthing exactly the same, viz., 15 feet at 1 foot and 12 feet at 5 feet, with boles of about 12 feet each. A splendid Scots fir stands near the mansion-house. It girths 10 feet 6 inches above the swell of the roots, and 9 feet 3 inches at 5 feet, with a fine clean bole of over 30 feet.

In the garden policies there is much to attract the attention of the visitor. One of the most notable objects of general interest is a large and uniquely-constructed heather tower at the east end of the pinetum. It is a very picturesque structure, and is the handiwork of Mr Wyber, who had been gardener at Cardross for about fifty years. It is built of larch and Scots fir pillars, sparred and thatched with heather outside, and lined with moss inside. The lower part is fitted up as a summer-house, and the tower reaches a height of 40 feet. It was built about thirty years ago with the object of affording a view of Stirling Castle and Abbey Craig, but from the luxuriant growth of the surrounding timber it has ceased to serve that purpose. The garden policies are further ornamented by a nice artificial pond, with a small island in the centre, and rockeries on either side. Three large and beautiful araucarias, the largest measuring 20 feet in height, spring from one of the rockeries; and surmounting the other are four gigantic Polynesian gods, made of wood, and a wooden figure of a New Zealand chief, brought to this country by Admiral Erskine. They are most grotesque and frightful-looking objects. So terrible, indeed, are they in appearance, that one which stood in another part of the property had to be removed on account of it frightening the cattle. This part of the policies was formerly known as the Ass Park, and was of a swampy character, overgrown with rushes, but it has since been drained, and laid out in such a way as to make it one of the most attractive parts of the policies. It is intersected by two fine grass walks, one of them lined on either side with beautiful Irish yews and *Cupressus Lawsoniana*. The most noteworthy tree in this neighbourhood is a great spreading oak, with a short bole but enormous branches. The girth of the bole at the narrowest part is 10 feet 10 inches, and the spread of branches is 74 feet.

The pinetum includes many rare and valuable specimens. Prominent amongst these is a grand *Wellingtonia gigantea* planted in 1861, in commemoration of Mr Erskine's marriage. It is about 40 feet high, and girths 8 feet at the ground.

There are also good specimens of *Cedrus Atlantica Pinus excelsa, Picea Nordmanniana, Pinus Pallasiana, Picea nobilis,* hemlock spruce, *Abies Menziesii,* Wyemouth pines, a very nice young snake-barked maple, a fine nettle-leaved elm, *Abies Douglasii,* &c. The latter promise to do particularly well, the best in the policies being about 60 or 70 feet high.

Some very fine trees of different varieties are also to be found in the Flower Garden. A *Cedrus Deodara* girths 6 feet 7 inches at the narrowest part of the bole. But the most notable tree in the policies is that known as the old staghorn elm, so-called from the staghorn-like appearance of some of its top branches. The girth at 1 foot from the ground is 18 feet 9 inches; and at 4 feet, the narrowest part of the bole, the girth is 18 feet 4 inches. It is altogether a fine massive old tree.

Another very interesting portion of the property, both from an arboricultural and a picturesque point of view, is "The Glen." "The Glen" is about one mile in length, and runs north and south along the eastern side of the policy park. On both sides the banks are very precipitous, and are loaded with many varieties of timber and shrubs, including oak, ash, scyamore, spruce, larch, birch, &c. A number of *Abies Douglasii* and *Picea Nordmanniana* have been planted here within recent years, and all of them are thriving splendidly. There are many grand larches, evidently belonging to the original stock, and are said to have been sent to Cardross from Dunkeld. Five of them measure as follows:—No. 1, at 1 foot from the ground, 10 feet 6 inches and 7 feet 6 inches at 5 feet, the height being 50 feet; No. 2, 11 feet 6 inches at 1 foot, 9 feet 3 inches at 5 feet, and 80 feet high; No. 3, 11 feet at 1 foot, 9 feet 2 inches at 5 feet, and 80 feet high; No. 4, 10 feet 6 inches at 1 foot, 9 feet 6 inches at 5 feet, and 85 feet high; and No. 5, 10 feet at 1 foot, 8 feet 9 inches at 5 feet, and 80 feet high.

Cardross has, from a very early period, been noted for its timber, and it still continues to be heavily wooded. Between twenty and thirty years ago this estate was considered to be one of the best-wooded properties, for its size, in Scotland, but about that period enormous quantities of excellent larch and

Scots fir trees were cut down. It is said, indeed, that the larch from Cardross supplied almost all the sleepers used in the construction of the Scottish Central Railway. Some idea of the immense quantity of timber taken off the property at that time may be gathered from the fact, that no less than thirty-five carts were constantly employed in driving timber for a couple of years. Since then the gaps have been largely filled up. Mr David Erskine, the grandfather of the present laird, was an extensive planter, and cut down very little timber. He planted the lower half of the Big Wood of Cardross, the largest on the property, and consisting of 400 acres. This enormous plantation lies along the whole of the south side of the Lake of Monteith, and contributes in no small degree to the beauty of the shores of this fine sheet of water. With the exception of some young wood to supply vacancies the whole of this plantation is loaded with mature timber—larch, spruce, Scots fir, and some natural birch trees. A plan of the estate, prepared in 1801, shows that only a small portion of the higher part of this plantation was then in existence, the whole of the lower part being then described as "heath pasture." The higher and older portion of the Big Wood was planted by Mr John Erskine, the author of the *Institutes*, who acquired the property in 1745, and died in 1768. The older portion was originally planted in rows, traces of this having been met with quite recently. The plantation is covered with knowes or mounds, giving it a more picturesque appearance. The soil is partly sandy and partly marshy. The wood is the home of an abundance of roe and capercailzie, the fox is frequently met with, and numerous otters dwell on the margin of the lake. Besides planting about the half of the Big Wood, Mr David Erskine laid down large plantations on both sides of the railway, and which consist of between 30 and 40 acres of larch, spruce, fir, and oak. He also planted a good deal over the property generally. The present laird continues to pursue this policy, planting principally with larch, Scots fir, and spruce. The soil, generally, is mixed loam and till, with a retentive subsoil, and timber of all descriptions thrive very well. *Abies Douglasii* is the most flourishing variety, rearing its head about 8 feet higher than any of the other kinds planted at the same time. The total extent of the property, is about 6200 acres. When first granted to the Erskine family, it was not so large, but several adjoining parts of the original estate, which had been feued from the Monastery, have since been acquired. Of the 6200 acres, 3200 acres are arable, 745 acres wood, and about 2255 acres moorland. Roadmaking and fencing have been executed to a large extent in recent years. Within the last five years the whole of the fences dividing Cardross parks, numbering seventeen in all, have been almost entirely renewed. A good many of the new fences are of iron, with $\frac{3}{4}$-inch rods for the top wire. The standards are $1\frac{3}{8}$ inch by $\frac{3}{8}$, with pronged feet,—every alternating standard being set in stone. We have stated that there are about 2255 acres of moorland. This is certainly not very profitable land, and nothing but the insuperable difficulties which stand in the way prevent it being turned to a more useful purpose. A number of years ago, when labour was cheaper, systematic efforts were made to remove the moss, its depth being from 8 to 16 feet. In two parts of the estate several families of Highlanders were allowed to settle on the borders of the moors, building for themselves houses of peat. They paid no rent for the land they cleared during their first lease of 19 years. For the next 19 years they paid a small rent for the land already brought under cultivation, and they were allowed to add as much as they pleased to their holdings by clearing the moss adjoining. At the end of this term one of these "colonies" came to an end, most of the reclaimed land being thrown into the two farms of South Flanders and Polden, comprising about 200 acres, now in the occupation of one tenant, —the original "Moss Lairds," as they were called, having died out, or parted with their holdings to their neighbours. The method adopted to remove the moss was very simple. A large central dam was constructed, from which the water was conveyed through the moss to the scene of the operations, and floated into the Forth. Like the ground underneath Blair-Drummond Moss, the subsoil here was the best of clay. The year following the removal of the moss, the land was levelled and drained, and sown with oats. Next year, it was limed and manured, and another crop of oats taken, varied occasionally by potatoes, until the land was fully reclaimed. Efforts are still being made to reclaim the land by a different method, and, although it is slower in operation, it is not to be despised. In the Collamoon district, the land is occupied by small crofters, with possessions extending from 6 to 16 acres each. They have good houses, and generally keep a horse and from three to five cows. They are, however, largely employed in the making of peat, and Port of Monteith Station being close at hand, the business is fairly remunerative—some families turning out about 100 tons annually, and finding a ready market in Glasgow for distillery purposes at 11s. per ton. The crofters are allowed to take as much of the peat as they like without charge, and whatever land is in this way reclaimed is cultivated by them, without rent, till the expiry of their nineteen years' lease, when a new contract is entered into.

Like most of our Scottish estates, the history of Cardross is somewhat chequered, and full of interest. From the foundation of the Monastery of Inchmahome it formed the principal part of its temporalities; and, as was sometimes the case in those days, the Commendatorship of the Monastery remained in the same family for several generations. In this way we find that the last Commendators (or Lay Superiors) of Inchmahome all belonged to the Erskine family, and at the suppression of the monasteries the Lordship and Barony of Cardross was erected by James VI. in favour of John, second Earl of Mar of the House of Erskine, by a charter dated 27th March 1604. It included all the lands, baronies, castles, towers, woods, parks, meadows, forests, fishings, &c., which formerly belonged to the Priory of Inchmahome and the Abbeys of Dryburgh and Cambuskenneth. This grant was made to enable the Earl the better to provide for his younger son by his second Countess, Lady Mary Stewart, a cousin of the King, of whom His Majesty took great care. There is a tradition that Lady Mary was in the practice of making grevious complaints to King James as to the state of dependence in which her second son, Harry Erskine, would be placed, as compared with his two elder brothers,—one of whom, by his father's first marriage would be Earl of Mar, the other, the eldest son of Lady Mary being Earl of Buchan,—and that the King assured her that her second son would be provided for, the fulfilment of this promise being the grant of the Lordship of Cardross. A very amusing tradition has been recorded as to how the Earl of Mar came to marry the King's cousin. The Earl had got his fortune told by some Italian conjuror, who showed him the limning of a lady whom, he declared, would be Mar's wife. The Earl fancied he recognised the same features in Lady Mary Stewart, and when he heard that the King had destined her hand for another, he wrote a plaintive letter to His Majesty, saying that his health was suffering from the dread of disappointment. The King thereupon visited the Earl, and assured him, "Ye shanna dee, Jock, for ony lass in a' the land," and immediately arranged the marriage. The Act for the erection of the Abbey and Priory lands into the temporal Lordship of Cardross, in favour of John, Earl of Mar, was passed on 9th July 1606, the Act narrating that this was done "in consideration of the good, true, and thankful services done by John, Earl of Mar, and his predecessors, in their great care, vigilance, and faithfulness in all things that tended to the advancement of His Majesty's honourable affairs," &c., special mention being made of the pains taken by the Earl and his late father in the education of the King, and "His Majesty's dearest son, the Prince." The first Lord assigned the Lordship and Peerage of Cardross to Henry, his second son by his marriage with Lady Mary Stewart. He did not, however include Cambuskenneth, which was assigned to Alexander Erskine, the third son of Lady Mary. Henry Erskine was thereafter styled Friar of Cardross, but having predeceased his father, he never enjoyed the Peerage of Cardross. Henry's only son, David Erskine, succeeded to the Peerage on the death of his grandfather, John, Earl of Mar, in 1634; and the Peerage of Cardross is now held as a male dignity by his collateral heir-male, the Earl of Buchan. The Erskines of Cardross have for ages occupied a prominent and honourable place in the country. No family has given more sons to the camp, or produced more eminent statesmen, distinguished lawyers, or useful works in the field of literature.

For hundreds of years they have been conspicuous in the affairs of the State, and have been honoured with the confidence of many Sovereigns, some of its members holding exalted positions around the throne. Some suppose the family to be of Danish origin, but it is certain that Henriens de Erskine was proprietor of the Barony of Erskine on the Clyde, in the reign of Alexander II., and it remained in the possession of the family for many ages. In the War of Independence, the Erskines were staunch adherents of the national party, being allied by marriage to the family of Bruce. Sir William Erskine was a man of exceptional bravery, and was the companion of the renowned Randolph, Earl of Moray, and the gallant Sir James Douglas. He accompanied the expedition to England in the year 1327, and was knighted under the royal banner as a reward for his valour. His eldest son, Sir Robert, was distinguished for his talents and accomplishments. He took a prominent part in the great events which make up the history of Scotland for the period, early displaying his opposition to the Baliol party. He was honoured with many high offices of State, and on several occasions distinguished himself as an Ambassador to the Courts of England and France. He had the interest of his Sovereign and his country so much at heart, that he gave his eldest son as a hostage for the payment of the ransom for the deliverance of King David after his capture at the Battle of Durham. His eldest son, Sir Thomas, was scarcely less distinguished as a statesman and a warrior than his father. His eldest son and successor, again, Sir Robert Erskine, first Lord Erskine, took a very prominent part in the Battle of Homildon, where he had the misfortune to be captured. Shortly after being released he was appointed one of the Commissioners to treat for the release of James I. in 1421, and three years afterwards he became one of the hostages for his ransom. He was released from captivity in 1425, and on the death of the Earl of Mar ten years later he succeeded in his claim to that Earldom. His son succeeded him in 1453, but in 1457 he was, by the Assize of Error, dispossessed of the Earldom, although he continued to take an important share in the national business, being one of the guarantees of a treaty with England, and took an active part in the cause of James III. His successor, Alexander, third Lord Erskine, was a man of great influence, and was entrusted with the charge of the youthful James IV., with whom he was ever afterwards a great favourite. The fourth Lord Erskine was killed at Flodden Field in 1513. The fifth Lord Erskine was also one of the distinguished men of his time, and was entrusted with the charge of his young Sovereign, James V. His eldest son was killed at the disastrous Battle of Pinkie, upon which John, his second son, who had been till then Commendator of Inchmahome, succeeded to the title. He was one of Scotland's most gifted sons, and his talents were freely used for the cause of his country, and the maintenance of Protestantism. On the advance of the English in 1560, the Queen Regent threw herself upon his Lordship's protection, and in consequence he assumed the charge of the youthful Queen Mary; and, in order to keep her in safety from the intrigues of the nobles, he sent her from Stirling Castle to the Priory of Inchmahome, and placed her under the care of his half-brother, David Erskine, who had succeeded him in the Commendatorship. There she remained, accompanied by her four Maries, till she went to France, whither Lord Erskine accompanied her. On the return of the young Queen Mary from France, he became a Privy Councillor, and was restored to the ancient title of Earl of Mar. The infant James VI. was committed to his keeping. It was through his prompt action that the King's party were saved from annihilation when the Regent Lennox was killed at Stirling, for which service he was chosen Regent of the Kingdom, special notice being taken of "his moderation, his humanity, and his disinterestedness." He strove to restore peace and prosperity to his country, but the ambition and selfishness of Morton preyed so acutely upon his mind, that he died of a broken heart in 1572. John, seventh Earl of Mar, was also a man of great talent, and was educated, along with James VI., by the celebrated George Buchanan. He frequently acted as an Ambassador, was a Privy Councillor, a Knight of the Garter, Secretary for Foreign Affairs, and High Treasurer of Scotland. James VI. created him Lord Cardross, with power to assign the title to any of his heirs-male as previously stated,—the record of the Parliament which met at Perth on the 19th July 1606, reading, "Act of erection of the Abbey of Dryburgh and Cambus-Kenneth and Priory of Inchmahome into a temporal lordship, called the Lordship of Cardross, in favour of the Earl of Mar—with the honour, estate, dignity, and pre-eminence of a free Lord of Parliament, to be called Lord Cardross in all time coming." It was he who, as already stated, married Lady Mary Stewart, and built the principal suite of apartments in Cardross House, where he entertained the King for several days with great magnificence. The third Lord Cardross was one of the many victims to adherence to Protestantism, having suffered the most cruel persecutions by fine and imprisonment. After remaining a fugitive in America for some time, he proceeded to Holland, accompanied the Prince of Orange to England, and was appointed to the command of a troop of dragoons. The fourth Lord Cardross also took an active part in the affairs of State, interesting himself deeply in the Hanoverian succession. In 1698 he succeeded to the Earldom of Buchan; and in 1745 the estate of Cardross was bought by a judicial sale from his Lordship's creditors by his cousin, John Erskine of Carnock, well known as the author of *Erskine's Institutes of the Law of Scotland*, and who was the first "Mr Erskine of Cardross." The succeeding members of the family have occupied honourable places both in the Church, the State, and in the profession of the law,—the present laird, as has been already mentioned, holding an important appointment in the House of Commons. The family has ever been distinguished for all that is good and noble, throwing the weight of its great influence on the side of those who struggled for the maintenance of the religion and liberties of the people of Scotland. Upon their own property they have always been held in the highest esteem, having been conspicuous for their desire to improve every part of their possessions, and to promote the welfare of all around them.

Having examined all the objects of interest at Cardross, a short walk brings us to Lochend, the residence of the genial Admiral John Elphinstone Erskine, author of *The Islands of the Western Pacific,* and formerly M.P. for Stirlingshire. This property is not very large, but it is charmingly situated on the shores of the Lake of Monteith, while the grounds surrounding the mansion are admirably laid out, and contain a number of magnificent trees. The house was built in 1715, but has since been considerably enlarged. All the trees appear to have been planted in accordance with a plan, when the mansion was erected. A very fine yew, with a bole of 6 feet, girths 7 feet 9 inches at the narrowest part. Another yew, with a bole of 12 feet, girths 8 feet 6 inches at 5 feet from the ground. Two of the beech trees have attained a respectable size,—the one having a girth of 15 feet 6 inches at 1 foot from the ground, and 12 feet 5 inches at 5 feet; and the other 15 feet 6 inches at 1 foot and 10 feet 6 inches at 5 feet,—both being fine clean trees, with boles of about 30 feet. There is also an elm with a girth of 10 feet 5 inches at 5 feet from the ground. At the Mill of Cardross, opposite the entrance gate to Lochend, there is a really splendid oak, but as it is situated in a low-lying place, its great proportions are somewhat obscured. At one foot from the ground the girth is 15 feet 8 inches, and at 5 feet the girth is 13 feet 5 inches—the bole being 35 feet, and the entire height of the tree 90 feet.

The Admiral kindly placed his boat at our disposal to explore the beauties of the lake, which is situated partly on the property of Cardross and partly on that belonging to the Duke of Montrose. The lake is a singularly beautiful expanse of water, rendered all the more attractive and picturesque by the romantic and tree-clad islands which rise from its bosom, with their still massive ruins of former greatness and splendour. The lake is circular in shape, about 7 miles in circumference, and in some places about 80 feet deep. Its waters are stored with several varieties of fish,—pike, however, predominating,—and numerous flocks of wild fowl and swans skim along the surface. The north, east, and south shores are embellished with a wealth of sylvan beauty to the water's edge, the boughs in some parts extending far over the margin and kissing the dark blue waters of the lake. Hills surround the lake in every direction, and to the north they reach to a considerable altitude. From Bendairg (the red hill), one of the northern hills, thirteen fine sheets of water may be seen, including the Lake of Monteith, Lochs Katrine, Rusky, Lubnaig, Achray, Vennachar, &c. North and west of Ben-dairg there is an extensive panorama of mountains and lakes. To the west, Ben Lomond rears his lofty head, in the centre of a family of great mountains. Southwards, the eye travels over the fertile plains of the Valley of Monteith, —from Drymen on the west to Stirling on the east,—the whole presenting one of the most gorgeous combinations of Highland and Lowland scenery to be met with in Scotland.

The Island of Inchmahome first claims special attention, not only on account of the remains of the celebrated Priory of Inchmahome, but because of its arboreal treasures. The former, however, are the more conspicuous, and at once recall the lines of Perthshire's most distinguished poet, Robert Nicoll :—

> "There stands the ruin'd Abbey's lonely tower,
> To speak of vanished pomp, exhausted power.
>
> . . . . .
>
> Decaying, roofless walls; and this is all
> That Desolation's blighting hand hath left
> Of tower and pinnacle, and gilded hall."

The Priory of Inchmahome was built by Walter Comyn, Lord of Badenoch, who married the eldest daughter of the Earl of Monteith, and who was a pious and patriotic Scotchman. The warrant for its erection was granted by Pope Gregory IX. on 16th June 1238, and is supposed to have taken the place of a Culdee cell. For upwards of three centuries it flourished as a religious house, till, at the Reformation, it shared the fate of almost all the ecclesiastical establishments, and was included in the grants of land which were assigned to the Barony of Cardross. It appears, however, to have been lost to that estate by proscription, probably during the years of trouble which befell the Cardross family, and it now forms part of the Monteith estates which were left by the last Earl of Monteith, who died in the year 1699, to the Duke of Montrose. The Priory was one of the order of St Augustine, and is the very earliest Augustinian Monastery in Scotland. It stands east and west, and the most interesting portion now remaining is the great west door, which formed the original entrance. The gable in which this gate is situated is 30 feet wide, and the doorway itself is a most beauliful specimen of Gothic architecture, 12 feet high and 6 feet wide at the ground. Notwithstanding its age of six centuries, and the neglect it has experienced during a great part of that time, it is still in tolerable preservation. The decay of ages has in some parts affected the grooving, but large portions still retain their pristine beauty. The choir of the church has long been used as a place of interment by the Earls of Monteith, as well as the Drummonds, who are closely connected with them. A very interesting monument to the memory of Walter Stewart, Earl of Monteith, and his Countess, who died in 1294 and 1286 respectively, is still preserved in the centre of the choir. The monument consists of two recumbent figures, 7 feet long, in full relief, and in loving embrace—

> "The steel-clad Stewart, Rèd Cross Knight,
> Monteith his Countess, fair and bright,
> Here live in sculptured stone."

A building, situated a few yards to the south of the Priory, is said to be the burial vault of the later Earls of Monteith and Airth. In the east gable of this building there are two windows, still entire,—one of three, and the other of two, arches. The former lighted the arched vault, and the latter the room above, still called Queen Mary's Room, from her having slept there during her stay in the island, to which we have already referred. Ash and other trees have grown up about the building, and now fill up this room, but parts of the iron and woodwork still remain.

The south-western portion of the island was marked off from the Priory as gardens and orchards for the Earls of Monteith, on the side adjacent to their castle on the island of Talla. It is in this part of the island that the garden and bower popularly known as Queen Mary's are situated, and which are now marked off by a neat paling. These are believed to have existed long before Queen Mary's visit, but their name is supposed to have been derived from the spot being a favourite retreat of the unfortunate Queen, during her temporary sojourn on the island. This historical garden is only about 35 feet square, and is surrounded by stone walls. A large boxwood tree stands in the centre, and is of more than ordinary interest, as it is reputed to have been planted by Queen Mary in remembrance of her visit. The tree is about 20 feet high, and measures at the ground fully 3 feet in circumference, and is still in a flourishing condition. Traces of the garden can be seen in the filberts and other fruit trees which still exist, and form an object of attraction and temptation to tourists. The bower, which is situated on a knoll at the west side of the garden, is a small spot, measuring only about 33 yards round the outside. It is said to have been originally adorned with a row of boxwood trees, and a thorn tree in the centre. Like the boxwood tree in the garden, these trees grew to a considerable size, but they have nearly all disappeared in satisfying the desire of tourists to become possessed of relics of Queen Mary. To those tourists who are afflicted with this relic mania, we would take the liberty of giving a gentle hint, attention to which may help to prevent the destruction of the plants at the bower, and save themselves from the ridicule of those who know better. Scarcely any of the boxwood which now adorns the bower was planted by Queen Mary, but was planted a little over twenty years ago by the late Mr Wyber, gardener at Cardross, to replace the trees which were destroyed. The cuttings, however, were taken from the original trees. They have now reached a considerable size, and the ignorant, in carrying off a cutting, might readily believe that they had a piece of the veritable tree planted by Scotland's most unfortunate Queen, instead of a plant obtained a comparatively few years ago from the gardens at Cardross. The pleasure-grounds of the ancient Earls were situated on the northern shore of the loch, and included the finely-wooded Hill of Cowden or Colden,

"Gay Coldon's feathered steep."

as well as part of the adjoining farm of Portend. On the south side of this hill, at the edge of the water, there is an excellent echo, words called out on the shore reverberating through the ruins of the Priory with remarkable distinctness. The most notable trees on the island are the old Spanish chestnuts, planted by the Augustine friars when they first settled here more than 600 years ago. Considering that the island is only about 5 acres in extent, these trees are remarkably numerous. The largest of them is that known as the "Antlered Chestnut," from the resemblance of the branches to the antlers of a deer. At 1 foot from the ground the girth is 20 feet 6 inches, and at 5 feet the girth is 17 feet 2 inches, with a bole of 25 feet, and a height of 80 feet—the tree being in very good health. Another chestnut, very much decayed, girths 16 feet 6 inches at the narrowest part of the bole. Another, in good health, girths 16 feet 6 inches at 1 foot from the ground, and 14 feet 6 inches at 5 feet, with a bole of 20 feet. The other varieties of trees upon the island include hazel in abundance, oak, ash, a few elder trees, sycamore, larch, a number of *Abies Douglasii*, and two young *Wellingtonias*. The largest of the oaks only girths 8 feet 4 inches, but one of the sycamores, with a bole of 15 feet, has a girth of 12 feet 8 inches at 1 foot and 11 feet at 5 feet from the ground. The whole island was cleaned and fenced a few years ago, and Admiral Erskine conducted extensive excavations round the Abbey and ruins about three years ago, which resulted in the discovery of about thirty human skulls.

On one of the islands, Inchtalla, there are the remains of the Castle of Talla, the principal residence of the Earls of Monteith of the Graham family from the time of Malise Graham, the first Earl, in 1427, to the last Earl, who died in 1694. No record exists to show who was the builder of the Castle, but it is considered probable that there had been a tower or keep on this island as early, if not earlier, than the Priory of Inchmahome, which, as we have stated, was erected by Walter Comyn, Earl of Monteith, in the year 1238, and that Malise Graham and his successors had erected additional buildings suitable for their residence in lieu of the Castle of Doune, the re-erection of the Earldom not including this stronghold, which had been converted into a royal residence. The Gaelic interpretation of Talla or Tulla is said to signify a hall or great man's house. The ruins, including a courtyard, occupy nearly the whole island, which is scarcely one acre and a half in extent. The buildings are almost completely enshrouded with luxuriant foliage, and such is the appearance of the island in summer, that it has been said to resemble an "umbrageous concrete." In consequence of almost the whole of the ground being required for the Castle, there is no old timber, the " umbrageous concrete " consisting of natural wood, mostly saplings of ash, sycamore, thorn, mountain ash, and a few hazel trees. A little to the west of the island of Talla is another small island called the Dog Island, only a few yards in circumference, the name being acquired from the Earls of Monteith having made it a dog-kennel, " when the island was less covered with water than it is now," according to the *Red Book of Monteith*.

We have said that the shores of the loch owe much of their beauty to the richness of their sylvan garment. This attraction is not confined to the Big Wood of Cardross. On the east side the shores are greatly embellished with timber planted by the late Lieut.-General Graham Stirling of Rednoch. This estate, now the property of the Rev. Henry A. G. Shepherd, carries a heavy load of fine mixed timber, which effectually conceals the mansion-house from the lake. The house is a modern erection, surrounded by extensive policies, containing a lot of well-grown mixed timber. The amenity of this portion of Monteith is considerably enhanced by the fir-clad hill to the west of Auchyle, and which is also on the Rednoch property, the hill being thickly clothed to the summit, a height of about 1500 feet above the level of the lake. To the east of Rednoch there is the exceedingly pretty and well-kept estate of Blairhoyle, the property of Mr A. H. Lee, who purchased it a few years ago. Since he became the proprietor, an addition has been made to the mansion-house, and extensive improvements have been made in fencing, draining, road-making, &c. Glenny, the property of Sheriff Grahame, stands on the hill to the west of the Lake of Monteith. The property occupies a charming situation, commanding a full view of the Lake and Valley of Monteith. The sheriff has recently planted soft wood very extensively upon the hillside. Next to Glenny is the property of Mondowie, belonging to Mr Robert M'Kechnie, now let, along with Glenny, to an agricultural tenant. Mondowie marches with the extensive estates of the Duke of Montrose.

Retracing our steps a little we come to Garden, which belongs to Mr James Stirling, and marches with Cardross on the south. This is one of the most beautiful and richly-cultivated properties in the district. The mansion stands upon an eminence at the foot of the sequestered Glen of Arngibbon, the lower portion of which is adorned with many fine trees. The old Castle of Garden stood a little to the north of the site of the present mansion, and was an important stronghold in its day. The policies at Garden are adorned with many grand old trees, the silver firs being the most noteworthy, six of these being 10 feet in circumference at 5 feet from the ground. An oak at the village of Arnprior, and near to the "Corner House," girths 12 feet at 10 feet from the ground, with a clean bole of 18 feet. Many large and good specimens of the ash are to be found on the property. The principal plantation, near Balgair Hill, contains a large quantity of very fine timber, mostly Scots fir, about 40 or 50 years old. The neighbouring estate of Gartmore, belonging to Mr W. C. G. Bontine, is also well wooded, and contains several very fine trees. The largest tree upon the property is a beech, girthing 14 feet at 5 feet from the ground. Other two beech trees girth 11 feet 3 inches and 11 feet respectively at 5 from the ground. The largest sycamore girths 12 feet 4 inches at 5 feet; the largest cedar of Lebanon girths 11 feet 8 inches at 5 feet; the largest elm, 11 feet; ash, 10 feet; oak, 9 feet 3 inches,—all at 5 feet from the ground. There are several very fine silver fir and larch trees. The two largest silver firs each girth 11 feet 6 inches; and the two largest larch trees girth 10 feet 6 inches and 10 feet 3 inches at 5 feet from the ground.

The whole district of Monteith, as we have endeavoured to show, is full of interest, and offers many inducements to visitors. Although difficult of access, owing to its distance from the great modern highways, there are few places more popular with tourists. Her Majesty the Queen sojourned in the district for several weeks during the autumn of 1869, and was greatly charmed with all she saw. This, as we have previously indicated, was by no means the first royal visit to Monteith. Royalty has frequently been attracted to the district from a variety of causes. King Robert the Bruce is known to have been at least on three occasions at Inchmahome, where he transacted national business. After James I. seized the Earldom, he and his immediate successors on the throne made the Castle of Doune one of the Royal residences; and Queen Margaret, wife of James IV., was provided with the Lordship of Monteith as part of her jointure lands. Her son, James V., made several visits to Monteith as the "Good Man of Ballengeich." Mary Queen of Scots, when a child of four years of age, here found safety and repose when these could not be afforded by the Royal palaces or fortresses. Her son, James VI., at Cardross, visited John, Earl of Mar, his former fellow-pupil under George Buchanan. Charles II. was also a visitor. At a later period, Prince Charles Edward, when prosecuting his father's pretensions to the throne, in 1745, remained here for some time, Doune Castle being the principal stronghold he had in Scotland. Royalty has thus shown a decided partiality for the district, and so long as its great natural attractions endure it is likely to do so.

XXVI.—STRATHALLAN.

An Arborous Sea—Strathallan Castle—The Drummonds of Strathallan—
The Policies — Triangular Group of Spanish Chestnuts — Other
Remarkable Trees—The Woods—Malloch's Oak—Tragic End of a
Greedy Meal Merchant—Birks of Tullibardine—*Abies Douglasii*—
Inexpensive and Effectual Method of Protecting Young Plants from
Rabbits—Tullibardine Chair Tree—Old Castle of Tullibardine—
Excavation Representing the Great "Michael" War-ship — The
Nursery—Result of Protecting Plants from Late Spring and Early
Autumn Frosts—An Ancient Target—Relic of Caledonian Forest—
Abies Douglasii v. Larch—Ravages of Recent Storms.

STRATHALLAN is one of the most richly-wooded properties in the county. Looking across the country from the admirable vantage-ground afforded by the tower of the fine mixed-baronial Castle, scarcely anything is seen up to the base of the encircling mountains, but one solid mass of dark green branches yielding to every breath of wind, and presenting a veritable arborous sea. Here and there an opening has been judiciously cut into the heart of the wood, which leads the eye to some patch of agricultural land, and at our feet lie extensive policies, with single trees standing boldly out in " pillared beauty," while a sparkling rivulet emerges " out of the forest dark and dread "— as is beautifully expressed in the last poem which Longfellow wrote—to enliven the precincts of the Castle, a duty which is shared by a fairy-like fountain playing on the lawn. These, however, only serve to bring out the prevailing feature of the landscape in stronger relief. It is the wealth of growing wood which first attracts the eye, and its seemingly endless extent which most powerfully impresses the mind. The western horizon, even viewed from the elevation of the Castle tower, is completely obscured by the dense mass of foliage, and only the merest peeps can be got of the mighty forms of Ben Vorlich,

Bed Ledi, and the Aberuchill Hills. The configuration of the ground in other directions, however, permits of a wider prospect. To the south and east the Ochils are seen rising out of a leafy bed, and their outline can be traced throughout the entire range, several of the more lofty of the Fifeshire hills being also distinctly visible. The pine-clad Hill of Moncreiffe is a conspicuous object in the distant east, and over the Gask and Dupplin woods are described the gigantic hills of the Stormont. To the north-west the view is broken by the mountains at the back of Crieff, while due north the Glenshee and Birnam Hills stand out most prominently.

The Castle of Strathallan, the seat of William Henry Drummond, sixth Viscount Strathallan, Lord Drummond of Cromlix, and Lord Maderty, is situated about a couple of miles from Tullibardine Station on the Crieff Junction Railway. It stands upon a site of some elevation, and is of a substantial and elegant design, in which the baronial style predominates. The exterior is comparatively new, but it encloses the old historic Castle, with which it is incorporated, the old house being in the centre of the new building. The first of this branch of the great house of Drummond was James Drummond, younger son of David, second Lord Drummond. He was educated with young King James V., who, in 1585, appointed him a Gentleman of his Bedchamber. He was one of the Commendators of the Abbey of Inchaffray prior to the Reformation. He was raised to the Peerage in 1609 under the title of Lord Maderty, and acquired the estate of Inchpeffray by his marriage to a daughter of Sir James Chisholme of Cromlix, who, through her mother, was heiress of that property. His eldest son, John, second Lord Maderty, joined Montrose after his great victory at Kilsyth, for which he was afterwards cast into prison, but liberated on the condition that he would never again oppose the Parliament. His son, who succeeded, lay under such suspicion of disaffection to the dominant party, that he was also placed in confinement. His own two sons having died in early youth, he resigned, in 1684, in favour of his brother, the Hon. William Drummond of Cromlix, who, accordingly, became fourth Lord Maderty. He was a devote adherent of the House of Stuart, and continued with the Royalists under Glencairn in the Highlands until they were dispersed in 1654. He afterwards found his way to Russia, and ultimately became a Lieut.-General in the army of the Czar. He was recalled after the Restoration by Charles II., who made him a Major-General of the Forces in Scotland, and a Lord of the Treasury. In 1686 he was created Viscount Strathallan and Lord Drummond. He was the author of the valuable *Genealogy of the Family of Drummond.* Upon the death of the third Viscount, the title devolved upon his cousin, William Drummond of Machany, descended from Sir James Drummond of Machany, second son of the first Lord Maderty. This branch of the family was also devoted to the Stuarts, and when the " Standard on the Braes of Mar " was raised in 1715, the new representative of the house of Strathallan was amongst the first to flock to it. He was captured at Sheriffmuir, but was discharged under the Act of Grace of 1717. This, however, did not cool his zeal on behalf of the House of Stuart, as when Prince Charles raised the rebel standard at Glenfinnan in 1745, he joined him within a fortnight, and was entrusted with the command of the Rebel troops in Scotland when the Prince marched into England. He was killed at Culloden. His Viscountess, a daughter of the Baroness Nairne, also suffered a great deal in the ill-fated cause, having been for about eight months a State prisoner in the Castle of Edinburgh. His eldest son, James, was attainted for his share in the '45. His son, again, Andrew John, became one of those Scottish military adventurers, who at this period distinguished themselves in many parts of the world. After serving with honour in America and on the Continent of Europe, he was, in 1810, appointed Governor of Dumbarton Castle, and, in 1812, became a General of the army. He vainly petitioned for the restoration of the forfeited titles of his family, and at his death in 1817, the forfeited estates, which had been bought back in 1755, fell to his cousin, James-Andrew-John-Lawrence-Charles Drummond, a grandson of the fourth Viscount. He was four times, twice unanimously, elected Member of Parliament for Perthshire,—first in 1811, then 1812, 1818, and 1820. In 1824 he was, by Act of Parliament, restored to the titles of Viscount Strathallan, Lord Maderty, and Lord Drummond of Cromlix. He was shortly afterwards elected a Scotch Representative Peer, and continued to be so till his death in 1851, when he was succeeded by the present Lord.

We have already incidentally alluded to the grandeur of the wood which adorns the policies. Within 100 yards of the Castle we are confronted with a numerous brotherhood of ancient and gigantic trees of several varieties. The most interesting of these is a triangular group of Spanish chestnuts of large size, the positions of the different trees in which is shown by the accompanying diagram. Whether the peculiar form in which

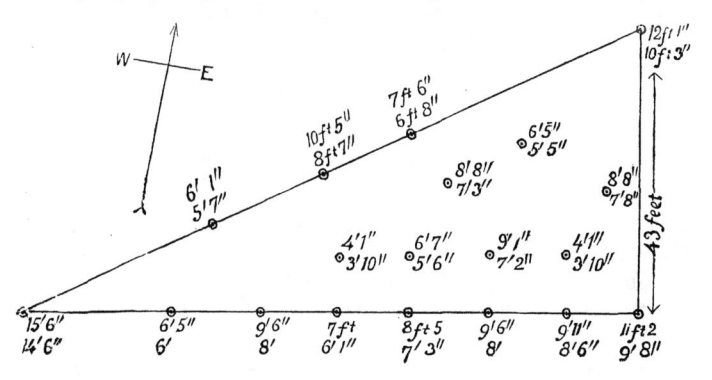

the group has been planted had originally any special meaning, cannot now be determined, but it is a singular and suggestive circumstance, that similar forms of grouping are to be met with in other localities, particularly in the southern counties of England. The trees are so arranged that from whatever point the group, or one of its individual members, is looked at in walking round, a triangle is presented. The sides of the outer triangle are 107 feet, 43 feet, and 99 feet long respectively, and the trees are so arranged that the triangles are numerous and varied—equilateral, scalene, obtuse, acute-angled, &c. The largest tree, which stands in the foreground of the group, measures as follows:—At 1 foot from the ground, 15 feet 6 inches in girth, and at 5 feet it is 14 feet 6 inches in girth, with a bole of nearly 40 feet, and a total height of 70 feet. The other trees girth as follows at 1 foot and 5 feet from the ground respectively:—12 feet 1 inch, 10 feet 3 inches; 11 feet 2 inches, 9 feet 8 inches—these two, together with the one already mentioned, forming the corner trees of the main or outer triangle. Proceeding along the line from the latter tree to the largest of the group, the girths are found to be 9 feet 11 inches, 8 feet 6 inches; 9 feet 6 inches, 8 feet; 8 feet 5 inches, 7 feet 3 inches; 7 feet, 6 feet 1 inch; 9 feet 6 inches, 8 feet; 6 feet 5 inches, 6 feet. Along the line of the triangle given above as 99 feet in length there are three trees girthing, at 1 foot and 5 feet from the ground, 6 feet 1 inch, 5 feet 7 inches; 10 feet 5 inches, 8 feet 7 inches; 7 feet 6 inches, 6 feet 8 inches. The inner trees girth respectively 9 feet 1 inch, 7 feet 2 inches; 8 feet 8 inches, 7 feet 8 inches; 8 feet 8 inches, 7 feet 3 inches; 6 feet 7 inches, 5 feet 6 inches; 6 feet 5 inches, 5 feet 5 inches—the remaining two girthing exactly the same, viz., 4 feet 1 inch at 1 foot from the ground, and 3 feet 10 inches at 5 feet. The site is level; the soil a dark loam, about 2 feet in depth, and the subsoil is a very hard retentive till. In one prominent part of the park, close to the castle, a very beautiful cedar of Lebanon " spreads his dark green layers of shade." There is only a bole of about 3 feet, but at the narrowest part the girth is 15 feet 5 inches. It is as a park tree, however, that it excels. A splendid elm also grows in front of the castle. It has a girth of 17 feet 9 inches at 1 foot from the ground, and 14 feet 6 inches at 5 feet, with measurable timber to a height of 35 feet. There is a much larger one in the avenue in front of the house, but it is very old and gnarled, and a considerable part of the trunk cannot be fairly measured as timber. Taking it as we find it, the girth at 1 foot from the ground is 20 feet 3 inches, and 15 feet 9 inches at 5 feet. An ash girths 16 feet 7 inches at 1 foot, and 12 feet 9 inches at 5 feet from the ground, with a fine round bole of 20 feet. An oak, with a bole of 25 feet, girths 15 feet at 1 foot, and 11 feet 4 inches at 5 feet from the ground. There is a very shapely beech with a girth of 19 feet 9 inches at 1 foot, and 12 feet at 5 feet, the bole being 12 feet in height.

Inside the garden fence there is a fine purple beech girthing 9 feet 2 inches at 1 foot from the ground, and 8 feet 1 inch at 5 feet, with a bole of 40 feet. There is also a very nice purple beech with a bole of 25 feet and a girth of 7 feet 10 inches at 1 foot and 7 feet 3 inches at 5 feet from the ground. The best spruce tree upon the property is also to be found in the Castle pleasure-grounds. It reaches a height of fully 90 feet, and has a girth of 14 feet at 1 foot and 10 feet at 5 feet from the ground. Beside it is another very good spruce girthing 13 feet 6 inches at 1 foot and 9 feet 4 inches at 5 feet from the ground. A few yards farther on we meet with the best specimen of *Abies Douglasii* at Strathallan. It was planted 25 years ago, and is exactly 65 feet in height, the girth at 1 foot from the ground being 7 feet, and at 5 feet 5 feet 7 inches. In the "Auld Boo" Park, a short distance from the Castle, there are also some very fine trees. An ash, with a bole of about 30 feet, and a total height of 90 feet, girths 15 feet 3 inches at 1 foot and 12 feet at 5 feet from the ground. There are a number of large sycamores, but the two following are perhaps the best:—No. 1, 14 feet 7 inches at 1 foot and 12 feet at 5 feet from the ground, with a bole of 20 feet; No. 2, close beside, girths 13 feet at 1 foot and 10 feet 4 inches at 5 feet, with a bole of 30 feet.

Attractive as the Castle pleasure-grounds undoubtedly are, they are no more interesting than the extensive woods—extending to nearly 2000 acres—by which they are surrounded on every side. It is not only in the abundance of saleable timber that they excel, for there are many trees in different parts of the woods which are quite as remarkable as those we have noticed in the policies. In the Castle Wood, for instance, there is an especially fine tree known as "Malloch's Oak," from a very tragic circumstance. During a great dearth of food at a remote period, there was a meal merchant of the name of Malloch who refused to sell provisions to the famishing people, in the expectation that the longer he held out the bigger price he would get. Having occasion to be in this neighbourhood one day, he was unceremoniously seized by the enraged inhabitants, and hanged upon one of the limbs of this tree. The branch upon which he is believed to have been hanged has rotted and fallen away, although the other branches remain entire. This tragic occurrence must have happened at a very early date, as documents in existence prove that the tree was called "Malloch's Oak" as far back as 200 years ago. The girth of the tree at 1 foot from the ground is 19 feet, and at 4 feet, where it can be most fairly measured, the girth is 15 feet 3 inches. One limb straight out from the trunk has the enormous length of 55 feet, and its great weight has to be supported by two heavy posts driven into the ground. The decayed limb is sheeted with iron to prevent water getting into the trunk. The tree is altogether a most interesting one, apart from the tragic circumstance to which it owes its name. In this wood there are some especially fine specimens of *Abies Douglasii*. They were planted on a clay-loamy soil in April 1866, when 3 feet high, and the best of those which we measured has a height of 48 feet exactly, and a girth of 4 feet 8 inches at 1 foot and 3 feet $7\frac{1}{2}$ inches at 5 feet from the ground. A beautifully compact *Cupressus Lawsoniana* has a circumference of 24 feet round the branches at 5 feet from the ground, and a height of 25 feet. This is really a splendid tree, but being in the heart of the wood, and close to the garden wall, it is not seen to proper advantage.

Our way next lay through the "Birks of Tullibardine," where we saw much that was interesting and instructive. So far as the "birks" are concerned, they are somewhat disappointing. The poet would not here meet with that

> "Wealth of birchen tresses
> Spread with diffusive leafy grace that knows
> No limit."

The "birks," in short, are conspicuous by their absence, although a few noble specimens survive,—one which we selected as a fair representative girthing 10 feet 4 inches at 5 feet from the ground. The Birks Wood is chiefly noteworthy for its promising crop of the newer coniferous trees, the chief of which are *Abies Douglasii*, of which there are between 7 and 8 acres without any admixture. Shortly after entering the "Birks," what appeared to be a peculiar fibre-like growth at the bottom of the young *Douglasii* trees attracted our attention, and forced us to ask for an explanation from the forester, Mr R. Maxtone. The result of our inquiries was the discovery of a very effectual and inexpensive method of preventing the destruction of the young trees by rabbits. The method adopted costs nothing but a trifling expenditure of labour, is perfectly effectual, and requires no further attention. The fibrous-like growths to which we have referred are simply a handful of spruce twigs, about 2 feet high, tied round the young tree when it is planted out about 3 feet high, any branches which obtrude being allowed to grow, as the rabbits can do no harm by eating them. The twigs are firmly stuck into the ground, and loosely tied round the tree with a piece of string. At first the twigs were planted round the tree in a circle, with a clear space round the stem, but it was found that the rabbits somehow got inside, and the precaution was thus of no avail. But by fastening the twigs close round the stem, tying them with a piece of string, a rabbit cannot possibly get its nose in. Even the string is scarcely necessary, at all events after a very few days, as the twigs cling so close to the trunk of the tree that a rabbit would find its task no easier by cutting the string. There is no pressure whatever round the stem of the tree, while the protection is so complete that the tree is absolutely impregnable to the attacks of this destructive vermin. The wood in which these young *Douglasii* trees are planted is a perfect rabbit warren, almost every footstep disturbing them in dozens, and sending them in haste to their burrows in the soft dry sand; and, although we searched most diligently, no trace whatever could be found of the slightest damage having been done. The rabbits, judging from one or two illustrations we met with, would appear to have a specially good appetite for *Picea Nordmanniana*, as the obtruding branches in several instances have been eaten to a height of 3 feet, although the bark was completely beyond their reach. We scarcely think that a cheaper or a more effectual means could be devised than that we have described for securing the young plants from the ravages of these woodland pests.

One of the most notable trees upon the property is to be found in the "Birks of Tullibardine." This is an oak known as the "Tullibardine Chair Tree." We have to trust to tradition for the origin of its name, but the tree itself, and the whole surroundings, corroborate the traditionary story. The old Castle of Tullibardine stood in a field about 150 yards west of the "Chair Tree," and it is said that round this oak the athletic sports of the muscular retainers were held in ancient times. The formation of the tree is such that a very commodious platform could be erected high among the boughs, and this, tradition says, was actually done to enable the lairds and their friends to have a complete view of the contests going on below. The tree is surrounded by the remains of a turf dyke, with a diameter of 25 yards, and within this circle the competitors had ample space to display their strength and skill. It is a very gnarled old tree,—probably about 600 or 700 years of age,—the whole trunk being thickly clothed with growing twigs—a sure indication of great age. The bole is about 40 feet, and is of the same girth for about 20 feet—the girth at 5 feet from the ground being 17 feet.

The old Castle of Tullibardine, whose occupants for generations must have gazed with veneration upon this aged king of the forest, was entirely demolished about fifty years ago, the materials being used for the building of steadings, &c., in the neighbourhood. Quite close to the site of the old Castle is another object of some interest, although it is not apparent to the casual visitor. When James IV., who had a penchant for shipbuilding, built the *Michael*, the largest war-ship of the period, he employed a skilful wright from Tullibardine to superintend the work. When the carpenter returned home, the people at the Castle were very anxious to obtain an adequate comprehension of the size of the great *Michael*, and with this view the carpenter was commissioned to have an excavation made near the Castle of the exact dimensions of the ship which had created so much sensation. The vessel was 240 feet in length; 36 feet within the sides, which were 10 feet thick; it took a year to build, and consumed all the oak wood of Fife, except Falkland. When ready for sea, she carried 1000 men, including 300 mariners. The Statistical Account makes no mention of this excavation representing the size of the *Michael*, but states that "her length and breadth is planted in hawthorn at Tullibardine by the wright that helped to make her." At the date of the last Statistical Account (1837) only three of the hawthorn trees survived, but they have since been entirely removed to facilitate agricultural operations. For the same reason the excavation has been almost completely filled in, the plough being now regularly worked over it. After it had been dug, it was filled with water, and continued to be an ornamental pond, in which aquatic plants and birds luxuriated until about twenty-five years ago. The shape of the vessel is still distinctly visible, and it will take a long time before it can be thoroughly obliterated. Within sight of this field, about 400 yards distant, stands the old chapel of the parish, burned by the Highlanders in the course of their retreat from Sheriffmuir. It has been restored, and is now used as a burial vault.

Proceeding through the home-farm parks, we come across a very nice nursery of 3 or 4 acres, recently formed by Mr Maxtone. All the ordinary varieties are reared, but spruce, Scots fir, larch, and *Abies Douglasii* predominate. The *Douglasii*, of which there are between 30,000 and 40,000 plants in the nursery, are now being raised from seed gathered in Strathallan Woods. The nursery has a fine southern exposure, and the site has been specially selected because it is so surrounded by trees as to prevent the young plants being damaged by the sun during the late spring and early autumn frosts. The effect of this protection is not only seen in the uniformly healthy condition of the young plants, but it may also be seen in at least one of the plantations, the success of which Mr Maxtone attributes to the impossibility of the sun striking the young trees before the frost has left their branches. This is at Blairskaith Plantation, where larch has been planted after a superior crop of the same variety. The young plants are from 10 to 12 or 15 feet high, and are thriving amazingly, with comparative freedom from disease. This, Mr Maxtone believes, is owing to the protection from the early sun afforded by the remains of the previous crop, of which fully 30 acres were blown down about fourteen years ago. In other plantations, where there is no such shelter, there is a good deal of disease, and even in virgin soil the crop is a comparative failure where it is exposed to the full sweep of the morning sun.

While inspecting the saw-mill, we were fortunate in seeing several objects of interest. An examination of an old green holly tree, recently blown down near the old Castle of Tullibardine, revealed numerous marks of arrow-shots in the outer bark, showing that it must have been used as a target for practising with the bow and arrow before Martini-Henry rifles entered into the imagination of man. Beside it, securely protected from the weather, was a quarter-section of a magnificent oak discovered four years ago, 4 or 5 feet underneath the moss at Tullibardine Muir, about 600 or 700 feet above the level of the sea. The length of the block was 16 feet, and the diameter of the section 4 feet, showing that the tree must have been about 20 feet in girth at a considerable height from the ground. This relic of the old Caledonian Forest has now been converted into picture frames, and other ornamental articles, the wood being firm, and of a beautiful black colour. At the mill a few *Douglasii* trees were being cut up, from which we could learn something of the great value of this variety as compared with larch. The trees were only thirteen years planted when they were blown down by the Tay Bridge gale, but there was as much red wood in the heart as in a larch of thirty years old.

Wherever we wandered about the property we found much that was interesting, and not a little, as we have endeavoured to show, that was instructive. The woods are so extensive that they seem to be interminable, and at almost every turn there is something to attract attention. No one could fail to be struck, especially, with the care with which the woods are tended. Drains are freely cut wherever necessary, and everything possible is done to promote the growth of the timber. Like all other places, Strathallan has suffered severely from the heavy gales of the past few years—the havoc, indeed, being so great as to be almost inconceivable. Enormous trees, and not unfrequently clumps of trees, have been turned over like nine-pins, their great roots rising from the ground like cliffs to a height of from 20 to 30 feet. These victims of the storms of winter certainly serve to make the plantations more picturesque and forest-like, but they involve an enormous loss to the proprietor.

XXVII.—DRUMMOND CASTLE.

The Attractions of Drummond Castle—The Founder of the House of Drummond—Origin of the Family Name—Glorious Death of the First of the Drummonds—History of the Family—Early Connection with Forestry—Queen Annabella—How Glenartney Forest was Acquired—The Builder of Drummond Castle—The Castle itself—The View from its Towers—The Gardens—Drummond Park and its Monster Trees—Remarkable Trees at Pitkellony, Muthill, &c.—A Curious Phenomenon—The Plantations—The Trossachs—Extent of the Property and General Improvements.

THE princely demesne of Drummond Castle, with its great natural and artificial beauties, its undying historic associations, and its priceless treasures, is one of the most notable places in the kingdom. The grandeur of its situation, and the fame of its gardens and richly-wooded policies, have attracted the attention of thousands of people from different lands, and their united testimony supports the presumption, that the gardens and grounds are, in the words of the Duchess of Sutherland, "unequalled in Europe, according to their scale." The beauties of Drummond Castle begin to dawn upon the visitor long before the entrance gate is reached. Leaving Crieff by the bridge which spans the Earn on the south-western side of the town, we very speedily obtain a foretaste of the joys which lie before us. For more than two miles from South Bridgend, the broad and well-formed public road runs through an avenue which has been described as "finer than the Italian Boulevards of Paris, and grander than the Linden Alley at Berlin." For the first three-quarters of a mile the avenue consists of beech trees, of noble proportions, averaging about 20 feet apart, with a few horse-chestnuts and limes intermixed, the branches beautifully interlacing in the embrace of more than a century. At Lagg Cottage, the avenue runs into lime trees,—which, in the words of

Pope, "part admit and part exclude the day"—for fully two and a half miles additional. The avenue is here and there backed with strips of mixed wood; and through the openings at intervals we can descry spacious parks studded with lordly trees, while beautiful vistas stretch far into the distance, and expose tracts of the fairest land. The entrances to the grounds are thoroughly in harmony with the character of the place. Gates of artistic design lead into alleys on the eastern and western sides of the public road, the eastern alley leading through the estate and rejoining the public road, and the western alley leading up to the ancient fortress of Drummond Castle. Passing through the Castle entrance-gate—a splendid piece of elaborately-decorated workmanship brought from Italy, and supposed to be about 300 years old—we enter the western alley, and are again confronted with an avenue—this time of enormous beech trees, although not the largest upon the property. The avenue is very narrow, being only 15 feet wide, with sidings to permit carriages to pass; but its very narrowness contributes to its unique character, and not a little to its beauty. The whole avenue is completely arched by the compact branches of the great beeches, and it is only occasionally that the eye can penetrate to the beautiful undulating country beyond, dotted with trees, and forming charming combinations of wood and water, hill and dale. The Castle comes upon the visitor somewhat abruptly, and immediately on passing through the archway into the courtyard we are involuntarily reminded of the poet's lines—

> "How many charms, by nature and by art,
> Do here combine to captivate the heart,"

Many eloquent pens have attempted to describe the beauties of Drummond Castle from this point, but none have been more eloquent or more accurate than MacCulloch. "Drummond Castle," says he, "is absolutely unrivalled in the low country and only exceeded in the Highlands by Dunkeld and Blair Placed in the most advantageous position to enjoy the magnificent and varied expanse around, it looks over scenery scarcely anywhere equalled. With ground of the most commanding and varied forms, including water and rock, and abrupt hill and dell, and gentle undulations, its extent is princely, and its aspect that of ancient wealth and ancient power. Noble avenues, profuse woods, a waste of lawn and pasture, an unrestrained scope, everything bespeaks the carelessness of liberality and extensive possessions; while the ancient Castle, its earliest part belonging to 1500, stamps on it that air of high and distant opulence which adds so deep a moral interest to the rural beauties of Baronial Britain."

The sight of the ancient stronghold of the Drummonds brings to recollection that it is the property of a noble family which has figured most prominently in the early history of Scotland, which gave us one of the best of the Scottish Queens, and which holds a distinguished place among the progenitors of the reigning Royal Family of Great Britain. The origin of the house of Drummond is traceable to the year 1067, when its founder landed upon the shores of Scotland, in times of great peril. In the reign of Malcolm II. of Scotland, and about the year 1013, the King of Denmark invaded England to revenge a massacre of the Danes, overcoming Ethelred in battle, driving him from his kingdom, and usurping his throne. Canute was the son and successor of this Danish King, but he was so strenuously opposed by Edmund, the Ironside, son of Ethelred, that it was mutually agreed to divide the kingdom between them. This treaty Edmund only survived one month, when the whole kingdom and his two sons, Edwin and Edward, fell into the hands of Canute, who was proclaimed sole Sovereign of England in 1017. To secure himself in full possession of the throne, Canute sent the two sons of Edmund to the Court of Sweden, on pretence of being educated, but charging the King to put them to death as soon as they arrived. The Swedish monarch disobeyed this savage mandate, and sent the young Princes to the King of Hungary, to be educated at his Court, where they were graciously received. Solomon, the King of Hungary, was so attached to the exiled princes that to Edwin he give his own sister in marriage, and to Edmund the sister of his Queen, Sophia, and daughter of the Emperor Henry II. Edwin died childless, but the wife of Edward bore Edgar Atheling, Margaret (afterwards Queen of Scotland), and Christina. On the death Canute in 1035, his son Harold succeeded to the throne, and when he died in 1039 he was succeeded by his brother, Hardicanute, who reigned only two years, and in whom ended the Danish usurpation. Edward, son of Ethelred, was unanimously called to the vacant throne, and as Edward the Confessor had no issue, he sent a deputation into Hungary, to invite over his nephew, Edward, son of his elder brother, Ironside, who was the only remaining heir of the Saxon line. That Prince accepted the invitation, and came to England with his wife and three children, together with a select train of Hungarian nobles, who were attached to his family by friendship and affection, and followed his fortunes as heir-apparent to the English throne. Edward died soon after his arrival, but his family continued to reside at the English Court till the death of the Confessor in 1065. Though young Edgar was now the rightful heir to the Crown, Harold, son of Earl Godwin, ascended the throne without opposition, but he did not long enjoy the fruits of his treachery. Confusions arose, which terminated in the Norman Conquest; and at length, on 14th October 1066, Harold encountered the invader at Hastings, and was slain, William the Conqueror being left sole master of the kingdom. All ranks of the nation at once submitted—even Edgar, who had just been declared King, resigning to William his pretensions to the Crown. As William's reign descended into oppression, multitudes consulted their safety by flight. Amongst the rest, Edgar Atheling, with his mother, sisters, and Hungarian friends, secretly set sail, directing their course to Hungary, but a furious storm overtaking them in the German Ocean, they were driven upon the Scottish coast, and came ashore at the haven since called St Margaret's Hope, in the Firth of Forth. At this time Malcolm Canmore, King of Scotland, was residing at Dunfermline, a few miles distant, and he generously invited the royal strangers to partake of the hospitality of the Scottish Court. To Edgar, Malcolm acknowledged the kindness of his relation, the Confessor, in sending 10,000 soldiers to his aid against Macbeth. He also received the friends who accompanied the Prince most generously, bestowing upon them offices and lands. Struck with the charms of Margaret, and the noble endowments of her mind, he took her for his Queen, celebrating the marriage with splendid magnificence. Amongst the distinguished strangers who landed at St Margaret's Hope with the royal voyagers, was the founder of the house of Drummond, Maurice, an Hungarian, long eminent for his faithful services. He is said to have been the son of George, a younger son of Andrew, King of Hungary. Educated at the Court with the Saxon Princes, his friendship for their family commenced early, and amidst all the vicissitudes of their fortunes, his attachment to their interests was firm and sincere. When accompanying the exiles driven on the Scottish coast, he is said to have commanded the vessel, and to his skilful management the safety of the Royal Family was attributed. It is believed, indeed, that it is from this circumstance that the family name is derived. Several derivations have been given of the word Drummond, the Greek, Celtic, and Latin being all laid under tribute. The word, *dromont* or *dromond* is of Greek origin, and was used, in different countries to denote a swift ship, the captains of which were called *dromonts* or *dromoners*. In the Celtic we have the word *drum*, a height, steep, or back, and *onde*, a wave—denoting a high wave; and to this correspond the bars called *unde* or *wavy*, as they are blazoned on the Drummond arms, which may thus have a reference to the storm by which they attained rank and fortune in Scotland. It is regarded, however, as more probable that the name is derived from the latin *dromo unda*, signifying a swift ship assailed by a tempest, and thus commemorating the storm which brought the founder of the name and the royal exiles to Scotland, and the appointment which he held as commander of the ship. It is thought quite probable, however, that the family name may have been derived from Drymen, the property in Stirlingshire forming part of the original grant of lands. On the marriage of Margaret to Malcolm, great honours and privileges were heaped upon Maurice. In addition to extensive grants of lands, he was appointed to high offices, and he had conferred upon him, as a lasting badge of honour, a noble coat-of-arms, representing his position and achievements. As a memorial of her Majesty's safe journey, the King assigned him for his armorial bearings, three bars wavy; and as the country where his property lay was full of rivers, woods, lakes, and mountains, these were emblematically expressed in the coat-of-arms. He continued to serve his new King most faithfully, and died gloriously in his cause. When William the Conqueror turned his arms against Malcolm, because he refused to deliver up Edgar Atheling, the first of the Drummonds fought with the Scotch until Edgar's safety was secured. When the successor of the Conqueror, William Rufus, invaded the dominions of Malcolm, in 1093, Maurice advanced with the King to repel the invader, and, on Malcolm falling by the treachery of Percy, Maurice hastened to revenge his death, but fell covered with wounds.

It is impossible, within the limits that can be conceded to this part of our subject, to give a complete outline of the family descent, with its innumerable branches. It will be of some interest, however, to trace the more prominent incidents in the family history, as then we shall be the better able to appreciate the rare attractions of Drummond Castle. We have seen how the founder of the family died gloriously in attempting to avenge the treacherous death of his Sovereign and benefactor; and we shall endeavour, as briefly as is compatible with clearness, to follow the career of the rest of the more prominent members of the family. The first of the family who was specially distinguished, after the founder, and the first of whom any written documents are preserved, was Malcolm Beg Drummond, the sixth Thane of Lennox—" Beg " being the Celtic for short, indicating that the Thane was below the ordinary stature. He succeeded about the year 1200, and died about 1259. He is represented as being held in high reputation among the Scottish nobles, and retaining all the lands given to his predecessors by Malcolm Canmore. The next Thane, also a Malcolm, is described as "a man of great respectability," and had possessions in land to an immense extent, which enabled him to give his younger children very considerable estates in different counties. The eighth Thane, Sir John Drummond, lived in the time of the unhappy confusions which arose at the death of Alexander III., but amidst the various contests for the succession to the Crown, he always stood forth as the defender of the liberties of his country. Sir Malcolm Drummond, the ninth Thane, flourished in the reign of King Robert Bruce, and was firmly attached to the interests of that great Prince, his undaunted bravery, and great military skill being conspicuous at the Battle of Bannockburn, his services there being rewarded by a grant of several lands in Perthshire. The tenth Thane of Lennox, Sir Malcolm Drummond, also distinguished himself in the service of the Bruce, and is described as "a worthy patriot, a steady loyalist, and deservedly esteemed for his merit and accomplishments." He is the ancestor of several branches of the family, including Concraig, Colquhalzie, Pitkellony, Megginch, Balloch, &c. His second son, Sir Maurice Drummond, was the first Knight of Concraig, and by his marriage with the daughter of Henrie, the Heritable Seneschal or Steward of Strathearn, he acquired a most intimate connection with the forests of the country. Besides the charter and confirmation he got of the office and lands belonging to Henrie, Robert, Earl of Strathearn, granted him a new gift of forestry to reach over all the forests in the country, making him Heritable Keeper. We are told that he was "a brave gentleman, and lived in great credit," from which we may infer that the forests did not suffer at his hands. The eleventh Thane of Lennox, Sir John Drummond, had what is appositely termed "an unhappy difference" with his neighbours, the Monteiths—a difference which, after forty years of "slaughters, robberies, and depredations," was settled by Royal Commission in May 1360. The lands of Roseneath were given to the Monteiths, but the King afterwards bestowed upon Sir John the baronies of Auchterarder, Cargill, and Kincardine. By this accession of lands, Sir John became one of the most opulent subjects in the kingdom, and from this time he changed both his residence and his designation. Unwilling to remain any longer in the neighbourhood of the hated Monteiths, he quitted the Lennox, with the Seneschalship thereof, of which the family had been the hereditary holders for three hundred years, and fixed his seat at Stobhall, which had been gifted to his grandfather by Robert Bruce, for services rendered on the field of Bannockburn. Sir John was the father of the Annabella Drummond who was married from Stobhall to Robert III., and became the mother of the hapless Duke of Rothesay, and James I., who was murdered in Perth. Queen Annabella was a lady of the most exquisite beauty and distinguished merit. She is celebrated as one of the best of the Scottish Queens, and her death, in 1401, was regarded as a great loss to the nation, having endeared herself to the people by the way in which she supported the dignity and splendour of the Court. Both of the succeeding lairds in the main line, Sir Malcolm (1373–1400) and Sir John (1400–1428), were notable men. It was in the reign of the latter that the splendid Forest of Glenartney—or Glen-Orkney, as it was then called—came into the family possessions. Sir John married Lady Elizabeth Sinclair, daughter of Henry, first Earl of Orkney, with whom he got the lands of Murthlow, in Banffshire, confirmed by a charter from Robert III., and the whole of the Forest of Glenartney. Whatever aspect the Forest presented in these days, it is now devoted to "the antler'd monarch of the waste," although a great deal has been done in recent years to adorn the ravines with sylvan beauty. The next laird, Sir Walter Drummond (1428-1445), was the ancestor of the families of Blair-Drummond, &c. The next notable member of the family was Sir John Drummond, the first Lord Drummond (1470-1519), who built "the strong castle of Drummond" in 1491, for which he was granted a special licence by James IV., given under his own hand and seal, whereupon his Lordship moved his residence from Stobhall to Strathearn. The Castle is built upon a picturesque rock, overlooking a great stretch of the most diversified and beautiful scenery. Entering the formidable donjon, with its ancient gate and guard-room still intact, we ascend the massive staircase to the balcony, from which the surrounding landscape is seen to the best advantage. Beneath are the fertile fields of Strathearn, with its many lordly mansions, stretching away to Invermay and Duncrub, while the position of Perth may be distinguished resting under the shadow of Kinnoull. Westwards, we look over "Glenartney's hazel shade," and gaze upon the pine-covered peak of Turlum. To the north, the thickly-wooded banks are seen to slope abruptly to the shores of a magnificent artificial lake, extending to nearly

100 acres, on which the swan, like that which Wordsworth pictures,

> "On still St Mary's Lake
> Floats double—swan and shadow!"

Beyond, the view is interrupted by the dark blue Grampians, many of the most prominent Bens being seen lifting their pointed tops far into ethereal space. Glancing at the Castle itself, we soon discover that it has been largely shorn of its original proportions. During the destructive campaign of Cromwell, whose army played great havoc upon the property, the fortress was nearly demolished, and it suffered again at the Revolution of 1688. In 1715 the remains were strengthened and garrisoned by the Royal troops; but in case it should ever again fall into the hands of the enemies of the House of Stuart, whose cause was warmly espoused by the Drummonds, Jane Gordon, Duchess of Perth, caused the greater part of the walls to be levelled to the foundation during the rising of 1745. From the remains of the old Castle which are still to be seen, there can be no doubt that the ancient fortress was of enormous strength. A portion of the old structure is at present fitted up as an armoury and art gallery, which of themselves, are of sufficient interest to form the subject of a long chapter. The armoury is richly stored with battle-axes, swords, targets, muskets, &c. Amongst the most notable of these implements of warfare is a ponderous two-edged sword which belonged to the Laird of Lundie, the valiant friend of King Robert the Bruce. The picture-gallery contains portraits of many of the distinguished characters who figured in Scottish and English history. Amongst the principal of these is that of King George III., by Gainsborough; Charles I., by Jamieson, and another by Vandyke; Charles II. and James II., by Lilley; the great Marquis of Montrose, by Honthuarst; Claverhouse, by Vanderhelst; Prince Charles Stuart; Henry, Cardinal York; Archibald, Earl of Argyle, who was beheaded by the Covenanters; George, Marquis of Huntly, the friend, companion, and fellow-sufferer with Montrose; and William Maitland of Lethington, Secretary of Scotland in the time of Queen Mary. There is also a cast of the head of Napoleon I., taken when his body was removed from St Helena.

The purely family portraits include several of the Earls of Perth; Lady Jane Gordon; James, fourth Earl of Perth and Lord-Chancellor of Scotland, the first Knight of the Thistle, when the Order was revived in 1687, and afterwards Duke of Perth; and Lord and Lady Perth, the grandfather and grandmother of the present Lady Willoughby. In this gallery are also stored many articles of great family interest, including the robes of the first Lord Drummond, the builder of the Castle, who was Lord Justice-General of Scotland; and the robes of the Earl of Perth, Lord-Chancellor of Scotland. We have brought our history of the Drummonds down to the builder of the Castle. The subsequent family history is conspicuously mixed up with the futile attempts of the Stuarts to regain the Crown, and the House of Drummond suffered severely in consequence. In 1715 James Lord Drummond was attainted, and his lands sequestrated to the Crown, but were restored, under certain conditions, in 1718. The same fate befell the house in 1745, but in 1784 the forfeited estates were restored to Captain James Drummond, who, in 1797, was created Lord Perth, Baron Drummond of Stobhall. He died in 1800, and the estates devolved upon his only surviving daughter, the Hon. Clementina Sarah Drummond, who, in 1807, married the Hon. Peter Robert Burrell, eldest son of Sir Peter Burrell, the first Lord Gwydyr, who assumed the surname and arms of Drummond. In 1820 he succeeded his father as Lord Gwydyr, and, in 1828, his mother, the Baroness Willoughby de Eresby, who was the sister and heir of the fourth Duke of Ancaster. He died in 1865, and was succeeded by his only son, the Hon. Alberic Drummond Willoughby, who died, without issue, in 1870, and was succeeded by his elder sister, the Dowager Lady Aveland, relict of the first Lord Aveland, and she is now the Baroness Willoughby de Eresby.

Having inspected the grand old Castle, the modern part of which is used as the autumnal residence of Lady Willoughby, our way lies directly through the far-famed flower garden. The garden is situated about 30 or 40 feet below the southern part of the Castle rock, and is reached from the Castle esplanade by a splendid flight of steps, communicating with three architectural terraces, artistically formed out of the steep bank, and tastefully adorned with shrubbery. The garden was originally laid out by John, the second Earl of Perth, whose death took place in 1662. The records of the time speak of him as a man of great integrity, a distinguished patron of literature, and himself learned in all the branches of knowledge becoming a nobleman. His younger days were spent in France, but on succeeding to the property he returned to Scotland, where he continued his pursuit of literature, and in the delightful retreat of Drummond Castle, he founded a thoroughly useful library of the most eminent ancient and modern authors. He paid great attention to the management of his estate, and conducted its affairs with marked prudence and economy. The garden is of a most unique design, combining the best features of the Dutch, Italian, and French styles of landscape decoration, which are made to harmonise, so as to present a magnificent whole. The garden is about 10 acres in extent, and oblong in form. Two broad walks of verdant sward run diagonally across the garden from the north-west to the south-east angle, and from the north-east to the south-west angle, intersecting each other in the centre, and thus throwing the general plan into a St Andrew's Cross. Each of the four sides is encompassed by a gravel walk, and three others run directly across the breadth, one passing through the centre. The whole is then divided into parterres, laid out with exquisite taste with the most choice flowers and shrubs, so arranged as to show the arms of the Drummond family, and to give the place almost as bright an appearance in winter as in summer. Everywhere the garden is studded with antique statues and vases, selected with great taste by the late Lord Willoughby, father of the present Lady Willoughby. The centre of the garden is adorned with a very fine sundial, about 15 feet in height, and containing about fifty faces, indicating the time in every direction. It was erected by the founder of the garden, whose arms, with the arms of his Countess, with a Latin eulogy in honour of his memory, though much defaced, are still traceable on the dial. At the top of the Broad Avenue in the park, but seen from the garden, there is a remarkably good marble statue of Dagon. During the memorable storm of 28th December 1879, this statue was knocked to the ground by a heavy limb which was broken off a huge silver fir in the neighbourhood, but although both statue and pedestal were thrown down, the figure sustained very little damage. Throughout the garden beautifully-trimmed yews, junipers, variegated oaks, and hollies are conspicuous. The walls of the terraces are hung with creepers, the most notable of which is the beautiful *Tropæolum speciosum*, which here flowered in the open air, for the first time in Scotland. On some of the specimens of *Citrus decumana* the fruit grows particularly large. The *Agave Americana*, or great American aloe, may also be included among the more notable plants. In the summer of 1832, one of these plants here reached the height of 23 feet; and in 1851, another reached to the height of nearly 30 feet. The whole of this Eden-like garden,—with its sharp contrasts of dark green foliage and gaily-coloured flowers, its bright marble statues, and its methodical outlines, its clean verdant walks, and its long broad vista, stretching up to the sky-line,—can all be taken in at a single glance, presenting to the beholder one of those glorious sights which are more often the products of fancy than of reality.

Leaving the gardens, our course next lies through Drummond Park, as the extensive policies surrounding the Castle are called. There are 511 acres in the Park, which is literally crowded with huge specimens of all the ordinary varieties of trees, and many that would be really remarkable specimens in other places are here completely dwarfed by their gigantic neighbours. It would be a tedious process to mention all of these in detail, and we shall, therefore, merely content ourselves with noting the more remarkable specimens of the different varieties, and pointing out a few of their peculiarities. Access can be had to the Park from different parts of the gardens, and, taking advantage of the way which is most convenient, we speedily find ourselves in the very heart of the policies, wandering amidst majestic trees that claim attention at almost every step, while new beauties are continually being disclosed—

> "Our journey lies through dell and dingle,
> Where the blythe fawn trips by its timid mother;
> Where the broad oak, with its intercepting boughs,
> Chequers the sunbeam in the greensward alley."

Glancing, first of all, at the "broad oak," whose "intercepting boughs" roused the enthusiasm of the poet of the Ettrick Forest, we find many specimens that might well inspire the worshipper of the muses to sing in their praise. The largest specimen grows by the side of the burn east from the Castle. The girth at 1 foot from the ground is 19 feet 6 inches, and at 5 feet from the ground 14 feet 8 inches, the length of the bole being 12 feet, the height of the tree 70 feet, and the spread of branches 114 feet. It is altogether a very noble tree. Two other large oaks grow close beside it. The girth of the one is 13 feet 4 inches at 1 foot from the ground, 10 feet 1 inch at 5 feet, with a bole of 14 feet, an entire height of 81 feet, and a spread of branches of 77 feet. The other girths 10 feet 10 inches at 1 foot from the ground, swelling out to 15 feet 8 inches at 5 feet, with a bole of 11 feet, and an entire height of 45 feet. Two very large oaks grow at the south corner of the pond, close by each other. The larger of the two girths 18 feet 7 inches at 1 foot, and 11 feet 10 inches at 5 feet, with a bole of 21 feet 6 inches, an entire height of 78 feet 6 inches, and a spread of branches of 100 feet. The girth of its companion at 1 foot is 14 feet 3 inches, and at 5 feet, 11 feet—the length of bole being 17 feet, the height of the tree 64 feet, and the spread of branches 73 feet. Another notable oak grows by the side of the walk circling round the south of the gardens, the girth at 1 foot being 14 feet 4 inches, and at 5 feet, 10 feet 9 inches, with a bole of 26 feet, and an entire height of 96 feet. A peculiar gnarled oak grows by the side of the burn east from the Castle, its peculiar formation giving it a strong resemblance to the Irish yew. The girth at 1 foot is 17 feet 4 inches, at 5 feet the girth is 17 feet 7 inches, the length of bole is 9 feet, and the entire height of the tree 66 feet. An ash, however, reaches a much greater size than any of the oaks, the girth at 1 foot being no less than 22 feet, and at 5 feet the girth is 14 feet 8 inches—the length of bole being 24 feet, the entire height 84 feet, and the spread of branches 75 feet. Large as these trees are, the giants of the Park are the beeches. The monarch of all is a monster on the east side of the Broad Avenue to the south of the garden. The girth at 1 foot is 29 feet, and at 5 feet 15 feet 11 inches—the length of bole being 9 feet, the entire height 71 feet, and the spread of branches 105 feet. We measured nine other beech trees of large dimensions and beautiful formations. The largest of these girths 23 feet 6 inches at 1 foot, and 14 feet 10 inches at 5 feet, with a bole of 12 feet and a height of 88 feet; and the smallest girths 15 feet at 1 foot, and 10 feet 3 inches at 5 feet, with a bole of 28 feet 5 inches, and an entire height of 100 feet. Several of those in this lot girth over 20 feet at 1 foot, and approach 15 feet in girth at 5 feet from the ground. One very beautiful beech, with eight large limbs, grows at the extreme east end of the walk east from the castle, and has a girth of 23 feet 5 inches at 1 foot, and 18 feet 2 inches at 5 feet, with a bole of 8 feet, and an entire height of 91 feet, and a spread of branches of 86 feet. There is a very beautiful purple beech, planted in 1842, when the Queen visited Drummond Castle. It has a girth of 4 feet 10 inches at 1 foot, a height of 56 feet, and a spread of branches of 39 feet. Lime trees are also conspicuous on account of their size and beauty. The largest of these are to be found on the west side of the garden, where there are three magnificent specimens within a very short distance of each other. The largest of these, a really splendid tree, girths 21 feet 7 inches at 1 foot and 14 feet 9 inches at 5 feet—the length of bole being 21 feet, the entire height of the tree 85 feet, and the spread of branches 77 feet. The next largest, also a most majestic tree, girths 19 feet at 1 foot and 15 feet 10 inches at 5 feet—the length of bole being 18 feet, the height of the tree 84 feet, and the spread of branches 83 feet. The smallest of the three girths 17 feet 10 inches at 1 foot and 13 feet at 5 feet—the length of bole being 21 feet 6 inches, and the entire height of the tree 91 feet 6 inches. There is a large collection of sycamores, and a good many of them reach respectable dimensions. The largest specimen grows by the walk along the west side of the garden, near to the lime trees already mentioned. The girth at 1 foot is 16 feet, and at 5 feet the girth is 11 feet 6 inches—the length of bole being 21 feet, and the height of the tree 90 feet. The second largest girths 15 feet at 1 foot and 12 feet at 5 feet—the length of bole being

30 feet 4 inches, and the height 86 feet. We measured several others ranging from a girth of 12 feet 8 inches to 14 feet 9 inches at 1 foot, and 10 feet to 13 feet 9 inches at 5 feet, with an average height of about 80 feet. The park is also adorned with several fine elm trees. The largest specimen girths 16 feet 2 inches at 1 foot, and 13 feet at 5 feet—the length of bole being 17 feet 6 inches, the height of the tree 71 feet 6 inches, and the spread of the branches 69 feet. The second largest girths 14 feet 2 inches at 1 foot, and 13 feet at 5 feet, with a bole of 7 feet 6 inches, a height of 50 feet, and a spread of 85 feet. There are a few Spanish chestnuts of fair proportions. The largest of these, situated a little to the west of the castle, has a girth of 20 feet 2 inches at 1 foot, and 15 feet 9 inches at 5 feet, with a bole of 27 feet 6 inches, a height of 68 feet, and a spread of 75 feet. The next largest, in the same locality, girths 20 feet 4 inches at 1 foot, and 14 feet at 5 feet, with a bole of 18 feet 4 inches, and a height of 64 feet. A horse chestnut on the north side of the alley girths 19 feet 3 inches at 1 foot, and 12 feet 6 inches at 5 feet, with a bole of 10 feet, a height of 65 feet, and a spread of branches of 80 feet. Several remarkable specimens of firs, and the less common varieties of park trees, are also to be met with in different parts of the grounds. Amongst the most prominent of these are silver firs, of which there are several very large specimens. Three enormous trees of this variety grow near each other along the side of the walk from the gardens. The largest of these girths 23 feet 9 inches at 1 foot and 17 feet 9 inches at 5 feet—the length of bole being 28 feet 6 inches, the height of the tree 101 feet 6 inches, and the spread of branches 80 feet. Three very large limbs spring from this tree like elbows. The first of these branches starts from the trunk about 6 feet from the ground, and takes a horizontal direction for about 5 feet, and then takes a sharp upright turn, growing straight almost to the top of the tree. The girth at the bend of this limb is exactly 10 feet 9 inches. The second largest silver fir girths 22 feet at 1 foot, and 16 feet 10 inches at 5 feet, with a bole of 32 feet, a height of 105 feet, and a spread of 80 feet. The other girths 18 feet 7 inches at 1 foot, and 15 feet 6 inches at 5 feet. The latter, a

remarkably fine timber tree, is about 25 feet up to the fork, and has an entire height of about 100 feet. A beautiful araucaria grows beside these silver firs, the girth at 1 foot being 3 feet 10 inches, and the height 34 feet. A *Douglasii* girths 7 feet 7 inches at 1 foot, has a height of 56 feet, and a spread of branches of 43 feet. There are many specimens of *Wellingtonia gigantea* deserving of notice. The girth of the largest at the ground is 12 feet, and at 1 foot 9 feet 10 inches—the height being 36 feet, and the spread 18 feet; a second girths 9 feet at 1 foot, reaches a height of 38 feet, and has a spread of 17 feet 6 inches; and a third has a girth, at 1 foot, of 7 feet, a height of 39 feet, and a spread of 14 feet. A Norway maple on the south side of the alley has a girth of 11 feet 2 inches at 1 foot and 10 feet 1 inch at 5 feet, with a bole of 11 feet, a height of 61 feet, and a spread of 75 feet. The only other tree in Drummond Park to which we shall allude is a very fine weeping birch near the roadside, but hid by a cluster of trees, and which has a girth at 1 foot of 5 feet 3 inches, and at 5 feet of 4 feet 5 inches—the height being 53 feet, and the spread of branches 31 feet. There are several very good yew trees a little to the west of the Castle, one of these girthing 8 feet 4 inches at 1 foot, and 7 feet 6 inches at 5 feet, with a bole of 15 feet.

The remarkable trees upon the property are not confined to Drummond Park. Some very fine specimens are to be found in other portions of the estate. At Pitkellony Park there are several large trees of different varieties. The largest is a Spanish chestnut, having a girth at 1 foot of 19 feet, and at 5 feet, 15 feet 9 inches,—the length of bole being 13 feet, the height of tree 67 feet, and the spread of branches 46 feet. An elm, a larger tree than any of the same variety at Drummond Park, girths 18 feet 5 inches at 1 foot and 13 feet 8 inches at 5 feet, —the length of bole being 19 feet, the height 62 feet, and the spread of branches 90 feet 6 inches. There is also another elm larger than any at Drummond Park, the girth at 1 foot being 17 feet 3 inches, and at 5 feet, 10 feet 4 inches—the bole being 15 feet, the height 71 feet, and the spread of branches 63 feet. A sycamore girths 13 feet at 1 foot and 12 feet 7 inches at 5 feet, with a bole of 15 feet, a height of 71 feet, and a spread of

64 feet. An oak girths 16 feet 2 inches at 1 foot, 9 feet 7 inches at 5 feet, with a bole of 13 feet, a height of 64 feet, and a spread of 77 feet. There is a very fine black Italian poplar, girthing 10 feet 6 inches at 5 feet, and having a height of 82 feet. There is also a clump of box tree in front of Pitkellony House, supposed to be the largest in Scotland, some of the trees being 30 feet high, with a girth of 16 inches. At Balloch, on the edge of the Balloch Road leading to Crieff, there is a grand natural weeping ash, having a girth of 11 feet at 1 foot, and 9 feet 7 inches at 5 feet,—the length of bole being 12 feet, the height 67 feet, and the spread of branches 68 feet. At Ballocharge Farm, on the same road, there are two fine ash trees of considerable size. The larger girths 17 feet at 1 foot, and 12 feet 5 inches at 5 feet, with a bole of 18 feet, and a height of 91 feet; and the other girths 16 feet 10 inches at 1 foot, and 13 feet 2 inches at 5 feet, with a bole of 14 feet, a height of 77 feet, and a spread of 94 feet. At Dargill Farm, on the north side of the alley at the East Lodge, there is a very good Scots fir, the girth at 1 foot being 10 feet 7 inches, and 9 feet 5 inches at 5 feet— the length of bole being 24 feet, and the height of the tree 66 feet. At Muthill there is a Spanish chestnut girthing 16 feet 10 inches at 1 foot, and 13 feet 6 inches at 5 feet, with a bole of 16 feet, and a height of 59 feet. In the garden at Balwharrie, near Muthill, a very curious phenomenon is to be observed, consisting of an oak tree growing out of the stem of a birch, and winding round it in the same way as ivy clings to its support. Both trees are healthy. At their junction, which is just above the ground, the birch is 6 feet 2 inches in girth, and the oak 16 inches.

The plantations at Drummond Castle are almost as full of interest as the policies, the area being extensive and the number of individual trees worthy of notice very large. The total number of acres under wood on this part of Lady Willoughby's property is 3965, including the 511 acres at the park. Templemill Wood is noted for its splendid Scots fir. One of these girths 10 feet 9 inches at 1 foot, and 8 feet 7 inches at 5 feet, with a bole of 28 feet, and a height of 90 feet. We took the measurements of other six trees with an average girth at 1 foot of 9 feet, and over 7 feet at 5 feet from the ground. In Drummond Wood, on the north side of the Castle, there are several very heavy trees of different kinds. A splendid silver fir grows on the edge of the walk leading from the Castle to the pond, the girth at 1 foot being 23 feet 8 inches, and at 5 feet, 14 feet 3 inches—the length of bole being 56 feet, the height of tree 98 feet 6 inches, and the spread of branches 60 feet 6 inches. A horse chestnut beside it girths 13 feet 9 inches at 1 foot, and 9 feet at 5 feet, with a bole of 11 feet, a height of 71 feet, and a spread of 74 feet. A Scots fir reaches close upon the size of the one at Templemill Wood. The largest of the larches girths 13 feet 5 inches at 1 foot, and 8 feet 4 inches at 5 feet, with a bole of 73 feet, and a height of 97 feet. Many large Scots fir and larch trees are also to be found at White Drums, although none of these reach the proportions of those already described. A great deal has been done in the planting of Glenartney Forest in recent years, there being no natural wood there. The principal varieties are spruce and silver fir, with a few Scots firs. The silver firs appear to be thriving best. The most interesting of all the plantations on Drummond estate is the splendid Hill of Turlum, rising to a height of 1260 feet above the level of the sea, and believed to be the finest wooded hill in Scotland. The extent of this plantation is 715 acres, and the wood has everywhere a most luxuriant appearance, and remarkably compact to the summit. The lower part of Turlum consists of dark Scots fir, now rather thin; then spruce; and afterwards larch, growing right up to the top. In some parts of the hill the wood is very dense, and numerous deer of great size find shelter in the thickets. The golden eagle, too, banished from almost every spot, here finds a congenial home. Recently one of the gamekeepers caught a splendid eagle in a trap, and by gentle treatment, and careful attention to its wounded leg, the enormous bird became almost as tame as a parrot, eating from the keeper's hand, and not attempting to molest him in the slightest. The bird measured 7 feet 10 inches from tip to tip of wing, and was in every respect a noble specimen. Many splendid capercailzie also inhabit the wood. A fine road leads from the Castle to the top of the hill, and here one may fully realise the delightful scene of which Nicoll sung—

> "Where forest paths and glades, and thickets green,
> Make up of flowers and leaves, a world serene."

To follow our Perthshire poet still further, the rambler may rest upon almost any spot

> "Upon the forest's velvet grass,
> And watch the fearful deer in distance pass."

The view from the top, we need hardly say, is most extensive, commanding the whole of the surrounding country, as we described it from the tower of Drummond Castle. The heaviest timber upon the hill consists of larch, the largest of this variety which we measured girthing 13 feet 8 inches at 1 foot, and 9 feet 3 inches at 5 feet, with a bole of 73 feet, and a total height of 98 feet. Many of them reach a girth of from 9 feet to 12 feet at 1 foot from the ground, and from 7 feet to 9 feet at 5 feet from the ground, and some reach a height of 100 feet. We noticed two very fine black Italian poplars. One girths 8 feet 2 inches at 1 foot and 6 feet 6 inches at 5 feet, with a bole of 57 feet, and a total height of 97 feet; the other girths 7 feet 5 inches at 1 foot and 6 feet 3 inches at 5 feet, with a bole of 50 feet, and a total height of 87 feet. A weeping birch girths 4 feet 7 inches at 1 foot, and rises to a height of 69 feet. The far-famed Trossachs also form part of the Drummond Castle property, and we do not require to point out how much of their beauty is due to sylvan adornment. In the general view, the Trossachs may be described as a contracted vale, whose sides are soaring eminences, wildly and irregularly feathered all over with hazels, oaks, birches, hawthorns, and mountain ashes, making up what Scott has described as the "scenery of a fairy dream."

The greater part of the planting, and the improving of the estate generally, has been done in comparatively recent years. A start appears to have been first made immediately after the Restoration of Charles II., between the time of the 3rd and 4th Earls of Perth. A good deal of planting was also done between 1785 and 1800, the Lord Perth of that period planting Turlum Hill, making the artificial pond, &c. A vast amount of planting has also been done by the present proprietrix. Lady Willoughby has not only sought to improve her wide possessions by judiciously conserving the woods, and planting where that was expedient, but she has done a great deal of work that is not so apparent to the casual observer. Before one can rightly comprehend the amount of work that has recently been accomplished, he must endeavour to realise the extent of the property which requires to be administered. The Drummond Castle estate, which is the principal Scotch estate belonging to her Ladyship, extends westwards, as the crow flies, from above 4 miles east of Muthill to Callander, a distance of 20 miles; from Callander it extends north-east to St Fillans, a distance of 15 miles; from St Fillans, it goes west to Ardveich, about 5 miles, running along the north shore of Loch Earn; from Ardveich, it goes north for 6 miles, and marches with the property of the Earl of Breadalbane; and from a point a little to the west of the Brig o' Turk the estate runs westwards along the north shores of Lochs Achray and Katrine, a distance of upwards of 10 miles. Within these boundaries are included a large portion of the town of Crieff, and the whole of the villages of Muthill, Callander, and St Fillans. This splendid estate consists of upwards of 10,000 acres of arable land, and 62,000 acres of hill land and plantation. It includes nearly the whole of the parish of Muthill, a large portion of the parishes of Comrie and Callander, and portions of the parishes of Crieff and Monzievaird. Indeed, the Drummond property, including that of Stobhall,—of which we have something to say afterwards,—extends over a large area of the fairest portions of the county of Perth, and includes some of the most magnificent scenery in hill and dale, in loch and river, that Scotland can boast of. Lady Willoughby, under the guidance of her factor, Mr Henry Curr, has within the past few years effected immense improvements of all kinds upon every part of the property. No less than £45,000 has recently been spent upon the erection of new farm buildings, or additions and alterations. Upwards of one hundred and sixty miles of fencing have recently been erected on the Drummond Castle estate, at a cost of over £16,000; and upwards of £8000 has been spent on drainage, while Lord

Aveland, her Ladyship's son, and tenant of the deer forest of Glenartney, recently erected about 16 miles of deer-fencing, involving an outlay of several thousand pounds. All over the property improvements have been recently carried out, and are still in progress, upon a magnificent scale, testifying to a most enlightened and judicious policy on the part of the noble proprietrix.

XXVIII.—STOBHALL.

Antiquity of the Drummonds of Stobhall—Castle of Stobhall—Extent of the Property—Great Height of Trees in Stobhall Wood—Strelitz Wood—Peculiar Origin of the Name—Ancient Religious Houses—The Monks as Foresters—Linn of Campsie—Sir Walter Scott's Allusions to the District—Antiquities.

As was indicated in our account of Drummond Castle, Stobhall is a very ancient seat of the Drummonds. Sir John Drummond, the eleventh chief of the House, first made Stobhall his residence in 1360, and even at that period it is described as "a part of the old inheritance of the family in Perthshire." It is believed, indeed, that Stobhall formed part of the lands granted to Maurice the Hungarian, and first of the Drummonds, by King Malcolm Canmore, as a reward for his distinguished services. Sir John considerably added to his property in this part of the country by his marriage with Mary Montifex, the eldest daughter and co-heiress of Sir William de Montifex, Justiciar of Scotland, and chief of a great and most ancient family, the portion given to Sir John with his wife consisting of "the baronies of Auchterarder, Cargill, and Kincardine." Stobhall continued to be the residence of the family until 1491, when the first Lord Drummond built "the strong castle of Drummond," to which he removed. The Castle of Stobhall is still standing, and, although it has not the same claim to be called a fortified residence as the one in whose favour it was deserted, it must, both on account of its situation and its massive walls, have been, in its day, a place of considerable importance from a military standpoint. It is impossible to tell when Stobhall Castle was originally built. The present building bears the date of 1578, but this is merely the year when the venerable pile underwent some important transformation, as the building is believed to have been erected fully 400 years previous to that date. The Castle occupies a most romantic situation on an elevated tongue of land on the north-east bank of the Tay, and commands a splendid view of the Stormont. The bank of the river is here lined with larch, sycamore, ash, poplar, gean, and walnut trees; and just beneath the ancient pile the trees are a little low, opening up a fine glimpse of the broad and gently-broken surface of the stream, which washes the edge of the property for five miles, the whole of this stretch of water affording the finest sport with the salmon to be had on the Tay. On the opposite bank we look down upon the beautiful and richly-wooded estate of Taymount, and away in the distance the view is only obstructed by the mountain barriers of the north. The Castle itself stands

> "Bosom'd high
> In Nature's sylvan majesty."

All around it there is a wealth of timber, both old and young, which adds immensely to its romantic beauty. Directly beneath its hoary walls there is a magnificent walnut tree of lordly proportions. At 1 foot from the ground the girth is 26 feet, and at 5 feet the girth is 21 feet. The length of the bole is 12 feet, the entire height of the tree 70 feet, and the spread of branches 99 feet. Another very fine walnut close beside it has a girth of 10 feet 8 inches at 1 foot and 9 feet at 5 feet, with a bole of 13 feet, a height of 70 feet, and a spread of branches of 66 feet. The Castle has a background to the south that is more than usually pretty. From a deep dell, with a tiny rivulet running through it, there springs some of the grandest timber in Stobhall Wood; and the walks are so ingeniously laid out that the pedestrian fancies he has walked far from the Castle, while he is never more than 100 yards from it. The walks here are the most beautiful in the country. The Den is well sheltered on all sides; and some of the trees thus protected, and with the water playing freely about the roots, are drawn up to a great height. There are spruce and silver fir trees, which the most careful measurement proved to be 120 feet in height, and one of the larches has a height of 110 feet. Before entering the massive portals of the Castle, we pass, first of all, a beautifully-variegated sycamore, and a nice purple beech; and then a beautiful reproduction of Drummond Gardens, in miniature. Passing through the main entrance, we reach a commodious court, in which are several buildings of more modern construction. The original Castle is still much in the position in which it was left. Within the inner entrance there can still be seen the ponderous bar of the old door, a block of heavy oak 5 inches square, so fitted that it can be rapidly inserted into a socket in the wall. The chapel has a very peculiarly-painted roof, representing the kings of the principal countries in the world. All the monarchs wear their crowns, and are mounted, one dusky potentate being seated on the back of an elephant, with two attendants on foot. The chapel has probably been an old dining-hall, which will account for the elaborate paintings representative of the mundane rulers. The paintings were only discovered about twenty-five years ago. The ceiling was at that time undergoing extensive repairs, and in the process of taking down the plaster it was found that there was another ceiling underneath ; and on traces of painting being found, it was cleaned, retouched, and allowed to remain, as a specimen of early Scottish art. The windows are now filled in with stained glass, bearing the arms of the Drummonds. The apartments in the Castle are generally small, and there are numerous recesses and queer corners which might serve for hiding-places. A press, with an opening leading into the rafters, is pointed out as the spot where one of the family remained in hiding during the troublous times of the Stuarts. There are also two very dismal-looking dungeons still intact—the one being 12 feet by 14 feet, and the other about 8 feet square, the height of the roof in both cases being about 6 feet.

In shape the estate of Stobhall bears a close resemblance to an egg, the Tay forming the north-west boundary, and the road from Perth to Coupar-Angus and the New Mill Burn the south-east boundary. It consists of 6791 acres of arable land, with about 900 acres under wood, and includes within its bounds the villages of Burrelton and Wolfhill. There are no policies, properly so-called, attached to the Castle, and almost all the timber worthy of notice is to be found in the woods. The most important of these is Stobhall Wood, to which we have already alluded. It is of considerable extent, and runs along the side of the Tay, giving to the river a beautiful sylvan aspect. Owing to the favourable situation, all the soft timber trees planted reach an unusual altitude, and are well developed, clean-grown trees. The spruce to which we referred as reaching a height of 120 feet, has a girth of 8 feet 1 inch at 1 foot from the ground and 7 feet 3 inches at 5 feet, with a bole of 80 feet, and a spread of branches of 36 feet. Another spruce has a height of 110 feet, with a girth of 10 feet 5 inches at 1 foot. At 5 feet there is a girth of 8 feet 7 inches—the bole being 80 feet, and the spread of branches 33 feet. One with a height of 105 feet has a larger girth, the circumference at 1 foot being 10 feet 9 inches, and at 5 feet 8 feet 9 inches—the bole being 75 feet, and the spread of branches 35 feet. We saw three splendid trees of the same variety which measured 100 feet in height. The first of these girthed 10 feet 3 inches at 1 foot and 8 feet 4 inches at 5 feet, with a bole of 70 feet, and a spread of branches of 36 feet; the second girthed 10 feet at 1 foot and 8 feet 2 inches at 5 feet, with a bole of 85 feet, and a spread of branches of 24 feet; and the third girthed 9 feet at 1 foot and 7 feet 6 inches at 5 feet, with a bole of 75 feet, and a spread of branches of 48 feet. There are a large number of spruce trees in this wood rising to a height of 90 feet, and much greater in girth than any of the higher ones. The girth of one of these at 1 foot is 11 feet 4 inches, and at 5 feet, 9 feet, with a bole of 70 feet, and a spread of branches of 42 feet. A large number of silver firs also reach to a great height, ranging even up to 120 feet. One which has attained this height has a girth of 10 feet 6 inches at 1 foot and 8 feet 7 inches at 5 feet, with a bole of 85 feet, and a spread of branches of 42 feet. Three of those measured have a height of 100 feet, all well-proportioned trees. The first has a girth of 13 feet 9 inches at 1 foot, and 10 feet 6 inches at 5 feet, with a bole of 60 feet, and a spread of branches of 60 feet; the second girths 12 feet 9 inches at 1 foot and 10 feet 9 inches at 5 feet, with a bole of 45 feet, and a spread of 51 feet; and the third has a girth of 11 feet 7 inches at 1 foot and 9 feet 2 inches at 5 feet, with a bole of 80 feet, and a spread of 42 feet. The larch has also many noble representatives. The tallest of this variety has a height of 110 feet, with a girth at 1 foot of 10 feet 2 inches, and at 5 feet of 7 feet 6 inches—the length of bole being 90 feet, and the spread of branches 30 feet. Another approaches to very nearly the same height, the exact altitude being 105 feet. The girth of this tree is 9 feet 7 inches at 1 foot from the ground, and 7 feet 4 inches at 5 feet—the bole being 90 feet, and the spread of branches 27 feet. There are several measuring fully 100 feet in height, and a good many reach to 90 feet. Scots firs are both numerous and of great size. The largest is 55 feet in height, having a girth of 8 feet 2 inches at 1 foot and 7 feet 3 inches at 5 feet, with a bole of 40 feet, and a spread of 30 feet. There is a larger one, however, in the Pleasance Wood, the height being 60 feet, the girth at 1 foot, 8 feet 6 inches, and at 5 feet, 7 feet 9 inches, with a bole of 22 feet, and a spread of 45 feet. In the same wood there is one girthing 10 feet 3 inches at 1 foot and 9 feet at 5 feet, with a bole of 12 feet, a height of 40 feet, and a spread of 60 feet. Amongst the hardwood trees in Stobhall Wood there are several very good specimens. A lime tree has a height of 90 feet, and a spread of branches of 60 feet—the girth at 1 foot being 10 feet 8 inches, and at 5 feet, 9 feet 6 inches, and the length of bole 20 feet. A poplar tree girths 11 feet 4 inches at 1 foot and 8 feet 11 inches at 5 feet, with a bole of 30 feet, a height of 80 feet, and a spread of 75 feet. There are many good beech, oak, and ash trees, but none of them are exceptionally large.

Although the heaviest timber is to be found in Stobhall Wood, it is not the most extensive of the plantations, the largest being Strelitz Wood, which consists of 274 acres of splendid larch, fir, and spruce, from 80 to 100 years old. The name of this wood has been derived from a singular circumstance. In 1763 the Commissioners of the Forfeited Estates, of which Stobhall was one, located at this place a colony of pensioners who had served in the German wars. They designated the township by the name of Strelitz, in honour of the Princess Charlotte, Queen of George III., who came from that place. There were altogether eighty of these old warriors provided for here, each having a neat dwelling-house, garden, and three acres of ground, fenced in with hedge and ditch, and sheltered by strips of trees. The holdings were erected in two rows, with a street or open space of 90 feet between them, and a life lease of the holdings was granted to the men at a nominal rent, as a reward for their martial services. They were notorious smugglers, and their carousals were frequently deep and protracted. To keep green the memory of their old campaigning days, they instituted an annual march through the parish to the strains of martial music, the demonstration generally ending in not a few bloody Fontenoys on a small scale. As the veterans bade farewell to all scenes of earthly strife, their possessions fell into ruins—no other cottars caring to live amongst them— and they gradually became extinct, leaving few traces behind them. The annual promenade was maintained for some years as a ploughman's festival, which ultimately merged into the Burrelton Market, which is also now almost extinct, being swallowed up by the Burrelton Games. The site of the old village of Strelitz is now wholly covered by the plantation of which we have been speaking, and the roe-deer, and other wild inhabitants of the woods, now take up their abode where formerly lived the heroes of a hundred fights.

Stobhall and district appears to have been well wooded from a very early period, as mention is more than once made of its woods in history. To the north of Stobhall, along the river side, is the historical Shortwood Shaw, where Wallace hid his booty after the successful raid upon Kinclaven Castle. In the middle of the 14th century the monks of Coupar-Angus had a religious house and chapel at the Linn of Campsie, near Stobhall Castle, dedicated to St Hannand. The monks belonging to the establishment acted as foresters to the abbey at Coupar-Angus, looked after the woods at Campsie, and sent supplies of timber for fuel and other purposes to Coupar. The road communicating between the two places is still called the Abbey Road. The monks were also proprietors of the fishings at the Linn. In those days the Linn deserved the name of a waterfall, as the rock extended across the whole breadth of the river, with a height of from 20 to 30 feet. The rock, on the Stobhall side, however, has since been blown with gunpowder to permit the fish the better to ascend to the higher reaches of the river, and also to allow of the floating down of the wood from the Highland forests. According to Sir Walter Scott, the "Fair Maid of Perth" and the Duchess of Rothesay made this religious house their retreat after the death of the Duke of Rothesay at Falkland Palace. The "Wizard of the North" still further throws a halo round those ancient ruins by making them the death-scene of Conacher or Eachin MacIan, the chief of the Clan Kay, after the sanguinary combat on the North Inch, of which he was the sole survivor. The chief of the Kays is described as having fled hither after the fight, and had a conference with Catherine Glover in the garden of the monastery immediately above the falls. Fearing capture by his enemies, he threw himself, in a fit of frenzy, headlong over the rock into the roaring waters, and was seen to rise no more.

There are many other matters of interest that might be mentioned in connection with Stobhall which are not properly within the scope of our work. Its relics of antiquity are specially valuable. Indications have been found of a religion reckoned to be older than the Druid era, and some very important discoveries of sculptured stones have been made by the present schoolmaster of Cargill, Mr Ferguson. Several traces of the Roman occupation can be pointed to, while its traditionary history is rich in stories of witchcraft.

A considerable number of new farm-houses and new steadings have been erected by the present proprietrix, the Baroness Willoughby de Eresby, at a cost of over £14,000, and upwards of £9000 have, we understand, been spent by the Baroness in draining the land.

XXIX.—THE CAIRNIES AND GLENALMOND.

Early Home of the Newer Coniferous Trees—The History of the Properties—Lord Provost Hay Marshall—The Policies at the Cairnies—The Finest Specimens of *Abies Albertiana* in the Country—How the *Albertiana* got its Name—Curious Place for a Tree—Progress of the Different Varieties of the Newer Conifers—"The Kirk of the Wood"—An Interesting Old Scots Fir—The Plantations.

THE properties of The Cairnies and Glenalmond have long been familiar to arborists as the early home of many of the rarest and most promising varieties of the newer conifers. Originally this part of the country was almost entirely destitute of timber, and it has only been within comparatively recent years that it has assumed that sylvan aspect which excites the admiration of all who pass through the district. Although larch and Scots fir existed to a certain extent previous to the time of the late Lord Justice-Clerk Patton, it is to him that we are entirely indebted for the magnificent collections of the newer conifers scattered over both properties for experimental purposes, and planted out in the woods as timber trees. It is impossible to estimate the value of the work which has thus been accomplished by the late laird, as his experiments being upon a most extensive scale, the country has the very best opportunity of judging of the value of the different varieties which have been introduced within the lifetime of many still living.

Looking, first of all, at the property of Glenalmond, we find that previous to the year 1805, it was in the possession of the Dukes of Athole. One of their titles is derived from this property, the third Earl of Athole having been created Viscount of Glenalmond in June 1703. In November 1805, however, John, the fourth Duke of Athole, sold a portion of Glenalmond to Thomas Hay Marshall, then Lord Provost of Perth, for the sum of £10,000 sterling. Lord Provost Hay Marshall is noted as the most enterprising citizen of Perth in his day, and as one of the most far-seeing and enlightened men who ever occupied the chief civic chair of the "Fair City." It was under his guidance that the excambion with the Earl of Kinnoull was carried out, by which the size of the North Inch was nearly doubled. By another excambion he was enabled to carry out the feuing of that fine street on the north side of the South Inch which worthily bears his name. He was the proprietor of the lands of Blackfriars, and he it was who laid out these grounds in their present form, the site for the Academy at Rose Terrace being the gift of the Provost. His memory has been honoured by the citizens by the erection of the handsome monument in George Street, in which are the Perth Library, and the Museum of the Literary and Antiquarian Society, and which bears this brief but expressive inscription :—" T. Hay Marshall, Cives Grati." In June 1808, Provost Marshall purchased a second portion of the Glenalmond estate from the Duke of Athole, at the price of £12,000. He was not destined to enjoy his acquisition for any length of time, as he died in the course of the same year in which he made the latter purchase. He left a deed of settlement in favour of Mr James Patton of The Cairnies, and Sheriff-Clerk of Perthshire, by which he became proprietor of the portion first purchased— Mr Patton also succeeding to the second portion as Provost Marshall's heir-at-law, the deed of settlement having been executed previous to the purchase of the second portion. In December 1808, Mr James Patton, the new laird, purchased the third and only remaining portion of Glenalmond from the Duke of Athole for the sum of £10,750. Mr Patton died in 1831, leaving one portion of Glenalmond to his eldest son, James Murray Patton, afterwards Sheriff-Clerk of Perthshire, and the other to his second son, Thomas Patton—his third son, Mr George Patton, getting The Cairnies. On the death of Mr James Murray Patton, in December 1853, Thomas, his immediate younger brother, succeeded to the whole of the Glenalmond property. Mr Thomas Patton died in 1869, leaving a deed of settlement conveying the property of Glenalmond to his brother, the Right Hon. George Patton, Lord Justice-Clerk of

Scotland. Coming to the Cairnies, we find that prior to the year 1691 the property belonged to the Arthurstone family, but in that year Mr John Murray of Arthurstone sold it to the Grahams of Balgowan for 9800 merks Scots. The estate was entailed by the Grahams in 1726, and in 1803 Colonel Thomas Graham, afterwards Lord Lynedoch, succeeded as heir-of-entail. In 1805 Lord Lynedoch obtained an Act of Parliament disentailing The Cairnies, and in the following year he sold the property to Mr James Patton, already referred to in connection with the property of Glenalmond. On his death in 1831, as already mentioned, he left a deed of settlement conveying The Cairnies to his youngest son, George, afterwards Lord Justice-Clerk. The Lord Justice-Clerk only held the united properties of The Cairnies and Glenalmond for a very few weeks before he met with his melancholy death, when the properties were again destined to be disunited. His widow, Mrs Malcolm Patton succeeded to Glenalmond in liferent, on the lapse of which the fee of the property was appointed to go to his sister, Miss Margaret Ann Patton. His sister also succeeded to the property of The Cairnies as her brother's heir-at-law, the house and policies being life-rented by Mrs Malcolm Patton, along with the estate of Glenalmond. Upon the death of Miss Patton in 1878, she left a deed of settlement conveying the fee of the property of Glenalmond to her nephew, Colonel Thomas Marshall Harris, until lately commander of the Royal Artillery at Gibraltar, the fee of The Cairnies being bequeathed to Colonel Harris' cousin, Lieutenant-Colonel Henry William Harris, formerly of the Bombay Staff Corps, but now, like his cousin, on the retired list.

While, for our present purpose, the two properties are practically treated as one, and spoken of almost indiscriminately, it will be understood that it is at The Cairnies where the grandest of those newer coniferous trees to which we alluded at the outset are situated. It was only for a very short time that the late Lord Justice-Clerk had an active interest in the Glenalmond property, although his influence was undoubtedly exercised for many years ; but from the time he succeeded to the Cairnies, in 1831, he displayed the spirit of an enthusiastic arborist, and spared neither trouble nor expense in securing specimens of the rarer varieties with the view of testing their adaptability to the climate of this country, either as ornamental or forest trees. Some of the finest of the trees we have to describe are, consequently, situated in the policy grounds surrounding the mansion-house at The Cairnies. Conspicuous amongst them all are the two magnificent specimens of the tree which bears the name of the late Prince Consort,—*Abies Albertiana*. The seeds of this tree were first sent home to this country by Jeffrey in his first Oregon expedition of 1850, and some of the earliest cones were sown by Mr Patton in the following year. The tree was at first known under a variety of names, but on a photograph of the two best ones at The Cairnies being shown to Her Majesty, after they were a few years old, she expressed a desire that the tree might be called the *Abies Albertiana*, in memory of the late Prince Albert,—a name by which the tree is now universally known. The two at The Cairnies are not only magnificent specimens, but they are acknowledged to be the grandest trees of the kind in the country. Growing at an altitude of 660 feet, on a poor thin moorish soil, with a subsoil of hard retentive clay, and with a southern exposure, they have thriven amazingly, withstanding all the ravages of the late very severe winters. The tallest one has reached a height of 54 feet, with a girth of 5 feet 8 inches at 1 foot, and 5 feet at 5 feet from the ground. The other has a height of 50 feet 6 inches, with a girth at 1 foot of 6 feet, and of 5 feet at 5 feet from the ground, one of the trees having a spread of branches measuring 42 feet in diameter. Amongst the other noteworthy trees in the policies at The Cairnies are the following:—*Picea Lowii*, 37 feet 6 inches high, with a girth of 5 feet 1 inch at 1 foot and 4 feet 1 inch at 5 feet; *Picea grandis*, 43 feet 6 inches high, with a girth of 4 feet 7 inches at 1 foot and 3 feet 6 inches at 5 feet; *Wellingtonia gigantea*, 38 feet high, with a girth of 8 feet 1 inch at 1 foot and 5 feet 3 inches at 5 feet; *Picea nobilis*, on the lawn in front of the mansion, 45 feet high, with a girth of 5 feet 10 inches at 1 foot and 4 feet 7 inches at 5 feet.

Entering Glenalmond on the east side of Buchanty

Bridge, a very few yards takes us into the midst of the woods. The first object to attract attention is an interesting larch growing on a rock projecting into the river, and from which the best view of the Buchanty Spout, a very picturesque waterfall, is obtained. It is somewhat surprising to find a large tree of any kind growing upon such a spot, as there are only a few inches of soil, but the story connected with the planting of the tree fully accounts for its presence upon such a peculiar site. The workmen engaged in planting this spot were one day enjoying a smoke after dinner, when one of them, who did not indulge in the weed, was seeking for recreation in some other form. Taking a spade in his hand, he cut out a piece of turf, and, laying it upon the rock, it occurred to him that he might stick in a plant, just for the fun of the thing, and it has been allowed to remain until it has become a tree containing about 70 or 80 cubic feet of timber, girthing 9 feet 2 inches at 1 foot from the ground and 7 feet at 5 feet up. It is marvellous to see the exertions the tree has put forth to draw sustenance for itself, the roots stretching over the rock in a most wonderful manner in order to fasten its radicles in the rich soil on the bank of the Almond. Near this place, a fine variegated spruce, a rare tree, was blown down by the Tay Bridge gale, very much to the regret of all who had seen it, the tree being well known throughout the country. It had attained a height of about 50 or 60 feet, and was a great ornament, especially at the fall of the year, when its magnificent foliage was seen to the best advantage. Mr M'Lagan, the forester, planted some seeds taken from cones grown on this tree, but as yet the young plants show no indications of that bright variegation of green and yellow which made the parent tree so beautiful. The tree was discovered about 1860, and the forester at that time, Mr A. Robson, grafted a few plants from it, but none of them retained the golden colour of the parent. As we proceed along the haugh below Buchanty, we begin to realise the great attraction which the newer coniferous trees must have had for the late Lord Justice-Clerk. Almost every tree belongs to one or other of the finer varieties of the fir tribe, and many of them being planted in the woods shortly after their introduction into this country, their value as timber trees is well illustrated. Amongst the best of these in the haugh is a magnificent *Pinus monticola*, which rises to a height of 69 feet, and has a girth of 4 feet 10 inches at 1 foot and 4 feet 4 inches at 5 feet. Surrounding this tree are some grand specimens of *Picea grandis, Abies Menziesii, Abies Douglasii, Picea Fraserii, Abies Albertiana, Pinus Cembra, Picea Cephalonica, Cedrus Deodara, Taxodium sempervirens, Picea Pinsapo, Picea nobilis*, &c. There are several specimens of a weeping variety of the silver fir (*Pectinata pendula*), the branches of which cling close to the stem, and have the pendulous peculiarity very distinct. The largest *Abies Douglasii* at the haugh girths 6 feet 10 inches at 1 foot from the ground and 5 feet 4 inches at 5 feet—the height being exactly the same as the *P. monticola*, 69 feet. There is a better specimen, one of the original trees, at the eastern entrance to Glenalmond House the girth being 7 feet 4 inches at 1 foot, and 5 feet 9 inches at 5 feet. It is scarcely so tall as the other, the height being 64 feet 6 inches. The neighbourhood of Glenalmond House is enriched by many fine specimens of the more valuable of the newer conifers. The first six pines were planted on the lawn in 1849, the plants being presented to Mr James Patton by the late Mr Turnbull of Bellwood. Directly in front of the house, which is an elegant and substantial mansion, there is a magnificent *Wellingtonia gigantea*, girthing 9 feet 9 inches at the ground and 5 feet 7 inches at 5 feet, with a height of 37 feet 4 inches. Another *Wellingtonia* has a height of 38 feet 6 inches, but the girth is not so good. A beautiful *Cedrus Deodara* girths 5 feet 1 inch at 1 foot and 4 feet 4 inches at 5 feet. The soil here is rich, inclined to sandy, and seems highly favourable for the finer varieties. A *Picea nobilis* at the garden has a height of 59 feet 6 inches, and girths 6 feet 4 inches at 1 foot and 5 feet at 5 feet from the ground. There are also a number of very fine araucarias. One of the original introductions of *Abies Menziesii* is growing at the east side of the garden, and is in every respect a grand specimen, although its progress is hampered by being too much confined. The girth at 1 foot from the ground is 11 feet 2 inches, and at 5 feet the girth is 9 feet. Its height is 56 feet. A beautiful specimen of *Picea grandis* has a height of 46 feet, and girths 5 feet 7 inches at 1 foot and 5 feet at 5 feet from the ground. The largest larch upon the property grows at the edge of the garden, and, along with a few others, is a relic of the original stock, the old stock having been cleared out to make way for oak at a time when oak bark was of considerably more value than it is at present. This relic of the Glenalmond larches has a girth of 10 feet at 1 foot from the ground, and 7 feet 6 inches at 5 feet.

One of the most interesting spots on the property, from an arborist's point of view, is the Pine Haugh at the Tulchan March of The Cairnies, on the south bank of the Almond. Here the late Lord Justice-Clerk laid out between 2 and 3 acres for the purpose of putting the different varieties of the newer conifers to a competitive trial. The pinetum is laid out in groups, thus affording a better test than single trees, and the present condition of the several varieties is of considerable value, as showing the extent to which they may be expected to become useful timber trees. *Abies Menziesii* has produced the most timber of all the varieties; the common spruce is doing very well, but is not equal to the *Menziesii* as a timber tree; *Picea nobilis* is doing very well; *Picea grandis* promises to be a grand tree, with magnificent foliage; *Abies Douglasii* is doing very well, but the gravelly subsoil is evidently too dry; the Weymouth pine is not doing very well, being affected by the frost; the same remark applies to *Pinus Laricio* and *Pinus Austriaca*, there being a good many deaths every year; *Pinus Cembra* is growing well, there being several very fine specimens in the group; *Pinus montana* is a first-class crop, but is being cut down to make way for trees that are more valuable for timber; *Abies Albertiana* grows splendidly, and seems quite at home; *Pinus monticola* has done very well up to a certain stage, when it seems to have received a check; *Picea Cephalonica*, *Picea Pichta*, and the araucarias are too much affected by the early frosts to be favourably reported on. There is also here a very fine *Pinus monticola*, 57 feet high, with a girth of 5 feet at 1 foot and 4 feet 4 inches at 5 feet. The haugh is about 10 feet above the level of the Almond, and is pretty well

exposed to the north and west. The ground rises abruptly to
the south, and, being heavily wooded, the sun is largely
excluded, although the pinetum has full advantage of the sun
in the afternoon. The soil is sandy, with a gravelly subsoil.
Several fine specimens are also to be found at the pinetum at
the quarry, including a *Picea nobilis*, 46 feet 3 inches in height,
with a girth of 5 feet 1 inch at 1 foot and 4 feet 4 inches at 5
feet; and a *Picea Nordmanniana* (a perfect model in size, form,
and foliage), 42 feet 3 inches high, with a girth of 4 feet 3
inches at 1 foot and 3 feet 4 inches at 5 feet. There is also a
very fine variegated *Abies Douglasii.*

In the Sma' Glen, at Conyachan, three or four miles west
from Newton Bridge, there is a solitary old Scots fir with an
interesting history, from which it may be inferred that it is the
sole survivor of a great wood. It stands near a spot known as
"The Kirk of the Wood," or "The Kirk of the Grove," and
close to the cave of the notorious Glenalmond robber, Alister
Bain, who paid the last penalty of the law at Perth. It is said
that at some remote period a rude chapel stood here, the remains
of which may still be traced in six huge boulders piled on each
other in a singular manner. There is also some evidence of a
rough foundation of stones enclosing a small bit of ground
immediately in front of the larger stones, but otherwise there is
no appearance of a burying-ground, or anything to lead one to
suppose that wood had once surrounded the place. It is
believed that the last solemnity celebrated in the chapel was the
taking of the Communion by a hundred and twenty young men
on their way to Culloden to fight for Prince Charlie. The fir
stands near the ruins of this old chapel, and in a superstitious
age, it was credited with possessing very extraordinary virtues,
it being said that if any person cut a branch from this tree he
died, while the branch lived. There is no record to show
whether it is entitled to this gruesome reputation, but as the
tree still stands the curious can test it for themselves. For our
own part, we preferred to permit the branches to fulfil their
natural functions. The height of the tree is about 50 feet, and
the girth is 9 feet 3 inches at 1 foot and 7 feet 2 inches at 5
feet. The tree is beginning to get very thin in the foliage, and

has a blasted appearance, showing unmistakably that age is telling on the solitary sentinel. Although it is not improbable that the tree may be the last of an ancient forest, it may be remarked that the Caledonians are said to have frequently planted a fir tree at the tomb of a warrior, so that the tree may in reality mark the last resting-place of an ancient hero :—" A tree stands alone on the hill, and marks the slumbering Connal."

The plantations extend, in all, to about 500 acres at Glenalmond and 300 acres at The Cairnies. The *Abies Douglasii* is very largely planted, but a good many of them have been blown down or thrown off their feet. They were from 3 to 4 feet high when planted, and the opinion of Mr M'Lagan regarding the difficulty of getting the trees to stand upon their own legs, as it were, is that they were too large when planted out. His experience is that small plants root better, and become more able to withstand the wind. However, large numbers of this variety are in a very healthy state, and growing vigorously, their heights ranging from 10 to 40 feet. The severe winter of 1880–81 injured a great many *Abies Albertiana*, and entirely killed a few. At all events, large numbers were observed in the following summer with their leaders dead, evidently from the effects of the severe frosts, indicating that this beautiful tree is not one of the hardiest additions to our coniferous sylva. Nurseries have been formed both at Glenalmond and at The Cairnies. The latter is the more important, and was formed two years ago by Mr M'Lagan near his own house. It is about one acre in extent, the exposure is inclined to the north-west, and the soil is loamy clay, with occasional bits of black soil. The nursery at Glenalmond was formed five years previously, and the stock of newer conifers at The Cairnies was transferred from there. While the larger portion of the nursery is occupied by larch, no inconsiderable space is devoted to the less common varieties, almost all of the more promising of the newer coniferous plants being largely represented. It may be said, indeed, that there is every prospect of the reputation which this part of the country has secured in the past as the early home of the most recent additions to our forest trees being fully maintained in the future.

XXX.—GORTHY.

State of the Property Sixty Years Ago—Antiquity of the Family of Mercer—Ancient Connection with Perth—"Merchant Princes"—A Famous Provost—Purchase of Gorthy—Remains of the Old Caledonian Forest—A Gigantic Sycamore—Other Notable Trees—The Plantations—Struggles with the Beetle.

THERE are very few properties in Perthshire which present a better example of what has been done in recent years to give to the land that smiling beauty which has such a charm for lovers of Nature than the estate of Gorthy, the property of Mr Græme Reid Mercer. Situated chiefly upon the high lands which overlook the romantic wilds of Glenalmond, the property, until half a century ago, almost entirely consisted of morass and bog, and had everywhere a cold, forbidding aspect—

> "Far as the eye could reach, no tree was seen,
> Earth, clad in russet, scorned the lively green."

A plan of the property, dated 1752, shows no wood whatever, with the exception of a few solitary trees on the margin of the Almond, where Tulchan House, the residence of the proprietor, still stands, but only a very few relics of those trees survive the blasts which periodically sweep down the glen with cyclonic force. At this date, almost the entire property was indicated as profitless moorland, although in early times it formed a part of the great Caledonian Forest, gigantic remains of which are frequently met with buried far down in the moss. The barren aspect of the country shown by this plan continued until the family now in possession acquired the property, in the year 1819, after which it underwent a complete transformation, and at present it is literally loaded with timber wherever the land is not sufficiently good for agricultural purposes.

Although the connection of the Mercers with this property is of very recent date, the family is of great antiquity, and has for

generations been closely associated with the county and city of
Perth. Of the Mercers of Perthshire we possess authentic
records from about the year 1300, but the surname was a
common one in England from a more remote period, and is
associated with several memorable events. The first ascertained
ancestor of the family was Aleumus de Mercer of Tillycoutry,—
born probably before 1200,—who in 1244 was one of the
Scottish magnates who signed the bond by Alexander II. of
Scotland to Henry III. of England, by which the Scots bound
themselves not to make war on, or annoy the English. His son,
Aleumus de Mercer of Tillycoutry, resigned his lands on the
19th of June 1261, and embarked in Continental mercantile
speculations. He is believed to have been the person whom family
tradition asserts to have been Treasurer to the Scottish host
which went under the Earls of Carrick and Athole to the Holy
Land in 1271. He appears to have had a son Thomas, who by
Edward III. is styled "Bonus Homo de Mercer," the "Bonus
Homo" being a man who held lands not *in capite* but of a
vassal. He having held lands in Perth under the Abbot of
Scone, was territorially designated "Bonus Homo" (or Good-
man). This burgess of the "Fair City" was the common progenitor of the family whose branches are represented at the
present day by the Marchioness of Lansdowne (in the female
line), the Mercers of Huntingtower, and the Mercers of Gorthy.
Contemporary with Thomas were Bernard le Mercier, whose
name appears among the burgesses of Perth who swore fealty to
Edward I. of England at Berwick, on 28th August 1296; Austin
Mercer, burgess of Roxburgh, and Walter Mercer, burgess of
Montrose, who also swore fealty in 1296; Stephen Mercer,
burgess of Berwick, who swore fealty in 1291; and Duncan
Mercer, who founded an altarage in the Church of St Nicholas,
Aberdeen, in 1292. It is supposed that ties of relationship
existed between all these Mercers. The early Mercers who were
burgesses of Perth were extensively engaged in mercantile
pursuits; and as Perth was at that time, being the capital of
Scotland, in the enjoyment of a large share of the national trade
and commerce, they came to be recognised as "merchant princes."
Thomas was succeeded by his son John, who, as a merchant of

Perth, was already on the road to eminence. He married the
Lady Ada Murray, daughter of Sir William Murray of Tully-
bardine, and granddaughter of Malise, Seneschal of Strathearn.
He was repeatedly Provost of Perth. He filled that office in
1356, when he was one of the deputies chosen from the four
towns of Edinburgh, Perth, Aberdeen, and Dundee, to treat with
the English Council at Berwick respecting the liberation, by
ransom, of David II. of Scotland, who had been a prisoner-of-
war in England since the Battle of Durham on 17th of October
1346. He also appears to have been very wealthy, as he advanced
loans of money for different royal purposes. After the death
of his father, he began to acquire various heritable possessions.
The lands of Kyncarrouchy passed into his possession by virtue
of an indenture with the Abbot of Scone, dated 22nd April
1358. In 1362 a charter of confirmation was granted at Perth
by Maurice de Drummond to John Mercer, burgess of Perth, of
the barony of Meikleour, with its appurtenances, in the Stormont,
which belonged to Allan of Kimbucke. In 1374, while John
Mercer occupied the position of Provost, his son, Andrew, was
First Bailie. From the important position which Perth occupied
in those days, John Mercer, as Provost, was called upon [to
discharge manifold public duties. He represented the city as
Commissioner in Parliament; and in 1366-7 was appointed an
Ambassador from Scotland to Flanders. At the Court of France
he was well known and highly esteemed, and his opinion on
commercial and political affairs was greatly valued. In 1376
John found occasion to visit Flanders on his own private affairs,
remaining in the Low Countries for some time. As the winter
drew on, he embarked with his merchandise in a homeward-
bound fleet of Scottish merchantmen. The vessels were wrecked
on the Northumbrian coast, and, although England and Scotland
were at truce, John Mercer was seized as a prisoner, and
immured in Scarborough Castle. When this became known in
Scotland, William, the tenth Earl of Douglas—from whom, as
Superior, Provost Mercer held the lands of Petlands of
Strathord—felt himself bound to interfere. His application had
the desired effect, although not without reluctance on the part
of the English, for we find Walsingham saying he was released

"to the great loss of the whole realm and people, for had he been held to ransom in the usual manner of prisoners of war, he would have enriched both the King and the kingdom by his vast wealth." No restitution was made of the goods plundered by the Northumbrian wreckers; but, from his influential position in the State, Mercer was soon enabled to enforce compensation. On his return to Scotland he was by Robert, Earl of Fife and Monteith (second son of Robert II.), appointed acting Chamberlain of Scotland, and in his accounts for that year recoups himself for his losses by shipwreck by deducting £2000 from the £4000 payable for the ransom of David II. The ransom money was carried to Berwick by Andrew, who, anticipating a rejection of his claim for the £2000, went accompanied by a fleet of Scotch, French, and Spanish vessels. Leaving these out of sight in the offing, he went with his own vessel into Berwick. The English Commissioners rejecting the claim, he rejoined his fleet, and swept the seas of the English. At length he appeared before Scarborough, the scene of his father's captivity. Mercer's squadron dashed in among the English vessels, ship after ship being taken to the number of sixteen, which he cut out and towed from their anchorage, in defiance of the onlooking garrison, who durst not descend from their rocky eyrie; so that the warlike Perth Bailie, without landing to assault the town, bore away with his prizes, trailing the English flag in the sea. He again scoured the coast, alarming the ports and capturing every English vessel he encountered. The Perth Bailie, however, did not long enjoy his triumph. While on his way homewards, and hampered by his prizes, he was attacked by a large English fleet sent in pursuit, under Philpot, the London merchant, and was defeated, after a stubborn engagement. Most of the Scottish rovers were taken, including Mercer's own ship, with himself on board. This occurred about 1377. Andrew might have been detained for ransom, but the English Council found it expedient to release him, and grant him a safe-conduct across the Border, which was dated 1st January 1378, and in which he is styled "Esquire of the King of Scotland." An embassy to England is the last trace of John Mercer, the Provost, whose busy and eventful life seems to have closed about the year

1380. The fortunes of Andrew continued to prosper. He inherited all his father's property,—new acquisitions being occasionally added,—and about the year 1384 he received the honour of knighthood. In the award granted by him in the dispute between Robert, Earl of Fife and Monteith and Logie of that Ilk, he styles himself "Dominus de Meiklour." This is admitted to be the oldest document extant in the Scottish vernacular, and is dated 14th of May 1385. It is not impossible that Sir Andrew fell with his liege Lord, the Earl of Douglas, at the Battle of Otterburn, in August 1388, as he appears from the Exchequer Rolls to have died in that year. He was succeeded by his son, Michael Mercer, who married Elizabeth, daughter of Sir Robert Steuart of Durrisdeer, a cousin of the Royal Family of Scotland. It is from these redoubtable merchants and patriots that the Mercers of Gorthy, and the others of the name we have mentioned, have sprung, and although the family pedigree contains many illustrious names, none have occupied a more prominent position in the country than the famous Perth Provost and his First Bailie.

The estate of Gorthy appears to have been held by the heralds of the Earl of Strathearn as their official fee, and passed from them to a junior branch of the Lundys of Lundy, and from them to Sir David Moray, a brother of Sir William Moray of Abercairny, and from him to George Moray, Bishop of Orkney, whose grandson, General Græme, dying in January 1797, the property was purchased by David Stewart Moncreiffe of Moredun, in Midlothian. The name of Moredun was originally Goodtrees or Gutters, but was called Moredun after the Hill of Moncreiffe, on the banks of the Tay, the property of Sir Thomas, his brother. After Sir David's death, his trustees executed an Instrument of Resignation of the lands and barony of Gorthy in favour of Sir Thomas Moncreiffe, a grand-nephew of the deceased. Sir Thomas married Lady Elizabeth Ramsay, daughter of George, ninth Earl of Dalhousie, and her dower was settled on the barony of Gorthy. Sir Thomas died on 26th March 1818, and a dispute arising as to his succession between Sir David, his successor, and George Augustus, second Earl of Bradford, who, on 5th March same year, had wedded Georgina, the only daughter of Sir Thomas, matters could only be adjusted by another sale of Gorthy. A private Act of Parliament was passed in 1819 providing for the sale, and in one of the Schedules appended the extent of the estate is given as 1330 acres Scots, and the rental at £1320. The lands and barony of Gorthy were then purchased by Mr George Mercer, father of the present laird, at the sum of £36,000. Mr George Mercer's descent was from the ancient house of Aldie and Meikleour. He was the eleventh in direct lineal descent from John Mercer, the celebrated Provost of Perth. Mr George Mercer was the youngest son of William Mercer of Pitteuchar and Potterhill, who held the office of Sheriff-Substitute of Perthshire. He had been a merchant in Calcutta, and was one of the original founders of the Colony of Victoria. He married in 1810, Frances Charlotte, daughter of John Reid, Esq., of the Bengal Medical Service. The children of this marriage were six sons and six daughters. In Mr Mercer's time—27th July 1846—an Act of Parliament was obtained repealing that of 9th October 1696, and appointing, as Commissioners for draining the Pow, the following Commissioners, viz. :—Lord Elibank ; Sir William Keith Murray of Ochtertyre ; William Moray, Esq. of Abercairny ; George Mercer, Esq. of Gorthy ; and Alexander Henry, Esq. of Woodend. Mr Mercer died on the 7th December 1853, and was succeeded by the present proprietor, Mr Græme Reid Mercer.

We have already alluded to the fact that the estate of Gorthy, although originally a part of the old Caledonian Forest, had degenerated to a howling wilderness before it came into the hands of the family presently possessing it. Remains of the old forest are frequently met with in the course of estate operations. Trunks of birch and oak trees have been got from time to time, buried six feet under the moss, and bearing unmistakable marks of the axe. The wood of the disembedded oak is remarkably hard and sound, and as black as jet—several articles of furniture which have been made of it having a very beautiful appearance.

Although there are not many trees of considerable size upon the property, there are a few that rank amongst the most remarkable of their kind. The most notable of all is a gigantic sycamore near the site of old Gorthy Castle, a stronghold of which not a vestige now remains. At 1 foot from the ground this tree girths 22 feet 4 inches, at 3 feet it girths 20 feet, and at 5 feet, which is the narrowest part of a bole of 17 feet, it girths 17 feet. The branches are numerous and heavy, giving the tree, which is in perfect health, a magnificent top. The two heaviest branches girth 10 feet 7 inches and 10 feet 3 inches respectively at 3 feet from the bole. The tree is slightly disfigured through two large branches being broken by people swinging upon them. The spread of branches is 70 feet, and the height of the tree about 70 feet. There is another very large sycamore in the cow park at Tulchan, the girth above the swell of the roots, 3 feet from the ground, being 22 feet 3 inches. It breaks into two stems, the one being 10 feet 9 inches in girth, and the other 10 feet—the boles being each about 20 feet. An oak close beside the first of these sycamores girths 14 feet 9 inches at 1 foot from the ground and 11 feet 4 inches at 5 feet, with a bole of 15 feet, and a height of about 70 feet. There is also, at old Gorthy Castle, a very fine tulip tree, girthing 9 feet 7 inches at 3 feet from the ground, being the narrowest part of a bole of 6 feet. It has been a good deal destroyed by having the bark peeled off by animals grazing in the park, but it has now been fenced in to prevent further destruction. There are some very beautiful trees in Gorthy Den, in the same neighbourhood. Amongst these, we measured the following:—beech, 12 feet 5 inches at 1 foot from the ground and 10 feet at 5 feet, with a bole of 20 feet; Scots fir, 8 feet 10 inches at 1 foot and 7 feet 4 inches at 5 feet; several oaks about 10 feet in girth, and lime trees about 8 feet in girth. There is a very nice oak tree in front of Tulchan House, having a girth of 10 feet 7 inches at 1 foot from the ground and 9 feet 7 inches at 5 feet, with a fine bole of about 30 feet. There is also here a beautiful wych elm, girthing 7 feet 11 inches at 1 foot from the ground and 6 feet 4 inches at 5 feet. There is a beech girthing 14 feet 4 inches at 1 foot from the ground and 11 feet 6 inches at 5 feet, with a bole of 20 feet. These are pretty large trees, considering that the soil is gravel on the top of red sandstone. One of the most interesting spots at this part of the property is the grass walk leading from the public road to the cemetery, and which was planted in April 1878, with an avenue of *Cupressus Nutkœnsis*. The trees are planted 20 feet apart, and the width of the avenue, from tree to tree, is fully 30 feet. The trees are now from 8 to 10 feet high, are fine healthy plants, and are carefully trained to run into single leaders. The old wood by which they are surrounded is cut down as the young trees grow up, so that in a few years the beauty of this avenue, with its fine beech-hedge background, will be fully disclosed. In planting these trees, the pits were dug 3 feet 6 inches across, and trenched about 3 feet,—a cartload of leaf mould being divided between every three plants. The upper part of the avenue consists of made-up soil, the remaining portion being a light loam on the top of gravel. The plants are found to thrive best where there are about 18 inches of black loam, with an open, gravelly bottom. An avenue to the west of Tulchan House contains some well-grown silver firs about 50 years of age.

In the plantations, we found a good deal that is interesting. These extend to about 550 acres in all, the large wood of Gorthy being about 400 acres and Tulchan Wood 150 acres. The young plants are now being largely supplied from a nursery of an acre and a half in extent, situated at the Home Farm at Glentulchan. It was laid out in 1878, and at present it is stocked with about 100,000 plants, principally Scots fir and spruce, with a few larches. All kinds are thriving very well. Several of the plantations have been terribly infested with *Hylobius Abietis*, and an examination of the damaged parts proved most instructive. The Splitters Wood, Glentulchan, about 60 acres in extent, was especially badly affected. In the summer of 1877, about 40 acres were cut, the refuse branches being burned in the fall of the same year. In the autumn of 1878 and spring of 1879 a start was made with the replanting of this portion with Scots fir, spruce, and larch,—a few of the newer coniferous trees being also introduced. Owing to the ravages of the beetle, considerable difficulty has been experienced in bringing up these plants. The insects, together with the severe weather of 1880–81, killed all the newer conifers, with a few exceptions. In the course of the season immediately preceding that during which the wood was replanted, clumps were planted at intervals to test whether there were beetles in the ground, when it was found that these pests were so abundant that one clump was entirely destroyed, and another clump injured, but not very seriously. It was thought, however, that the damage was not so serious as to prevent replanting, but in the course of the same spring—that of 1880—the beetles again made their appearance where the clumps had been previously destroyed, and have since spread over the greater part of the plantation. It is worthy of note that where the young plants have had the shelter of the old wood, comparatively little damage has been done, but in the open the plants have been almost entirely eaten up. A portion was again replanted in the spring of 1881, and is doing well. We were a little surprised to find upon an examination of the plants destroyed in the spring of 1880, that they bore fresh marks of the beetle, thus showing that these voracious insects eat the dead wood as readily as the living. In another part of this wood a number of *Abies Menziesii* have been planted upon deep moss, and are thriving splendidly. They show a great superiority of growth to the spruce and Scots fir, but they have the advantage of being planted along the top of a ditch, so that they did not require to struggle with the heather, and were, besides, larger trees when planted. But, making full allowance for these advantages, there is a marked superiority of growth.

In Gorthy Wood the beetle has also made considerable havoc. In the year 1873, 45 acres of this wood were cut down, the branches being all burned off the ground in the same year. The land was afterwards well grazed with cattle. In the last week of April 1878 a few clumps were planted with the view of testing the ground as to the presence of the beetle, but these pests, notwithstanding the measures adopted to eradicate them, still existed in such numbers that every plant was totally destroyed after being two weeks in the ground. In the spring of the following year the ground was replanted—the plants again meeting with the same fate, with one or two exceptions.

In the month of May 1880 the whole of the 45 acres were burned, and in the following year experimental clumps were again planted, and it is of importance to know that, as yet, there is no appearance of the return of the beetle. The wood consists of larch, spruce, and Scots fir, nearly the whole of which is ready for the axe. The timber is of fine quality, the trees being nearly all straight-grown and clean. The amenities of the property have been further enhanced by the planting, in the spring of 1879, of Newton Den, with hardwood trees, larch, and a few *Abies Douglasii*, which, in a few years, will give the Den a very ornamental appearance, and afford much-needed shelter.

The whole of the property, as may be judged from the improvements which have been effected during the past half century, is managed with great care. The arable land is highly cultivated, as far as the soil and climate will permit; while the moorland, wherever practicable, is planted with the most suitable softwood trees. Fencing and draining have also been extensively carried out, and farm and other buildings kept in first-rate order. Altogether, as we stated at the outset, there are few places where more really good work has been done in the same period of time, the effect of these improvements being that since the present proprietor came into possession the rental has been trebled.

XXXI.—MEIKLEOUR.

Meikleour House—The Mercers of Meikleour and Aldie—The Policies—
Remarkable Trees—Kinclaven Castle—Village of Meikleour—The
Great Beech Hedge—The Plantations.

MEIKLEOUR HOUSE, the seat of the Dowager-Marchioness of Lansdowne, occupies a delightful situation on the north-east bank of the Tay, about half a mile above the place where the Isla mingles its waters with those of the more rapid and more majestic river. Shrouded by many stately trees, its great proportions and elegant design are not fully revealed until we stand directly in front of it, when the beauty of its architectural style, and its commanding position, are abundantly apparent. The plans were supplied by the late Mr Bryce, Edinburgh, and the architecture is of the mixed French style, the principal features being a magnificent circular staircase, and a commodious and tasteful balcony. To make the surroundings suitable for so noble a mansion, considerable alterations were effected on the policies. The garden was removed from a position close to the house to a spot about 500 yards to the west. Here a very extensive range of vineries, extending to upwards of 300 feet in length, was erected by Mr Scrimgeour, the well known hothouse builder at Methven, the vineries being built on the unique curvilinear style designed by the late Mr William Gorrie, the celebrated landscape gardener, Edinburgh. Several new drives were also formed, the stables were remodelled, and a new dairy and byre erected in a style harmonising with the main edifice. While these improvements were being effected upon the mansion and its surroundings, the other parts of the property were not neglected, as during the past twelve years almost the whole of the steadings have been renewed, and a large extent of fencing erected.

Although the family of Lansdowne derives its name and principal honours from Ireland, it represents, in the female line, one of the oldest of our Perthshire families—the Mercers of Meikleour and Aldie. We have already traced the history of the Mercers in the chapter on Gorthy, and it only remains now to show their connection with the noble family which presently holds the property of Meikleour. In 1362, as we formerly pointed out, a charter of confirmation was granted at Perth by Maurice de Drummond to John Mercer, burgess of Perth, of the barony of Meikleour, with its appurtenances in the Stormont, which belonged to Allan of Kimbuckie. The oldest document extant in the Scottish vernacular, which is dated 14th May 1385, is one in which Andrew Mercer, whom we have described as the "warlike Perth Bailie," styles himself "Dominus de Meikleour." The present proprietress of Meikleour is lineally descended from Sir Laurence Mercer of Melgins, who married Margaret Mercer of Aldie. Marriage also gave the Mercers of Meikleour and Aldie the blood of the two noble houses of Nairne and Elphinstone. Their connection with the famous Jacobite house of Nairne was brought about by the marriage of Jean Mercer, heiress of Aldie, with the Hon. Robert Nairne, second son of William, second Lord Nairne, who had been forfeited for his share in the Rebellion of 1715. Robert Nairne fell at Culloden, and by his death the estate of Aldie was saved from forfeiture. His eldest son, William, who succeeded, entered the army, and rose to the rank of Colonel. He, again, married Margaret Murray, heiress of Pitkeathly, and died in 1790, leaving three daughters, the eldest of whom, Jane, became the wife of the Hon. George Keith Elphinstone, second son of Charles, tenth Lord Elphinstone. This marriage also allied the Mercers to the two noble Jacobite houses of Fleming and Keith-Marischal, the mother of George Keith Elphinstone being Clementina Fleming, only surviving child and heiress of John, sixth Earl of Wigton; while her mother, again, was Lady Mary Keith, eldest daughter of William, ninth Earl of Marischal. George Keith Elphinstone afterwards became the celebrated Admiral Sir George Keith Elphinstone, K.B., Viscount Keith, G.C.B., K.C., F.R.S., &c. He purchased the estate of Tulliallan, where he carried out gigantic improvements, but frequently resided at Meikleour, his only child and daughter, Margaret Mercer Elphinstone, being the heiress of Meikleour and Aldie. In 1817 the heiress of

Meikleour married Count de Flauhault de la Bellarderie, *aide-de-camp* to Napoleon Bonaparte, and latterly French Ambassador at the Court of St James. We have heard a very good story regarding Napoleon's *aide-de-camp*, when he was first introduced at Meikleour. His lady, according to custom, instructed the piper to play round the room after dinner, but the Frenchman found the music of "the great Highland bagpipe" to be so excruciating, that he peremptorily ordered the piper to stop, upon which Dougal retorted, " I suppose she'll have heard too much of that music at Waterloo!" On the death of Admiral Keith in 1823, his daughter succeeded him as Baroness Keith of Stonehaven —Marischal in the Peerage of Ireland—and as Baroness Keith of Banheath in the Peerage of the United Kingdom. On the death of the last Lord Nairne, in 1837, she claimed the title of Baroness Nairne, and her claim was sustained by the House of Lords. The Baroness and Count had five daughters, and in November 1843, their eldest daughter, the Hon. Emily Jane Mercer Elphinstone de Flauhault (Baroness Nairne in her own right) married Henry, fourth Marquis of Lansdowne, and their son is the fifth and present Marquis of Lansdowne, who is well known as having held several important offices of State, and is at present Governor-General of Canada. It may be added that the proprietors of Meikleour are by an old Act of Parliament bound to adopt the name of Mercer.

The sight of the new mansion-house of Meikleour, to which we have already adverted, has been admirably selected so as to take full advantage of one of the chief arboreal features of the policies—a grand sylvan Gothic arch, formed by a double row of sycamore, beech, and lime trees. The centre of the house is directly opposite this unique and beautiful feature, and through the opening the eye is carried along a splendid avenue directly to the top of Dunsinane Hill, which is clearly discernible in the distance. Looking up from the bottom of the avenue, the view is even more pleasing, as the house is viewed at the top of the arch as if it were through a gigantic telescope. The arch reaches a height of from 70 to 80 feet, and is 80 yards in length. Several of the trees, particularly the beeches, which form the avenue terminating in this arch are of gigantic size,—the largest of the beeches girthing 20 feet at 1 foot and 13 feet 9 inches at 5 feet from the ground. It is regularly pruned on the inside so as to maintain the view from the mansion, but, notwithstanding this, the branches of the largest tree have a spread of 86 feet in diameter.

The grounds immediately surrounding the mansion are studded with several trees of more than average proportion. One of the first to attract attention is a superb silver fir fully 90 feet high, and having a diameter of branches of 60 feet, sweeping the ground all round the tree. Beside it is a rare golden American oak, which in spring sparkles like a ball of fire, and imparts a brilliant glow to the whole neighbourhood. It is about 40 feet high, with a fine round bole of 12 feet, and has a girth at 1 foot of 10 feet 3 inches and at 5 feet of 8 feet 6 inches. In the same grove there is also a very fine larch girthing 12 feet at 1 foot from the ground and 10 feet 4 inches at 5 feet. Travelling further afield, we meet with many trees of respectable size in the parks. A very fine ash, for instance, has a girth of 15 feet 5 inches at 1 foot and 13 feet 7 inches at 5 feet; and within a short distance of it there is a vigorous old Spanish chestnut, girthing 14 feet 4 inches at the narrowest part of the bole, and sending forth such tremendous branches that all the lower ones have to be supported by heavy props. A fine lime, a very nice park tree, but with rather a short bole, girths 13 feet 9 inches at the narrowest part, and another, having nearly the same girth, spreads its branches to a diameter of 70 feet. Several very fine trees also grow on the south bank of the river—one of the grandest of which, a silver fir, grows close beside the ferry landing-place. Along with a number of other giant firs, it shoots its top to a height of fully 120 feet, and has a girth of 15 feet at 1 foot from the ground and 14 feet at 5 feet.

On our way to Kinclaven Manse, where there is a really good sycamore, we pass the ancient Castle of Kinclaven, the history of which we have already narrated in describing Methven. It was this castle which Wallace attacked with a handful of men, putting the entire garrison to the sword, including Sir James Butler, the governor. This exploit is commemorated at the castle by an iron plate bearing the following inscription:—" Wallace took this fort in the year 1299. Placed A.D. 1869." In its palmy days the castle must have been a place of much importance, as the ruins show it to have been a place of great size and strength. It was erected by Malcom Canmore (1057–1093), and for centuries it was a favourite residence of the monarchs of Scotland. In its decay the spacious court has been turned into an orchard, and its walls give support to innumerable creepers, which give a touch of the picturesque to the extensive ruins. The sycamore to which we have referred as growing at Kinclaven Manse reaches a height of fully 80 feet, and has a grand bole of 25 feet. The girth above the swell of the roots is 18 feet, and at 5 feet from the ground the girth is 13 feet. Two old lime trees growing at the gate leading to the walk from the ferry to the church are worthy of being noted, the girth of each of them being 13 feet 4 inches at 1 foot and 12 feet 7 inches and 11 feet 7 inches at 5 feet respectively.

Leaving the policies on our way to the famous beech hedge on the Perth and Blairgowrie Road, we pass through the interesting village of Meikleour, nestling amidst the woods a little to the north-east of the mansion. In ancient times the village seems to have been a centre of rural merchandise, as the Tron for weighing wool and other bulky articles still stands in the old market-place, with the jougs attached to which offenders used to be fastened. The ancient cross, bearing the date 1698 (the second last figure, however, now broken off), also stands in the centre of the village square, with a neat iron railing round it. The village itself is one of the most beautiful in Perthshire, and its proximity to the great beech hedge, which may justly be described as one of the arboreal wonders of the world, gives it an air of some little importance. This hedge is so remarkable that parties have been known to travel from the most remote parts of the country for the express purpose of seeing it. The trees composing the hedge, which is neatly trimmed, reach an average height of about 80 feet, and the length of the hedge is 580 yards, or exactly one-third of a mile. Throughout its entire length the hedge is as compact and perfect as it could possibly be, and looking down upon it from the corner of the road leading to the village, it appears as a great arborous wall, shutting out everything behind it, and extending as far as the eye can reach. The hedge is believed to have been planted in the year 1746. An old inhabitant, who died about 1878, remembered of his father telling him about the building of the dyke in front of the hedge, which was impressed upon his memory by the fact of a number of the men engaged in the work having left to take part in the Battle of Culloden, which was fought in 1745. The dyke was completed that year, and, in accordance with the usual custom, the hedge was planted in the following year. The hedge bears evidence of having been annually pruned and trimmed for the first forty or fifty years. For about twenty years afterwards the trees

THE GREAT BEECH HEDGE AT MEIKLEOUR.

appear to have been allowed to grow wild, until their wide-spreading branches became an obstruction on the highway. For the last fifty years or more the hedge has been regularly cut on the side next the road, but the side next the wood has not been trimmed for thirty or forty years, there being no necessity for interfering with that side. It is absolutely necessary, however, that the side next the road be regularly pruned, and the result of this is not only to keep the road clear, but to give to the row a truly hedge-like character. The operation of pruning the hedge is carried out every five or six years, and, from its great height and extent, the work is one of no ordinary labour and difficulty. The operation, however, is now much facilitated by the adoption of an ingenious, but simple, method, which the Dowager Marchioness of Lansdowne observed in use upon the tall lime hedges at Versailles. Her Ladyship was so much struck by the machine in use there, that she procured a plan of it, and had one constructed by Mr Scrimgeour, Methven, under the superintendence of Mr Matheson, the land steward. The machine is simply a double ladder, 30 feet high, supported on four wheels, and by means of which the heavier branches are cut with a single stroke saw. The twigs are then cut by an averuncator to the height of 45 feet. Up to this point the work is comparatively easy, but beyond that height there is not a little difficulty and danger. One of the foresters climbs as near to the top as possible, fastening himself to the strongest branch he can get, by means of a belt round his waist, and holding on to another branch by his left hand. With his right hand he wields a light billhook reaching to a height of 75 feet, which is as high as can be trimmed with safety, and which is quite high enough, as the uppermost branches gradually taper towards the apex of the trees. The hedge is, altogether, the most wonderful thing of the kind in existence. Every tree throughout its entire length is in perfect health, and the height throughout is as uniform as it well could be, the hedge having every appearance of being trimmed at the top as well as on the side.

The plantations extend to 1050 acres, and consist chiefly of oak, beech, Scots fir, larch, and lime. One of the finest of the plantations is that at the Court Hill,—a very fine oak forest of between 300 and 400 acres. Very few of the newer coniferous trees have been planted in the woods, but a number of thriving young specimens are scattered throughout the policies, the most notable of which are *Abies Albertiana*, the best specimen being a promising young plant 20 feet high. The other varieties include *Abies Menziesii*, *Abies Douglasii*, *Picea nobilis*, *Picea Nordmanniana*, and *Wellingtonia gigantea*.

XXXII.—TAYMOUTH.

The Character and Extent of the Property — Poetic Allusions — The Campbells of Breadalbane — The Approaches to Taymouth — The Castle—The Policies — The Early Planters — Remarkable Trees—Trees Planted by Royalty—Drummond Hill—The Great Vine at Kinnell—Plantations in other parts of the Property.

THE Breadalbane country is so extensive and so varied that there is not one of those characteristics for which our Scottish scenery is famous which it does not embrace. In Loch Tay, the salmon loch, *par excellence*, of Scotland,—we have one of the finest sheets of water in the country, while its outlet gives rise to our grandest river. In fertile plain and wooded height its charms are inexhaustible, and these are heightened in no small degree by an abundance of moor and mountain, which, while not shutting out the Lowland beauties, give the district an essentially Highland character. In extent the property stretches from Aberfeldy to the Atlantic Ocean,—a distance of upwards of one hundred miles, making it the longest property, we believe, in Britain. Within its extensive bounds there is, indeed, everything which can gratify the lover of Nature, and fire the imagination of the poet. The extent to which it has touched the poetic chord can never be known, as even the effusions which have seen the light are almost endless. In the presence of the Muse of our National Bard, however, all others sink into insignificance, especially as the impromptu which he left in pencil over the chimney-piece of Kenmore Hotel in 1787, is as expressive as it is beautiful:—

> " Admiring Nature in her wildest grace,
> These northern scenes with weary feet I trace ;
> O'er many a winding dale and painful steep,
> Th' abodes of covey'd grouse and timid sheep,
> My savage journey, curious, I pursue,
> 'Till famed Breadalbane opens to my view—

> The meeting cliffs each deep-sunk glen divides,
> The woods, wide scattered, clothe their ample sides.
> Th' outstretching lake, embosom'd 'mong the hills,
> The eye with wonder and amazement fills;
> The Tay, meand'ring, sweet in infant pride,
> The palace, rising on its verdant side;
> The lawns, wood-fring'd in Nature's native taste;
> The hillocks, dropt in Nature's careless haste;
> The arches, striding o'er the new-born stream;
> The village, glittering in the noon-tide beam—
>
> . . . .
> Poetic ardours in my bosom swell,
> Lone wandering by the hermit's mossy cell;
> The sweeping theatre of hanging woods;
> Th' incessant roar of headlong tumbling floods—
>
> . . . .
> Here Poesy might wake her heaven-taught lyre,
> And look thro' Nature with creative fire;
> Here, to the wrongs of Fate, half-reconcil'd,
> Misfortune's lightened steps might wander wild;
> And, Disappointment, in these lonely bounds,
> Find balm to soothe her bitter-rankling wounds;
> Here heart-struck Grief might heavenward stretch her scan,
> And injured Worth forget and pardon man."

Professor John Stuart Blackie is one of the most recent of those whose depth of feeling upon seeing Taymouth has found vent in verse, and he gives us his impressions as follows:—

> " A piece of England, ramparted around
>   With strength of Highland Ben and heather brae,
>   Art thou, fair Taymouth, with thy long array
> Of stately trees and stretch of grassy ground,
>   Soft with ancestral moss, and thy fair flow
> Of the clear amber stream in winding reaches,
> That laves the roots of smooth wide-branching beeches,
>   With swirl of waters murmuring deep and low.
> Who lauds not these is blind."

Neither prose nor poetry, however, can adequately express the beauties of Taymouth. These are simply ineffable, and must be seen to be realised.

Taymouth Castle is the seat of the Earl of Breadalbane, who belongs to a branch of the Argyle stock, the Campbells of Breadalbane being descended from Duncan, first Lord Campbell of Lochow, and ancestor of the Dukes of Argyle. The family can boast of having royal blood in its veins, Duncan's wife having been Marjory Stewart, daughter of Robert, Duke of Albany, Regent of Scotland, and brother of King Robert II. The chiefs of the house of Breadalbane embrace seven earls, six knights, four baronets, and two marquises. The first knight was Sir Colin, third son of Duncan, Lord of Lochow, who settled Glenorchy upon the new knight. James III. afterwards bestowed the barony of Lawers upon Sir Colin, for his success in bringing the murderers of James I. to justice. The Castle of Kilchurn was built by him in the year 1440. The second knight, Sir Duncan, was made Hereditary Bailie of the King's lands of Discher, Foyer, and Glenlyon, and received charters of the King's lands of port of Loch Tay, of Glenlyon, of Finlarig, and other places in Perthshire. Sir Colin, the third knight, had strong ecclesiastical inclinations. He is noted as a reformer, and sat in the Parliament of 1560, by which Protestantism was established. He was the builder of the Chapel of Finlarig, to be " ane burial for himself and his posteritie." It was him also who, in 1580, built the old Castle of Balloch, now Taymouth, and regarding whom the familiar story is told about " birzing yont." The Castle was built close to the eastern boundary of the property, and on being asked why he erected it at the extremity of his lands, he replied " We'll birze yont," meaning that he intended that the Castle would stand nearer the centre of his territory. The extensions, however, have been in the opposite direction from Aberfeldy. He also built the Castle of Edinample, a seat of the family on the side of Loch Earn. His eldest son, Sir Duncan, was the first baronet of the family, and was a great favourite of James VI. and Charles I. Both he and the second baronet, Sir Colin, did a vast amount of work in improving the property, arboriculture and agriculture owing much to their skill and exertions. The third baronet, Sir Robert, is chiefly remembered for his sufferings and resolute adherence to the Covenanting cause. The family was ennobled in the person of Sir John, the only son of the fourth baronet, born about 1635, who was created Earl of Breadalbane and Holland, Viscount of Tay and Pentland, Lord Glenorchy, Benderaloch, Ormelie, and Weik. The second Earl, John, Lord Glenorchy, was Lord-Lieutenant of Perthshire, and was repeatedly selected as one of the sixteen representatives of the Scottish Peerage. The third Earl distinguished himself at a very early age, having been sent as an Ambassador to the Court of Denmark when only twenty-two years old. He held various other important offices and represented Saltash, in Cornwall, in the Parliaments of 1724 and 1734, and Orford in Oxford, in the Parliament of 1741. John, the eldest son of Colin Campbell of Carwhin, succeeded as fourth Earl of Breadalbane, and sat as a Representative Peer from 1784 to 1806. He was afterwards made a Peer of the United Kingdom by the title of Baron Breadalbane, and was created a Marquis at the Coronation of William IV., in 1831, by the title of Marquis of Breadalbane and Earl of Ormelie. It was he who raised the Breadalbane Fencibles. When a battalion of these was enrolled as the 116th Regiment in the regular service, he was appointed its Colonel, and in 1809 he was raised to the rank of Major-General. He was more distinguished, however, as an improving landlord than as a statesman or soldier, many of the most extensive and tasteful of the improvements upon the policies of Taymouth having been effected by him. His only son became the second Marquis, succeeding his father in the year 1834. He was chosen Lord Rector of the University of Glasgow in 1831; was Member of Parliament for Perthshire from December 1832 to May 1834; and was made a Knight of the Thistle in 1838. When Her Majesty and Prince Albert visited Taymouth in 1842, he gave them a most magnificent reception, Her Majesty stating in her *Journal* that she "never saw anything so fairy-like." When he died in 1862, the titles of Baron Breadalbane and Marquis of Breadalbane and Earl of Ormelie became extinct. He was succeeded, as Earl of Breadalbane, by John Alexander Campbell, Esquire of Glenfalloch, whose son, Gavin Campbell, is the present Earl.

The various approaches to Taymouth Castle are singularly beautiful. The road from Aberfeldy is for several miles almost completely under a thick leafy shade, with openings here and there, revealing the clear, sparkling waters of the Tay, cultivated fields and grazings, the tree-clad heights of Weem and Drummond Hills, and the dark blue mountains lying against the sky in the distance. A fine row of oak trees on the south side of the road was planted by the late Marquis in 1842, on the suggestion of Her Majesty. Many of the trees which skirt the road are of considerable size. At Croftmoraig a very complete Druidical circle is easily seen from the road. The approach from Kenmore is even more attractive. The broad expanse of Loch Tay, with its picturesque isle, on which are some ecclesiastical ruins, is spread out before us, and on crossing the bridge, the pretty village, with its tastefully-kept flower-plots, powerfully strikes the imagination. A splendid view of the Castle is obtained from the road between Aberfeldy and Kenmore, but it

TAYMOUTH CASTLE.

requires a nearer view fully to realise the princely magnificence of the pile. It is built of a fine blue stone found on the estate, and consists of four storeys, with round towers at the angles, an airy central pavilion, and a quadrangular tower rising in the centre to a considerable height above the main building, carrying up a spacious and ornate staircase. The interior—particularly the Queen's Room, the Baron's Hall, the Chinese Room, the drawing-room, and the dining-room—is gorgeously fitted up, and contains paintings by Titian, Rubens, Salvator Rosa,

Carracci, Teniers, Vandyke, and other great masters. The Library is another magnificent apartment, and contains many very rare and valuable works. Picturesquely situated in a beautiful valley, the Castle is surrounded by a landscape of exquisite beauty. One writer, accurately describing it, says :—
" The vale is not spacious enough to admit of that apparently boundless contiguity of park which constitutes such a charm around the baronial residences of England ; but the very hills which confine it may be said to present a still superior charm. These hills are abrupt, luxuriantly wooded, and broken into every sort of picturesque and varied outline; while the level alluvial space below is as green as the gown of Spring herself, and at the same time adorned in the richest manner by fine old trees." The bold rock of Drummond Hill, with its steep and richly-wooded sides, presents a grand background to the Castle, and a glance around reveals the proximity of Schiehallion, Ben Lawers, Ben More, and many other mountain Goliaths. The view from the Castle is greatly enlivened by a glimpse of the blue waters of Loch Tay, and a view of Kenmore, nestling among groves of stately trees.

The policies are remarkable for the size and number of their gigantic old trees. The newer conifers have not yet found much favour, although specimens are to be met with occasionally which promise to become as noteworthy as the trees which now excite our admiration. The first systematic attempt at planting the Breadalbane property, and perhaps the first systematic attempt at planting in Scotland, was made by the first baronet, Sir Duncan Campbell, known as "Black Duncan," who succeeded in 1583 and died in 1631. During this period of forty-eight years he did an immense amount of planting. Drummond Hill was partly planted by him, and it is to his forethought that the policies are indebted for the grandest of their trees. It was him to whom we referred as being held in great esteem by James VI. and Charles I., and amongst the favours he received from the former monarch, was an appointment in 1617 as Heritable Keeper of the Forests of Mamlorn, Bendaskerlie, &c. The third baronet also appears to have taken some interest in the planting of the property, as Dr Walker notes in his essays that the lime tree was planted at Taymouth in 1664. The third Earl planted a large portion of Drummond Hill, extending, as is believed, from the east and west to above Taymouth Gardens. John, the fourth Earl of Breadalbane, who succeeded in 1782, was distinguished as one of the three largest planters of the period, sharing this honour with the Duke of Athole and Sir J. Grant of Strathspey. The work of the fourth Earl attracted so much attention at the time that in 1805 he was awarded the gold medal of the Society of Arts. The policies were at this time greatly improved, and the plantations much extended, including the completion of Drummond Hill to the west end. The north end was planted by the late Marquis.

The largest of the trees upon the property are all within an easy walk of the Castle, and chiefly in the deer park. The largest of all is an enormous beech, which girths 27 feet at 1 foot from the ground before one of its great limbs was broken off about ten years ago. Even at present the girth round the top of the roots is 24 feet 6 inches, and 21 feet at 5 feet from the ground. It has a fine bole of about 15 feet, and an entire height of about 60 feet. There are many fine lime trees. The largest of these girths 19 feet 4 inches at 1 foot and 18 feet 6 inches at 5 feet from the ground, after which it swells out to a still greater girth. It has a bole of 12 feet, and an immense spread of branches. Another lime close beside it, although not so large in girth, has a much greater spread of branches, and presents a grander appearance. The girth at 1 foot is 18 feet 8 inches, and 17 feet at 5 feet from the ground, with a bole also of 12 feet. The tree, which is about 90 feet high, breaks out into a great many huge branches, richly loaded with foliage— the diameter of the tree being about 100 feet, and the circumference 288 feet. Another very fine lime has a girth of 15 feet 6 inches at one foot and 14 feet 8 inches at 5 feet, with a bole of 12 feet. Opposite the Castle there is a grand old ash with a girth of 19 feet 8 inches at 1 foot and 16 feet 10 inches at 5 feet from the ground—the bole being 20 feet, and the entire height about 90 feet. At the Rail Bridge, which spans the Saw-mill Burn, in front of the Castle, there are four famous Spanish chestnuts. They are situated between the wooden and the stone bridges, and, counting from the former, the girths of the four are as follows :—No. 1, 18 feet 10 inches at 1 foot, and the same at 5 feet from the ground ; No. 2, 17 feet 8 inches at 1 foot, and 16 feet 4 inches at 5 feet ; No. 3, 14 feet at 1 foot, and 13 feet 7 inches at 5 feet ; No. 4, 17 feet at 1 foot, and 16 feet 5 inches at 5 feet. The height of these trees is from 70 to 80 feet, and each contains about 460 cubic feet of timber. At Newhall, where a handsome iron bridge has been erected by the present Earl, there is a nice sycamore, girthing 14 feet 7 inches at 1 foot, and 13 feet at 5 feet ; and an old and picturesque elm, close by, girths 17 feet at the narrowest part of the bole. Immediately to the south, and overlooking the Castle, is a fine walk cut out of the face of the hill below the fort, and known as the Surprise Walk. On the north side of this walk is a row of splendid larches, about the oldest in Taymouth park, and containing from 80 to 100 cubic feet of timber. The following are the girths of the three largest, the height in each case being about 100 feet :—No. 1, 17 feet 4 inches at 1 foot, and 11 feet 6 inches at 5 feet from the ground ; No. 2, 15 feet 3 inches at 1 foot, and 13 feet at 5 feet ; and No. 3, 15 feet at 1 foot, and 13 feet at 5 feet. In the Flower Garden, close to the Castle, are four trees of more than ordinary interest, having been planted by royal hands in 1842—

> "When Scotland's loved and loving Queen
> Was gay in Taymouth's gayest hall."

They consist of two oaks and two Scots firs, one of each variety having been planted by the Queen and by Prince Albert. They are all thriving very well. Both of the oaks are about 25 feet high. The Scots fir planted by the Queen is the taller of the two, the height of the one being 40 feet and of the other 30 feet. To the south of these trees are other two, also an oak and a Scots fir, planted by the Empress Eugenie when on a visit to Taymouth about twenty years ago. They are also thriving well, and are each about 15 feet high. Behind the Castle there is a double row of grand lime trees, forming an avenue in the shape of the letter D. The places of a number of those which have been blown down from time to time have been replaced by saplings of the same variety.

Crossing the river by the Chinese Bridge immediately behind the Castle, we reach the Hill of Drummond. Starting near the junction of the Tay and Lyon, about a mile from the Castle, this hill extends westwards to the hamlet of Fernan, a distance of 7 miles, with about 1 mile in breadth, and the greater part of it is clothed with a heavy crop of valuable timber,—principally larch and Scots fir,—the lower portion of the hill containing a mixture of hardwood trees, including oak, beech, elm, and sycamore. Ascending a steep passage from the river side, we reach the North Terrace Walk, a splendid promenade 3 miles in length, and 50 feet in width, —the walk extending all the way from Kenmore Bridge to the junction of the Tay and Lyon. The entire north side of this beautiful walk is planted with a line of beech trees, standing about 10 or 12 feet apart, and girthing from 6 feet to 9 feet, the height of the trees being from 70 to 80 feet. The south side is planted with a strip of mixed wood. The two heaviest larches on Drummond Hill girth as follows:—23 feet 6 inches at 1 foot, and 16 feet 3 inches at 5 feet; and 19 feet 10 inches at 1 foot, and 13 feet 9 inches at 5 feet. They are both grand trees, and contain about 90 feet of measurable timber, both trees being clean bole to the top. What may be accepted as a fair specimen of the Scots firs girths 11 feet at 1 foot and 10 feet at 5 feet. The spruce trees girth from 12 to 15 feet, the oaks from 8 to 10 feet, beeches from 11 to 12 feet, and there are several Spanish chestnuts about the same size as those at the Rail Bridge. There is a lime tree girthing 17 feet at 1 foot and 15 feet at 5 feet from the ground. The hill contains altogether about 4000 acres of woodland, the age of the crop varying from 80 to 160 years. The trees are generally well matured, and on the west shoulder beginning to deteriorate in some instances. At the old house of Kinnell, Auchmore, there is a vine celebrated for its great size, filling as it does a glass house 170 feet long. It is about fifty years old, and still in fine bearing condition. When Professor Blackie saw this tree he was so affected that he has written, "I made a vow on the spot, whenever I might be troubled with low and vulgar imaginations, to think upon this vine." He also wrote the following lines as a memorial of so elevating a spectacle:—

LINES TO THE KINNELL VINE, AUCHMORE.

" Come hither all who love to feed your eyes
   On goodly sights, and join your joy with mine,
Beholding, with wide look of glad surprise,
   The many-branching glory of this vine,
Pride of Kinnell! The eye will have its due,
   And God provides rich banquet, amply spread,
From star-lit cope to huge Bens swathed in blue,
   And this empurpled growth that overhead
Vaults us with pendent fruit. Oh, I would take
   This lordly vine, and hang it for a sign
Even in my front of estimate, and make
   Its presence teach me with a voice divine—
     Go hence, and in sure memory keep with thee,
     To shame all paltry thoughts, this noble tree !"

The plantations at other parts of the property are very extensive, being, in fact, as extensive as any to be met with in Scotland. There are also several noteworthy specimens of different varieties of trees in other parts of the property. In the grounds at Moness House, near Aberfeldy, there is a fine group of larches—one specimen girthing 18 feet at 1 foot from the ground and 11 feet 6 inches at 5 feet; and another 15 feet at 1 foot and 10 feet 6 inches at 5 feet. They vary in height from 94 feet to 100 feet, and are very symmetrical throughout. On the road leading to the celebrated Falls of Moness, the scene of Burns' song "The Birks of Aberfeldy," there is a splendid cluster of Scots firs, of large size. Within the "Birks of Aberfeldy" there is also some magnificent timber. Groups of spruce trees are met with varying from 11 feet 10 inches at 1 foot, and 9 feet 2 inches at 5 feet, to 12 feet 6 inches at 1 foot, and 9 feet 8 inches at 5 feet. There are also many stately larches, girthing close upon 10 feet at 5 feet from the ground. Ash, elm, and beeches of large dimensions are also to be met with. In the Killin district there are several large plantations of great value. The soil, almost everywhere, is congenial to the growth of timber, and especially so with regard to larch, which grows more luxuriantly than any other variety. Extensive as the plantations are,—reaching close on 11,000 acres,—much yet remains to be done, and the present Earl contemplates planting extensively. His Lordship has already done much to beautify and enhance the property by having considerable tracts planted under the superintendence of Mr William Dunn, the land

UPPER FALL OF MONESS.

steward. A nursery has been formed for the rearing of forest trees, and these are being planted out at intervals. Many of the barer parts of the property are thus being clothed gradually, to the immense improvement of the landscape, and the ultimate benefit of the proprietor.

XXXIII.—CASTLE MENZIES.

The Present and Former Castles—The Policies—Remarkable Chestnut, Sycamore, Beech, Oak, and Elm Trees—The Newer Conifers—The Largest *Wellingtonia gigantea* in Britain—Comparison between the Growth of the *Wellingtonia* and the *Cedrus Deodara*—The Castle Menzies Gean Trees—Weem Rock and Woods—The " Black Watch " —St David's Well—The Plantations—Spring *versus* Autumn Planting —Great Improvements at Rannoch.

CASTLE MENZIES has long been justly famed for the magnificent timber which surrounds its ancient Walls. Indeed, there is no place in Scotland, far less Perthshire, where so many gigantic specimens of most of the ordinary varieties of trees are to be seen in more luxuriant growth and beauty. All the larger specimens are to be met with in the rich and extensive lawn at the foot of Weem Rock, under whose shelter the castle is built. The present castle is a very old structure, but it is not the first edifice which has been built upon these grounds. The original residence of the Menzies family in this part of the country was at a Castle built on the Rock of Weem, north-west from the present one. There was also, as one of the residences of the family, Comrie Castle, on the banks of the Lyon, the name being derived from the lands upon which the Castle was built. The latter was destroyed by fire in 1487 ; after which, the then baronet, Sir Robert de Menyrs, built the old Castle of Weem, Comrie Castle being repaired as a residence for the second son. At this time the laird of Weem had a new grant of his whole lands from the Sovereign, the property being erected into a free barony, under the title of the Barony of Menzies. In consequence of a family quarrel about the lands of Rannoch, the old Castle of Weem was burned by a relative, Neil Stewart of Fortingall in 1502 ; and in return for this the Menzieses burned the Stewarts of Fortingall's stronghold at Garth Castle. These fires proved to be a great misfortune to the family, as many valuable private papers, which threw much light upon the history of the Clan, were consumed. The existing Castle appears, from the date upon it, to have been built in 1571, and is a fine old Scottish baronial edifice. Extensive additions were made to it in 1840, in keeping with the original design, rendering it one of the most commodious and picturesque castles in Perthshire. The present Castle is also the centre of some interesting historic incidents. The Clan Menzies took the side of Parliament and the people in the Civil War of the seventeenth century. When the Marquis of Montrose began his march from Athole into the Lowlands in 1644, he was the cause of some trouble to the Menzieses. The Marquis sent a trumpeter with a friendly message to the Chief of the Clan, intimating that he purposed passing through the Menzies lands with his army on his way south. The Menzieses, with that defiant spirit which characterised the Highlanders of the age, ill-treated the messenger of Montrose, hung on the rear of the Rebels, and did all in their power to harass them. This was a game, however, at which two could play, and the Rebels retaliated by plundering the country, destroying the crops, and giving the roofs of the inhabitants to the flames. The Clan Menzies fought on the side of General Mackay at Killiecrankie, and after his defeat the General found refuge within the walls of Castle Menzies. A good story is told regarding the fidelity of the individual members of Clan Menzies, in connection with the Rebellion of the next century. The Chief, at this time, Sir R. Menzies, was lame, but he sent 200 of the Clan to fight for Prince Charlie, and these were led by Menzies of Shian, being present at Culloden. One of the Clan, Menzies of Meggernie, having been pardoned for his share in the rising of 1715, was restrained from personally taking part in the Rebellion of 1745, but he showed his sympathy with "bonnie Prince Charlie" by sending him a splendid charger as a present. The servant entrusted with the conveyance of the horse had the misfortune to be apprehended, but, with the noblest fidelity to his Chief, he preferred to die rather than betray his master. Castle Menzies can boast of having been dwelt in by Mary, Queen of Scots, who came here after the hunting excursion with the Duke of Athole, when she lived in a lodge put up for the occasion at Loch Long, of which all traces have now disappeared. In 1745 Prince Charlie spent a night in the Castle.

Many of the more remarkable trees which surround the castle were evidently planted about the date of the building of the original castle, and it is rather marvellous how they escaped destruction during the distracted period which supervened. Having survived the ravages of war, and eluded the attentions of the Vandal, they have been able to reap the full advantage of their fine situation and the rich soil in which they luxuriate, living to reach a green old age, and attain dimensions that provoke the wonder of all who see them. The lawn upon which the Castle and its arboreal treasures stand is about 300 feet above the level of the sea. It is finely sheltered on the north by the Rock of Weem, which presents a most magnificent background, the many-tinted woods rising boldly out of its steepy slopes, with huge masses of grey rock peeping out at intervals, and contributing to the embellishment of the sylvan picture.

Leaving the hospitable roof of the Castle, a walk of a very few yards brings us into contact with the monarchs of the forest. The first tree to which we directed our attention is a grand old Spanish chestnut, situated in the washing-green. The tree is much decayed, and carries a good many "growths," but it is noteworthy from its extraordinary girth. At 6 inches from the ground it girths 26 feet 6 inches; at 2 feet, 22 feet 2 inches; at 3 feet, 20 feet 6 inches; at 5 feet, 19 feet 2 inches; at 7 feet, 18 feet 4 inches. It has a bole of about 12 feet, and an entire height of 60 feet. At the top of the bole it breaks into two very large limbs, each of which is equal in size and girth to ordinary well-developed trees of this species. Behind the garden there is another large Spanish chestnut, girthing 19 feet 2 inches at 2 feet from the ground and 15 feet 10 inches at 5 feet from the ground. It has a good bole of about 20 feet, and an entire height of about 80 feet. An additional charm is lent to its hoary, massive trunk by one side of the bole being beautifully and completely covered with a profusion of the pretty fern, *Polypodium vulgare*. From a record, dated 1867, we find the size of this tree stated as :—Height, 74 feet ; girth at 6 feet from the ground, 13 feet 10 inches ; diameter of branches, 42 feet ; contents, 297 cubic feet. The soil is a rich sandy loam, the exposure southern, and the tree is well sheltered by others which surround it. Several large branches have been cut off.

Although no traces of the old Castle of Weem are now to be found, there are not wanting indications pointing to the site upon which it stood. These consist of the remains of the avenues which evidently led to the front entrance, and many of the largest of the trees were originally units in these avenues. Some of the trees in the avenues are not so large as might be expected from their great age, having been allowed to grow too close, but wherever the trees have had full sway, they have grown into veritable giants. A little to the west of the Castle there is a very fine sycamore tree, which, although now somewhat isolated, was at one time within the avenue leading to the Castle which was burned by Neil Stewart of Fortingall. The girth at 1 foot from the ground is 23 feet 11 inches, and at 5 feet 17 feet 11 inches. It is 80 feet in height, and has a splendid bole of 40 feet. There are a few branches striking out from the main trunk, but they do not interfere with the value of the bole. A great limb was broken off during the memorable gale of 28th December 1879, but the tree is very little disfigured in consequence. The spread of branches is 85 feet, and it contains upwards of 1000 cubic feet of timber. A short distance from this sycamore we find the avenue, of which it originally formed part, so far complete, but the trees are not nearly so large, on account of their having been allowed to grow only about 6 feet apart. About 100 yards east from the Castle there is a sycamore of even larger dimensions than the one already mentioned. The girth at 1 foot from the ground is no less than 28 feet 6 inches, and at 5 feet from the ground the girth is 19 feet 3 inches. It is a most symmetrical and picturesque tree, but, like many of the others, it is rather decrepit, and pretty knotty. It also appears to have formed part of an avenue leading to the old Castle at the foot of the rock. The trees in this avenue have been planted wider, and they are consequently of larger size and more uniform growth. The sycamore adjoining the tree we have just mentioned girths 19 feet 3 inches at 3 feet above the ground, being immediately above the roots. The whole of the trees in this avenue are about the same size, with an average height of 80 feet, and are noble and symmetrical specimens of this fine park tree, which is among the first to cheer the spring with its vivid green.

SYCAMORE AT CASTLE MENZIES.

The beech is represented by several remarkable specimens. Some very good ones are to be seen quite close to the Castle,— one of them having a very peculiar growth, being, in point of fact, a vegetable "Siamese Twins." The roots are completely joined together, but about 1 foot above the ground the trunk separates for $3\frac{1}{2}$ feet, and then unites again, leaving an opening through which an ordinary-sized person might pass. One of the separate limbs girths 9 feet 10 inches, and the other girths 7 feet 3 inches. At the point where the two limbs reunite the girth is 14 feet 2 inches. It would seem as if two trees had been planted and had agreed to amalgamate the better to resist the storms of life, in accordance with the old motto, " Union is strength." An adjoining tree has grown up in much the same way. Although there is virtually one trunk, there can be little doubt that there were originally two trees, as the connection is perfectly visible. At 1 foot from the ground this double tree girths 16 feet 3 inches, and at 5 feet from the ground the girth is 13 feet 2 inches, with a fine bole of 20 feet, after which the tree splits into two branches. There is another very fine beech girthing 16 feet 6 inches at 1 foot from the ground and 12 feet 8 inches at 5 feet from the ground, with a grand bole of 18 or 20 feet.

There are several very good oaks. One of these, growing close to the public road between Weem and Fortingall, girths 15 feet at 1 foot from the ground and 12 feet 6 inches at 5 feet from the ground. There are also a number of very fine elms on the property. One of these, near Camserney Mill, has a girth of 22 feet at 1 foot from the ground and 15 feet 2 inches at 5 feet. An ash in the policies girths 17 feet 5 inches at 1 foot and 13 feet 10 inches at 5 feet, and another ash at Boltachan has a girth of 18 feet 3 inches at 1 foot and 13 feet at 5 feet up. A larch at the west end of the garden girths 16 feet at 1 foot and 10 feet 6 inches at 5 feet from the ground.

Having detailed the more remarkable specimens of the common hardwood trees at Castle Menzies, we shall now proceed to describe the rarer and newer varieties of timber for which this property is no less distinguished. Amongst the more prominent of these is a splendid *Wellingtonia gigantea*, believed to be about the largest specimen in Great Britain. It was amongst the first batch sent out from London, and, although it was then no bigger than a man's finger, it cost the extravagant sum of £3, 3s. It was kept in a pot for a couple of years, and planted out in the garden in 1858. When measured in May 1879 it had a girth at the ground of 11 feet 1 inch. In September 1880 the girth was 12 feet. In May 1883 the girth at the ground was 13 feet 7 inches, at 1 foot from the ground the girth was 12 feet 10 inches, and at 5 feet from the ground the girth was 9 feet 3 inches. The height was 44 feet, and the spread of branches 25 feet. There are many splendid specimens of *Cedrus Deodara*, and, as they are about the same age as the *Wellingtonia*, we have a good opportunity of contrasting them with the more rapid-growing tree. There is a very fine avenue of the *Cedrus Deodara* lining a grass walk a little distance from the Castle, and planted about twenty-five years ago to commemorate the natal day of Mr Neil Menzies, younger of Menzies. One of these trees, taken as a fair sample of the lot, girths 4 feet 6 inches—a very striking contrast to the girth of the *Wellingtonia* of much the same age. The trees in this avenue range from 30 to 35 feet in height. They are all exceedingly beautiful specimens, although one or two have lost their leaders. About two years ago a quantity of *Cedrus Deodara* seed was obtained from India, and planted throughout the garden. The seed has sprung up beautifully, and now there are about 500 thriving young specimens planted out. There is also a very good specimen of *Cryptomeria Japonica* of the same age as the *Wellingtonia*, but it is not making much progress, although it is in perfect health. One of the most curious trees in the grounds is a specimen of *Arbutus Menziesii*, planted about twelve years ago. It is a large shrub, the leaves of which have a strong resemblance to the laurel. It casts its bark,—which is only the thickness of writing paper,—exposing underneath a very fine quality of wood, of a beautiful clean cinnamon colour. The height is about 12 feet, and the plant is well spread. It does not seem to suffer in any way from the weather, but as yet it has yielded no fruit. It is a very ornamental plant, and has a somewhat tropical appearance. *Abies Albertiana* promises to do very well in this part of the country. One specimen in the policies—raised from seed, and planted by Sir Robert Menzies about thirty years ago—has attained a height of 60 feet. A *Picea grandis* close beside it has also proved to be a fast-growing and very promising tree. It was planted about ten years ago, and has been known to add three feet in the course of one year. There is also a very shapely specimen of *Picea Lowii*. We next directed our attention to a large and handsome tulip tree. At 1 foot from the ground the girth is 10 feet 10 inches, and at 5 feet from the ground the girth is 10 feet 1 inch. The bole is fully 8 feet, —the tree breaking out into two limbs, which grow wide apart and have to be bound by an iron loop. It is believed to be the largest specimen in Scotland. A hornbeam close beside the one just mentioned, is also a notable tree. It has a bole of 25 feet, and the girth at 1 foot from the ground is 8 feet 6 inches ; and at 5 feet from the ground the girth is 8 feet 4 inches. *Abies Menziesii* is also found to luxuriate in the deep rich soil and sheltered situation at Castle Menzies. One of the best specimens is fully 70 feet high, with a girth of 10 feet 6 inches at 3 feet from the ground. The branches are very dense, and so close to the ground that there is some difficulty in measuring the tree. It is altogether a most shapely plant, and as good a specimen as can be seen anywhere. There are several remarkably good specimens of the black Italian poplar. What is understood to be the best of these rises to a height of about 90 feet, and has a splendid bole of 40 feet. At 1 foot from the ground, it girths 12 feet 5 inches, and at 5 feet from the ground the girth is 11 feet. There are several others that may claim to be almost equally good. There are a number of fine specimens of *Picea Cephalonica*, *Pinus Nordmanniana*, *Abies Douglasii*, *Pinus Benthamiana*, and *Pinus ponderosa*. A large variety of the newer conifers, principally from Japan, have been introduced in recent years, and, so far, are thriving as well as could be wished. These newer importations include *Picea polita*, a beautiful ornamental tree ; *Abies orientalis*, a very handsome variety of the spruce ; *Abies Alcoquiana*, which attains a height of 100 feet under favourable conditions ; and *Picea bifida*, also a very promising variety. Sir Robert does not possess what can be called a pinetum, the exotic conifers being planted all over the property, and in all sorts of situations as to soil and climate. This, we need hardly say, would have been a very expensive proceeding if the plants had been procured from nursery grounds in the ordinary way. Sir Robert, however, has been enabled to adopt this plan from having subscribed to two Associations, organised for the purpose of sending out properly qualified men to California and British Columbia as seed-collectors, and in this way he received a share of all the new plants imported by Jeffrey and Browne. In addition to the *Douglasii*, the *Menziesii*, and the *Albertiana*—which Jeffrey sent home as *Mertensiana*—and Browne's *grandis*, which are all proving useful additions to the timber trees of Scotland, we found that the *Thuja Craigiana* sent home by Jeffrey, and otherwise known as the *Thuja gigantea* and *Libocedrus decurrens*, is also doing very well indeed, growing quite as fast, and thriving quite as favourably, as any of the others. If the wood should ultimately prove to be good and useful, this variety will have to be added to the others that will be extensively planted in the future.

Throughout the whole policies, something attractive may be seen at every turn. Here we meet with an old avenue of elm or sycamore, reminding us of the more ancient Castle ; and there we pause to admire some of those newer varieties which will undoubtedly become the timber trees of the future. The rich loamy soil of the extensive lawn surrounding the Castle, and the splendid shelter which is afforded by the Rock of Weem, present a combination which is most favourable to the growth not only of all varieties of timber which can resist the baneful influences of the climate, but many plants thrive here in a way which they can do in few other places in Scotland. All the fruits capable of being reared in the open air grow in abundance, and more than one variety can be seen in perfection. From a very early date the Castle Menzies gean trees obtained celebrity all over the country. So famous, indeed, were these trees at one time, that their seed was sought for from all parts of Great Britain, and large quantities were even sent abroad by request.

Leaving the lawn, of which we have had so much to say, we bend our steps towards Weem Rock, which rises to a height of 600 feet immediately behind the castle. Its bold sides are richly covered with a good deal of old and young timber, presenting a very picturesque object in the landscape. From the summit a most extensive and charming prospect is opened up, the eye looking down upon

> " A blending of all beauties—streams and dells,
> Fruit, foliage, crag, wood, cornfield, mountain, vine.

Straight beneath us lie the romantic old castle and the peaceful village of Weem. At one glance the eye takes in almost the whole of Strathtay, with its highly-cultivated fields, tastefully scored with lines of well-nourished trees, principally oak. We trace the Tay from its source, and mark the lines of birch and oak, with an occasional fir, which grace its banks, and then direct attention to the items which go to make up the magnificent scene. Two miles distant we have a full view of the pretty village of Aberfeldy. A little higher up the slopes of the south side of the Tay we see the famous "Birks of Aberfeldy," immortalised by Burns. Not the least interesting item in the scene immediately before us is the substantial stone bridge, with its four large obelisks, built by General Wade in 1733, as part of his great military road from Stirling to Inverness. At the point where the bridge crosses the Tay, the river is no inconsiderable size, and is frequently so much flooded that the adjoining fields are sometimes under water to the depth of several feet. A little to the north of the bridge we see the field on which, in 1739, the famous "Black Watch" was first embodied, this title being acquired from their dark tartan uniform. Although the regiment was embodied here in 1739, half a dozen companies were raised some years previously. The independent companies known as the "Black Watch" came by two different routes to the "Loblen Mor," or the large field north from Tay Bridge;—those from Braemar *via* Kirkmichael and Pitlochry; and those from Lochaber and Badenoch by Dalnacardoch and Coshieville, and having been formed there into a regiment, they marched across Tay Bridge, and encamped on the field to the north of it at Aberfeldy, where they remained for ten months. In May 1740 the regiment was raised to an aggregate of over 1000 men, who were quartered on this field for nearly eighteen months. At that period the regiment was known in the Line as the 43rd; but on the reduction of the original 42nd Regiment in 1749, they were gazetted as the Royal 42nd Highlanders. Many of the men, we are told, who composed this fine Regiment, were drawn from a higher station in society than is ordinarily the case with soldiers, being principally connected with the families of gentlemen and farmers, which may have not a little to do with the distinction which they gained in their earlier battlefields, and which has stimulated them ever since. Casting our eyes westwards from the historic field of which we have been speaking, we obtain a view of the eastern end of Loch Tay, its blue waters sparkling in the sunshine, and looking in the distance, with the great hills on either side more like a winding river than a great lake. At the western end we see the mighty Ben More frowning down upon the bright expansive waters, as if it were commissioned to confine them to their present basin. There, too, is the huge Ben Lawers, and the pine-capped Drummond Hill. Taymouth Castle we cannot see—embowered, as it is, amidst umbrageous trees. On the south side of the Tay, the ground gradually rises from the river side, until it attains an almost mountainous height. The land is cultivated for a considerable distance up the slopes, varied here and there with battalion-like strips of timber, and a grand fir plantation along the crest of the hill directly opposite the rock from which we look upon this grand panorama, reminding one forcibly of the lines in Goldsmith's *Traveller*—

> "Its uplands sloping deck the mountain's side,
> Woods over woods in gay theatric pride."

In exploring Weem Rock we find a good deal more that is interesting than we could anticipate. Disturbing in our passage the capercalzie, the hawk, and the roe-deer, by which it is much frequented, in addition to the more familiar inhabitants of the woods, we proceed to examine the nature of the wood which clothes its slopes and imparts so much that is beautiful to the surrounding landscape. About 150 or 200 feet above the Castle we come across an avenue of large-sized beeches. One which we measured as a fair sample of the lot girthed 13 feet 5 inches at the narrowest part of the trunk. The avenue is of considerable length, but is broken here and there by trees that have fallen and been removed. This avenue, like some of those we noticed on the lawn, has evidently had some connection with the old Castle. The timber upon the Rock consists principally of oak, beech, larch, and Scots fir. Some of the hardwood trees throughout the wood are of large size, particularly the beeches. A very fine one which we measured girths 14 feet 7 inches at the narrowest part, being 3 feet from the ground, and in every respect a very shapely tree. An oak girths 13 feet 2 inches at 5 feet from the ground. A larch at 1 foot from the ground girths 13 feet 11 inches, and at 5 feet from the ground 10 feet 2 inches. The bole tapers very little until it reaches a height of 35 feet,—there being altogether a splendid trunk up to the top of the tree, which is about 100 feet in height. There are a number of very good Weymouth pines. One of these girths 11 feet 10 inches at 1 foot from the ground and 10 feet 1 inch at 5 feet from the ground. Another, close beside the one just referred to, girths 10 feet 11 inches at 1 foot from the ground and 9 feet 8 inches at 5 feet. A good many of these trees immediately at the back of the garden wall have boles averaging 25 feet. There are a few Scots firs of fair proportions. One which we measured girths 7 feet 3 inches at 5 feet from the ground, tapering very slightly until the bole reaches a height of about 40 feet. A spruce close beside the latter girths 7 feet 3 inches at 5 feet from the ground, the bole being quite as good as that of the fir. Another which we saw was much larger, the girth at 1 foot from the ground being 11 feet 4 inches, and at 5 feet 10 feet 8 inches. There are also a few very fine silver firs,—one near the garden girthing 14 feet 11 inches at 1 foot from the ground and 12 feet 9 inches at 5 feet. A good many of the newer coniferous trees have been planted throughout the Rock, including *Pinus Austriaca, Abies Douglasii, Abies Albertiana*, &c. A large quantity of rhododendrons have been planted over the Rock, principally as cover for game. Although the soil in many parts is not very deep, and the trees have but a slight grip of the ground, there are some portions where the soil is of a very superior quality ; and all along the brow of the hill, the land, which is a light loam, is under cultivation, and capable of producing very fine crops. At some places the Rock is marked by dangerous-looking fissures, and occasionally great masses of stone break away and rush down the slopes with terrific force.

One of the great attractions to ordinary visitors is the spring known as St David's Well. This is one of those ancient "wishing wells" into which the devotees used to drop money or valuables, and invoke the blessing of the patron saint, and which proved a rich mine to the priests in the district. The well occupies a most romantic situation, and one which would readily stir the superstitious feelings of the ignorant devotees. The water, cool and bright, percolates through the rock in one of the corners of a large cave, capable of accommodating fifty people under the immense overhanging rock. Some time ago the ground in front of this great hollow was dug up, with the result that a quantity of bones was discovered.

When we come to consider the plantations, we find that the lairds of Castle Menzies are entitled to rank amongst the great planters of their day. A good deal of waste land was planted during the time of the late Sir Neil Menzies, Sir Robert's father, but by far the larger portion of those miles of woodlands which now adorn the Menzies property was the work of the present laird. Since Sir Robert came into possession in 1845, he has planted no fewer than two millions of trees; and the great improvement which this has effected upon the appearance of a country that was previously bleak, barren, and cold-looking, can scarcely be imagined. The first work of this nature which he undertook was the planting of the sides of the Hill of Tgarmuchd, which marches with the Wood of Dull on the west. The Wood of Dull itself consists of about 300 acres of mature wood, principally Scots fir. The hill of Tgarmuchd was planted in 1847, and extends to about 100 acres—about 230,000 trees being utilised in the planting. The wood here is mostly Scots fir and larch, intermixed with hardwood, principally oak, beech, and ash. The wood has not grown so well here, as in some other parts of the property. At Foss a large acreage was also planted about thirty years ago, Scots fir and larch predominating. Within the last twelve years a very fine plantation has been laid down above Weem Rock. The wood here is also principally larch and Scots fir, the plants being obtained from Perth Nursery, Dundee, and Aberdeen. There are altogether about one million trees in this plantation. Taken as a whole, the young trees are thriving vigorously, and promise to yield a very fine crop.

The hand of the improver is perhaps most apparent at Rannoch. Even when the present baronet entered into possession, there was scarcely a tree to be seen throughout this extensive part of the property, almost the only relief to the dull monotony of great mountains, acres of stone,—"the riddlings of creation,"—and the blue waters of the loch, being the famous "Black Wood of Rannoch," on the Dall property. Even at present, the district of Rannoch has a wild and barren aspect; but it is still a veritable "Garden of Eden," compared with what it was thirty-six years ago, when Sir Walter Scott's description of another part of Scotland was equally applicable here—

> "By my halidome,
> A scene so rude, so wild as this,
> Yet so sublime in barrenness,
> Ne'er did my wandering footsteps press,
>   Where'er I happ'd to roam.
>
> . . . . . .
>
> Above, around, below,
> On mountain or in glen,
> Nor tree, nor shrub, nor plant, nor flower,
> Nor ought of vegetative power,
>   The weary eye may ken.
>
> For all is rocks at random thrown,
> Black waves, bare crags, and banks of stone.
>   As if were here denied
> The summer sun, the Spring's sweet dew,
> That clothe with many a varied hue
>   The bleakest mountain-side."

Indeed, "The Wizard of the North" had a keen perception of the barren loneliness of Rannoch in his day, and embodied his observations in the opening stanzas of the Fourth Canto of *The Lord of the Isles*—

> "Stranger, if e'er thine ardent steps hath traced
> The northern realms of ancient Caledon,
> Where the proud Queen of Wilderness hath placed,
> By lake and cataract, her lonely throne;
> Sublime, but sad delight thy soul hath known,
> Gazing on pathless glen and mountain high,
> Listing where from the cliffs the torrents thrown
> Mingle their echoes with the eagle's cry,
> And with the sounding lake, and with the moaning sky.

> Yes? 'twas sublime, but sad.—The loneliness
> Loaded thy heart, the desert fired thine eye;
> And strange and awful fears began to press
> Thy bosom with a stern solemnity.
>
> .   .   .   .
>
> Such are the scenes where savage grandeur wakes
> An awful thrill that softens into sighs;
> Such feelings rouse them by dim Rannoch's lakes."

Now, however, such gloomy thoughts are hardly likely to arise on visiting Rannoch, where, although the railway has not penetrated within twenty miles of its beautiful loch, large numbers of tourists, in search of health and of the "savage grandeur" in Nature, annually find their way to the foot of Schehallion, and traverse the shores of Loch Rannoch. All over this extensive track of almost barren country,—where only the deer, the grouse, and, in some parts, the hardy blackfaced sheep or Highland cattle, can find a picking—several miles of woodlands have been laid down at intervals during comparatively recent years, with the view of at once improving the aspect of the country, and turning the more barren spots to a profitable purpose. In some parts, the planters have evidently been actuated more by a love of the beautiful than by pecuniary considerations. Along the sides of the loch, and by the banks of the burns and streams which flow into it, Sir Robert has added greatly to the beauty of the landscape by planting birch very extensively,—a tree which Coleridge has described as the

> "Most beautiful
> Of forest trees—the lady of the woods."

All the plantations at Rannoch appear to have been laid down in the course of the present century. At Aulich, 15 acres of Scots fir were planted in 1802 by Sir John Menzies, who also planted 10 acres of the same variety at Killiechonan in 1806. In 1809, Sir Robert Menzies planted 40 acres of Scots fir, larch, oak, and ash at Kinaclachair; and his successor, Sir Neil Menzies, father of the present proprietor, extended this wood in 1816, by adding other 60 acres. All of these older plantations have thriven remarkably well, and now contain very fine trees for their age. At Kinaclachair, the wood suffered severely by the Tay Bridge gale. No planting was done between the years 1816 and 1845, when the present proprietor came into possession, and commenced his extensive improvements. The first of Sir Robert's plantings took place at Craiganour, where, in the autumn of 1845, 70 acres of Scots fir and larch, with a few oaks, were laid down. This plantation has also thriven very well, and in 1860 it was extended by about 25 acres of the same varieties of timber. Sir Robert brooked no delay in the work of improvement, and in the spring of 1846, 120 acres were planted at Tirchardie, the wood being of the same varieties as planted at Craiganour, in the autumn of 1845. By laying down the one plantation in autumn and the other in spring, an opportunity has been presented of contrasting the merits of the two systems. In both cases, the woodlands are at the lochside, and are exactly the same as to soil and climate, and yet we find that the spring plantation is apparently ten years behind the one planted in autumn. In 1873, 60 acres of fir and larch were planted at Culvullin, and are doing very well. In the later plantations— those laid down since 1845—some of the larches have not done so well; but still there are many fine specimens throughout the woods. Other plantations, more or less extensive, have been laid down during the past few years.

It is almost impossible to overestimate the value of such work as this. Since the year 1802 about 500 acres of plantations have been laid down in the Rannoch district, and of these more than 350 acres have been planted by the present Sir Robert Menzies. Within the last thirty-eight years, then, an alteration has been made upon the face of this part of the country of which the younger generation can have but a very slight conception, and which will become even more marked in the future. Where formerly nothing was seen but unproductive moorland, we have now the waving pine tree, the hardy fir, and the graceful birch, adding a new beauty to the face of the earth, and yielding a rich return to the planter. Experiments are also being made with varieties of the newer conifers, which may serve as guides for future planters in such unpromising situations. *Picea grandis* is planted throughout the woods, and is thriving very well. *Abies Albertiana* is very extensively planted at Rannoch, and is also growing in a most satisfactory way.

Although Sir Robert takes a very great interest in the woodlands, and personally directs their management, he is equally solicitous regarding the improvement of his agricultural holdings. Everywhere, throughout his vast domains, the hand of the improver is visible; and although, from the general character of a great part of the country, he has had enormous difficulties to overcome, he has been rewarded with such a measure of success as to encourage him to persevere in the same commendable direction.

XXXIV.—THE BLACK WOOD OF RANNOCH.

Origin of Name—Relic of the Original Forests of Scotland—Age of the Trees—Buried Remains of the Ancient Forest—A Forest Fire—Destruction of a Portion of the Wood by a Public Company—Difficulties Connected with the Removal of the Timber—Scarcity of Food at the Time—Natural History Items—Rannoch Fir as an Illuminator—Remarkable Trees.

ALLUSION has already been made to the picturesque effect of the Black Wood of Rannoch upon the southern shores of the magnificent loch which takes its name from the district. This wood is certainly a most conspicuous object in the landscape of a district which cannot boast of a superabundant wealth of timber, although, as we have previously shown, improvements of considerable magnitude have already been effected, and are still in progress in Rannoch. The Black Wood, in all probability, derives its name from its peculiar sombre appearance; but this is one of those characteristics which contribute to its charm, its sable shade presenting a perpetual contrast to the lighter tints of the surrounding woods, as they change their hues with the returning seasons. But while its waving branches and gloomy aspect enrich the landscape with a beautiful and imposing picture, there is another circumstance which increases our interest in this celebrated wood. The Black Wood of Rannoch is a valuable relic of the original forests of Scotland, being the only important remains of the great Caledonian Forest. From it we gather some faint idea of what Rannoch may have been like when the whole surface of the country was one unbroken stretch of woodlands, the haunt of the hungry wolf and the ferocious wild boar, of which many traditional stories are still related in the district; or, more recently, when it was the refuge of the unsubjugated freebooter.

The bulk of the present growth of timber does not appear to be particularly old, although its age cannot be determined with anything like certainty. It has evidently, however, grown up in succession to a wood which had been almost consumed by fire at a period not very remote; at least, at a time long subsequent to that when the woods of the country generally are supposed to have been cleared by the same agency. Whether oak or fir was most plentiful in the ancient forests, it is now very difficult to say, but tracts of both may have predominated in the same district at intervals according to the nature of the land. The remains of fir, however, are chiefly turned up in the mosses of Rannoch, the boglands being largely studded with the stocks of fir embedded in cinders, the trunks being of somewhat similar dimensions to their modern representatives. But the remains of oak are not unfrequently discovered. A few years ago a fine log of oak, measuring 5 feet 9 inches in diameter, was dug up in this wood.

The Black Wood of Rannoch is on the estate of Dall, the property of Mr T. V. Wentworth, an enthusiastic arboriculturist. Mr Wentworth has displayed commendable taste in planting very extensively all over an already well-wooded estate, and in the transplantation of trees of various sizes from the native stock. Within the last twenty years Mr Wentworth has advantageously put about 200 acres under plantation, the wood being confined to the proprietor's favourite, the old Scots fir—

" The pine which for ages has shed a bright halo."

In a short time this will very considerably enlarge the aspect of the Black Wood, which, at the same time, carries on its own work of propagation and extension. A few years ago, a very promising plantation of about 80 acres was almost destroyed by fire, which seized the heather among which the trees were planted. The fire, unfortunately, speedily obtained a strong hold upon the wood, and, although many willing hands used their utmost efforts to preserve it the fire was not extinguished until the greater part of the plantation was destroyed. The trees which were saved are of fine growth and great beauty, and as the trees consumed have been carefully and successfully replaced by wholesale transplantation from the Black Wood, the ravages of this great fire will, within a short time, scarcely be visible.

The area of the Black Wood is not quite so large as it must have been in the last century, but it is still of considerable size, its total extent being about three miles long by two miles broad. In addition to this, there are many detached clumps scattered round about, but not connected with the main thicket.

Oak, ash, and other native varieties, are interspersed all over the Dall estate, while the indigenous birch flourishes in great abundance. The old knotted birch, unrivalled for cabinet work, is very abundant; while the weeping variety, with its fine trailing locks, is also plentiful, and meets with much merited admiration in the balmy days of summer, when this tree appears in all its graceful beauty. We have mentioned that the area of the Black Wood is not at present so extensive as it was during last century. This is greatly to be regretted, but it is matter for profound satisfaction that the famous wood was not completely obliterated, as at one time was being rapidly accomplished by contract. About the beginning of the present century, a portion of the wood was sold to a Public Company belonging to the south, who felled a great part of it, and would have carried off considerably more timber, but for the lucky circumstance, at least for the country, that the enterprise failed to prove remunerative owing to the difficulty of transit, &c. Although the Company did not prosper, it was the means of doing some little good to the district while the operations lasted. At that time the country was in a very bad state, bordering almost on famine. Meal could scarcely be procured at any price, work was scarce, and labourers only too plentiful. The cutting down of the Black Wood was, accordingly, looked upon at the time as a providential circumstance, and proved a great blessing to many of the poor people in the district, as the Company imported provisions from the south to supply the workers in a way that was impossible for the natives to do without extraneous aid. Indeed, so great was the poverty at this time, and the scarcity of oatmeal,—the staple food of the people of this and other parts of Scotland,—that no more than one half-

THE BLACK WOOD OF RANNOCH. 413

boll of meal was distributed at a time to a single family, the rest of their supplies having to be made up with sea biscuits.

The wood actually borders on the loch, but, owing to its great extent inland, the Company was compelled to adopt a rather ingenious method for conveying the timber to its expansive waters, thence to be carried out to sea. Canals were formed along the brow of the hill, with locks or basins at fixed intervals on each level, into which the timber was floated for collection. About one mile from the loch, a long sluice was constructed, down which the heavy trees were slipped one by one, a man being perched upon a tree to signal by a loud whistle when the log was launched. The trees sometimes came down this great incline with such force that they broke into splinters on the passage, many went into the loch with such tremendous precipitation that they stuck as fast as stakes in deep water, while the sheer weight of not a few caused them to sink to the bottom, never to rise again. The trees which found their way in safety into the loch, were bound into rafts, to undergo a dangerous passage down the Tummel and the Tay. Even some of the rafts which reached the Firth of Tay in safety, floated away to the German Ocean and were lost, some of them being found stranded on the shores of Holland. With such enormous difficulties to contend against, it is not surprising that the Company became seriously embarrassed, and had to give up the undertaking. It was thus that the famous Black Wood of Rannoch has been preserved till this day.

The state of the wood is now considerably changed, although the canals are still to be seen in green or heather-covered ditches. Beautiful drives are now opened up in every direction, and a finer drive or promenade cannot be enjoyed in this country than here in the bracing days of autumn, amidst the sweet-scented heather and the fragrant birks, and under the shadow of the old Scots fir. It is worthy of note that this natural wood has not destroyed the heather, as plantations have a tendency to do. Indeed, the wood rather seem to have stimulated the heather, for its strength and luxuriance of growth is such as can be excelled nowhere in Scotland. No trace has been found of any habitation having ever been attempted within the wood, although there are not awanting evidences of the time when seclusion and quiet were highly necessary conditions for distilling a "good drappie." In the centre of the wood there is a fine green opening of about 25 acres, called Inniscalden. This was once under cultivation, but it is now given up to the use of game, with which the wood necessarily abounds. The strong red-deer of the forest seeks shelter here, and the spot is the favourite haunt of the roe. Grouse are plentiful on the higher parts of the wood. There are also plenty of blackgame, while the capercailzie are so much at home, and have become so numerous, that they seem to some extent to have driven the blackgame outside. A great number of these fine birds, of enormous weight and splendid plumage, are shot every season. Hares and the ubiquitous rabbits also roam over the entire wood in great numbers. The industrious ants, too, are seldom seen to greater advantage than in the Black Wood of Rannoch, where they construct pyramidical mounds rising to a height of four or five feet. So great is their rapacity, that a single community will consume a rabbit amongst them in a night, and polish its skeleton to a nicety. One of the most interesting features connected with natural history in the Black Wood, is the large numbers of rare lepidopterous insects to be found, in search of which entomologists annually come from all parts of the country. Their exertions are frequently rewarded by the capture of some of the most valuable flies, and specimens have been occasionally secured that were hitherto unknown. We have heard of specimens being captured so rare that their value was set down at from £25 to £30.

The timber of the Black Wood is so rich in resin and turpentine that before the days of paraffin oil, the whole of Northern Perthshire depended on the stocks of the Rannoch fir alone for the illumination of their dwellings. Many thousand tons of these stocks were regularly sent to the different fairs in the county for sale, the purchasers breaking up the stocks into splinters, which were used not as substitutes for the candle, but as the indispensable domestic light. One of the most noticeable characteristics of the Rannoch firs is the enormous weight of the limbs in proportion to the body of the trees, and their general uniformity. Taking, as most convenient, two fine trees at the curve of the approach west of the mansion-house, we found that they were almost identical in size and structure. The first one is exactly 14 feet in girth at 1 foot, and 10 feet at 10 feet from the ground, continuing at this girth for about 20 feet up. The tree then divides into several ponderous branches, from which less heavy, but by no means light, branches again break out— the top of the tree being rich in that bushy finish peculiar to the native fir. The second tree is almost similar to the first. The girth at 1 foot from the ground is exactly the same, but at 5 feet the girth is $11\frac{1}{2}$ feet, this measurement being obtainable up to a height of about 30 feet. The diameter of the ground covered by these two trees is no less than 50 and 51 feet respectively. Along the loch side, near the public road, there are a few trees worthy of being tested with the tape line. About one mile west from Dall Bridge, where the denser part of the wood begins, there are three very remarkable "sister trees." At a little distance they present the appearance of having sprung from the same stem, but a closer inspection demonstrates their individuality. They are so near to the lake that their roots spread into the water, and draw their nourishment directly from the loch. They are apparently of the same age as the majority of the trees in the wood. There is nothing particularly striking about them as individual specimens, but it is very unusual to meet with three such fine trees so close to each other in a wood of Nature's own planting. On being measured, they were each found to girth 9 feet at about 5 feet from the ground, this girth being maintained up till a height of about 25 feet from the ground. They have a very beautiful palm-like appearance. About half a mile farther up the loch side, and close to the margin of the water, there is a magnificent sample of the oldest stock of trees in the wood, although it has reached no great height. It is situated a little below the private road leading to Inniscalden opening, and is easily reached. Its gigantic limbs and beautiful symmetry combine to render it a grand specimen of the native fir. At 1 foot from the ground it girths 16 feet, at 16 feet from the ground it girths 12 feet, and at 20 feet high the girth is 9 feet. The first branch, which breaks out at 8 feet from the ground, girths 5 feet close to the trunk. Before the tree reaches more than 20 feet in height, there are as many as ten similar branches, varying only one or two inches in girth. The lowest branches, too, are by no means the heaviest. Up to the height of 30 feet there are no fewer than twenty strong branches. Above this height the leader almost loses its individuality, and combines with the other branches in forming a magnificent top—the whole tree presenting the appearance of a gigantic beehive. On leaving this tree, and proceeding to the middle of the wood, we come across another fir of a most unique appearance. In some aspects it presents a strong similarity to the Irish yew. Starting out about 3 feet from the ground, the branches, within a couple of feet from the leader, turn upwards and inwards, causing the tree to assume a form which we can only compare to the ordinary paraffin-lamp glass. The shade of this tree is more than usually dark, although the foliage is similar to the rest of the firs in the wood. Cones from this tree have been looked for, but none have yet been procured ; so there has been no way of testing whether the tree would reproduce its own kind, or whether its unique formation and peculiar singularities are due to one of those " freaks of nature," which are as common in the vegetable as in the animal kingdom. Although all the fir in the Black Wood is undoubtedly native, there is apparently more than one variety. Whether this difference, which is shown by the length and colour of the foliage, is owing to such accidental circumstances as variation in the character of the soil, and diversities in the age or condition of the trees, we cannot determine, but the separate shades are quite easily distinguished.

The measurements of other trees might be given, but we have said enough to show the general character of the timber which prevails throughout almost the entire wood. The whole of this remarkable native wood is thriving most luxuriantly, and has every appearance of continuing to remain and increase in all its pristine glory, and be the mother of forests for ages yet to come.

XXXV.—DUNALASTAIR.

The Amenity of the Property—The Mansion—Dunalastair Wood—
"Macgregor's Cave"—Crossmount—Relics of King Robert The
Bruce—Notable Trees—Plantations—General Improvements—The
Macdonalds of Dunalastair.

DUNALASTAIR, in many respects, presents a striking contrast to the surrounding country at Rannoch. It may be described as a veritable oasis in the wilderness. Not that Rannoch is devoid of those qualities which give richness to a country and beauty to a landscape, but the richness and the beauty consist chiefly of examples of Nature in her wilder and bleaker moods—

> "Forlorn hill slopes, and grey, without a tree;
> And at their base a waste of stony lea."

Heather and stones are the characteristic features of a great part of Rannoch, whether we roam by the valleys or climb the mountains, but at Dunalastair the wildness of Nature is completely subdued by Art, or blended so as to harmonise most effectually with its milder surroundings. The cold and cheerless aspect of an almost treeless district gives place, at Dunalastair, to a landscape singularly warm and bright—the lower lands yielding average crops of golden grain, and the higher parts clothed with a "silent sea of pines." Until about eighty years ago, Dunalastair shared in the general bleakness of Rannoch, since the destruction of its natural woods; but the improving spirit which was then moving across the country breathed upon the blasted heath, and its influence has never ceased to operate. Since that period, improvement has followed improvement, until at the present day this is one of the most delightful Highland properties in Scotland. The beauties of the situation are greatly enhanced by their unexpected development. Immediately after leaving Struan Station, the traveller enters the heather region, diversified by glimpses of the foaming Errochy, as it rushes down its rocky channel, or flows through a narrow strip of meadow land, while here and there a belt or clump of plantation breaks the monotony of the scene. At Auchleeks we have an agreeable variety in the umbrageous surroundings of this fine shooting lodge, but afterwards the road lies directly

"O'er the moor amang the heather,"

until we catch sight of the verdant fields of the Vale of Rannoch, and the bright sparkle of the Tummel winding its way from the loch between gigantic hills of deepest blue. Shortly before entering the policies,—which we do by a splendid new road made at the sole expense of the proprietor, and extending for several miles,—we pass through a deep gorge, from which nothing is for some time visible but the "bonnie blooming heather," and the cone of quartz which tops Schiehallion. Then we come to rich green parks, and dark, distant woods, relieved by glimpses of the glittering waters of Loch Rannoch. The approach is beautifully lined with thriving beech, oak, and larch trees; and the fields on either side are fringed with fine young timber, and surrounded by romantic crags. The house is reached somewhat suddenly—a circumstance which materially heightens the effect of the scene, and fixes it indelibly upon the memory.

The site of the mansion is particularly well chosen. It is sheltered from the north by Ben Hualach, and looks straight across to Schiehallion. Brightness is the prevailing characteristic. The east wind does not reach it disagreeably, and the sun beats pleasantly upon the south windows of the house, receiving in winter reflected help from the glacial snows of Schiehallion. The climate is further modified by extensive drainage,—the air being free from dampness, and most invigorating. The particular spot upon which the mansion is built is exceedingly dry, and mostly of rock. The house—which was erected in 1859 by the late General Sir John Macdonald from designs by Mr Heiton, Perth—is Scotch baronial in style, and is built of a beautiful whinstone, with Dunmore freestone facings. It presents a highly-decorated and imposing exterior, while the interior is comfortable and modern in arrangement. The terrace in front of the garden is one of the prettiest spots imaginable, commanding magnificent views in every direction. All around are waving woods of spruce, larch, and fir, which the hilly character of the country shows to great advantage. Both east and west charming peeps are got of the Tummel, with its boiling rapids and deep, still pools. Looking westwards, Loch Rannoch is the focus of a splendid picture, backed by the Black Mount, and the Glen Etive and Glencoe mountains. The chief scenic attraction, however, is the view towards the south, embracing the great mountain of Schiehallion, rising symmetrically to a height of 3547 feet above the level of the sea, and belonging to the Dunalastair estate. Its precipitous face is scored with many deep gorges, down which the rain rushes in torrents, and forms numerous pretty cascades. The lower hills round the foot of the mountain are well clothed with natural timber of fair size, which contributes a good deal to the general effect. In wandering over the policies we learned that there are several rings—or " castles," as they are called in the neighbourhood—about which very little seems to be known. One near the home-farm is 60 feet in diameter, and looks as if it had been the site of some ancient sentinel station. It is at present used as a burial-place for favourite animals.

Where so much has been done in comparatively recent years to improve and beautify the property, it is not disappointing to find that gigantic trees are conspicuous by their absence. The most of the timber is within 100 years old, but, for trees of this age, many magnificent specimens are to be met with. The best of the timber adorns Dunalastair Wood, on the top of the Black Rock—a really beautiful crag, commanding a view of the whole of the glorious scenery of the district from a large stone table, erected for the convenience of visitors. This wood was planted about eighty years ago by Colonel Alexander Robertson of Struan, whose family then possessed the property. The Scots firs predominate, and, as these have been raised from Rannoch seed, their excellence as to size and quality need not be enlarged upon. There are also several oak, ash, and beech trees of respectable dimensions. Shortly after entering this wood our way lies along the banks of the Tummel, by a fine new road constructed in 1880 by the proprietor, Major-General

Alastair Macdonald. At a prominent part of the road the following inscription is carved in large letters upon a huge boulder, additional prominence being given to the notice by the letters being painted black :—" This road was made by Major-General Alastair Macdonald. 1880. Distance to East Lodge, 1884 yards." Some of the best of the trees on the Dunalastair property are in this neighbourhood. At the river side, near the bottom of the garden, there are some very nice beech trees, with an average girth of 8 feet at 5 feet from the ground. Round the Silver Well, a fountain of the brightest and coolest of waters, there is a beautiful clump of good ash and gean trees. There are several clumps of aged yew trees, probably about 200 years old, with a girth of from 5 to 7 feet. A number of larches reach about 100 feet in height, and several of the spruce trees have boles of 80 feet, one of the latter which we measured girthing 9 feet 3 inches at 1 foot and 7 feet 6 inches at 5 feet from the ground.

At the west end of the river side drive an elegant and substantial iron girder bridge—12 feet wide, and nearly 100 feet long—was erected by General Macdonald in 1876, in order to connect Dunalastair with his property of Crossmount, on the opposite bank of the Tummel, and which was purchased from Colonel Macdonald of St Martins in 1875. The bridge must have cost a large sum of money, as an immense rock had to be blasted on the Dunalastair side before a start could be made with the bridge itself. The road leads to Aberfeldy, Kinloch-Rannoch, and to Craig-an-Tuathanich, and "Macgregor's Cave," by what is perhaps the most beautiful walk upon the property. Here and there are glimpses of the river, fine trees, open spaces, views of the house and mountains, glades and rocks, distant peeps of water and valley, and ultimately of Loch Rannoch and its weird background. At one point the road is continued over a ravine by a novel bridge or conduit, made with enormous labour by the water being allowed to pass through masses of stone, covered with earth to the height of about 40 feet, giving the whole the appearance of an ordinary road, with a tunnel for the passage of the burn below. On the south side of the burn there is a very pretty waterfall, and the burn itself runs in a north-easterly direction through a dell of great beauty, the water being completely hid from view by over-hanging woods of natural birch and oak. Continuing this road for a few hundred yards further, "Macgregor's Cave," the historical retreat of the outlaw, is reached. Here a neat house, shielded by rocks, so as to be scarcely visible, has been built at considerable expense, and is of some utility for pic-nicing purposes. Everything has been done at this point to heighten the scenic effect,

QUEEN'S VIEW AT LOCH TUMMEL.

the view being almost unrivalled for variety and extent. Two hundred feet beneath, the Tummel surges and roars as it passes through " Macgregor's Leap," another of those places where the celebrated outlaw is said to have jumped the river when hotly pursued. The ledge on which he is believed to have alighted has been cut away to provide for the better drainage of the Rannoch Valley, but the height from which he is said to have leaped is from 20 to 30 feet above the stream. The whole

Valley of Rannoch, with the giant hills surrounding it—from the top of the loch to the "Queen's View" at Loch Tummell—is seen at a glance from the seat at "Macgregor's Cave," and almost the entire extent of the Dunalastair property is spread out before us—the mansion appearing as a gem in the most gorgeous setting. The road to "Macgregor's Cave" was only made in 1878, so that it is quite a new attraction to the property. Although the whole course of the Tummel is sufficiently beautiful to make it one of the most notable rivers in Perthshire, the famous Falls of Tummel, as well as the

FALLS OF TUMMEL.

"Queen's View" of the loch, are best known. The falls, which are a few miles below the loch, are ranked amongst the most picturesque in the country, and are a favourite resort with tourists.

At Crossmount—which, as we have stated, was acquired by purchase in 1875—there is a wood of historic interest, it having afforded a safe hiding-place to King Robert the Bruce when he was a fugitive in Rannoch. The names of several places about Dunalastair, Dalchosnie, and Crossmount—which are all united in the one property—bear testimony to the district being a regular resort of Bruce and his followers. The "Queen's Pool," in the Tummel, a short distance from Dunalastair House, takes its name from its having been a ferry used by the Queen of Robert the Bruce, when she, along with her Royal husband, were secreted in a hut erected in a gully just opposite the pool, and to which they fled after the King had slain Red Comyn at Dumfries, and after his defeat at the Battle of Methven. The remains of the hut are still visible, and the Gaelic name by which it is known ("Seomar-an-righ") signifies the "King's Hall." In another portion of the estate, the remains of a rude circle are to be seen, supposed to have been the ruins of the Castle of Bruce, or some of his fortifications. General Macdonald came upon this ancient fortification by accident, and has had considerable excavations carried out about the ring, but without any special results. The place appears to have been of some strategetic importance in defending Bruce from surprise in his place of concealment by the river side. On one occasion Bruce gained a notable victory over the English and renegade Scots at Dalchosnie, or about three miles from Dunalastair. Indeed, the name of Dalchosnie is derived from this circumstance, signifying as it does "The Field of Victory," in the same way as Innerhadden signifies the point where the battle began; and Glen Sassein, the glen by which the English approached to the scene of conflict. Wallace was also a frequenter of the district, but no place seems to be specially identified with his exploits. The timber at Crossmount is much superior to that in other parts of the property,—the larches, in particular, being exceptionally good. Outside the garden wall there are several above the ordinary size. The largest has a girth of 10 feet 3 inches at 1 foot and 8 feet 3 inches at 5 feet, with a height of 120 feet. Another girths 11 feet at 1 foot from the ground and 9 feet 3 inches at 5 feet from the ground, with a clean stem of about 90 feet. In the policies there are many beautiful park trees, although none of them are particularly heavy. The largest of the beeches girths 11 feet 3 inches at 1 foot from the ground and 9 feet 3 inches at 5 feet. Ash, sycamore, and Scot fir trees have a circumference of about 8 feet; oaks, about 9 feet; and silver firs, about 7 feet. In front of the house there is a nice purple beech, and a splendid *Abies Menziesii* about 40 feet high. The plantations extend to about 900 acres, and are in a most thriving condition.

Besides the extensive improvements in the way of road-making, a large outlay has also been incurred by the present proprietor in fencing. During the last five years between seven and eight miles of Morton's patent wire-fences, with straining-posts, have been erected. Several thousands of pounds have

KINLOCH-RANNOCH.

also been spent in drainage since General Macdonald came into possession; and, in addition, a large sum of money has been expended upon river embankments, &c. In the village of Kinloch-Rannoch,—a very quiet and pretty summer resort,—of which General Macdonald is proprietor, a very extensive addition has been made to the Macdonald Arms Hotel, and many other improvements are still in progress or in contemplation.

All of these improvements are being carried out at a time when the military duties of General Macdonald are exceedingly arduous, he being at present the Commander of the Forces in Scotland. The General, we may add, belongs to a family which has long been distinguished for the gallant military services of its members, being descended, through the Macdonalds of Keppoch, from Alexander, son of John, Lord of the Isles, by the Lady Margaret, daughter of King Robert II. of Scotland. Alexander Macdonald of Dalchosnie, the great-great-grandfather of the present representative of the family, fell at Culloden. One of his sons was also killed in action; another, John Macdonald, was an officer in the Highland Army of 1745-6; and a third, Donald Macdonald, distinguished himself at Waterloo as the officer commanding the 92nd Highlanders. The father of the present laird, Sir John Macdonald, K.C.B., of Dalchosnie and Dunalastair, was a distinguished General in the Army, and Colonel of the 92nd Highlanders. The present laird himself has had a distinguished military career. When barely sixteen years of age he received his commission as Ensign in the 92nd Highlanders; became a Lieutenant in the same regiment in the following year, 1847; and in March 1848, was appointed aide-de-camp to his father, a position which he held till 1854. In the latter year he was appointed aide-de-camp to General Sir John Pennefather, and served with him in the Crimean campaign; was present at the Battle of Alma, where he was wounded; and at the battle of Inkermann, where he was again severely wounded; and at the siege and fall of Sebastopol. He was appointed Major of the Rifle Depot Battalion at Winchester, and afterwards became its Lieutenant-Colonel. He was Assistant Adjutant-General at Dover, and afterwards aide-de-camp to H.R.H. the Duke of Cambridge, until his promotion to the rank of Major-General in 1877. He has the Crimean medal, with clasps for Alma, Inkermann, and Sebastopol, and the fifth clasp of the Medjidie and Turkish medal. He is now Major-General Commanding the Forces in Scotland, and, as such, took a prominent part in organising the Royal Review in August 1881, when he received the very highest compliments for the way in which the troops were handled; and had the honour of entertaining the Duke of Cambridge for several days at the beautiful place which we have been endeavouring to describe. His Royal Highness visited the village of Kinloch-Rannoch and the mausoleum of his comrade-in-arms, the late Sir John Macdonald, and sailed to the head of Loch Rannoch, the little village being decorated with flags, and the royal salutes being fired from the quarries above the pier in a most effective manner. The Royal Standard floated on the Rannoch waters for the first time on the occasion of this visit, which was the first that one of the members of the reigning Royal Family has paid to this beautiful valley.

XXXVI.—GLENLYON.

Garth—The Ancient Castle—" The Wolf of Badenoch "—Fortingall—
The Birthplace of Pontius Pilate—The Roman Camp—The Fortingall
Yew, the oldest tree in Europe—Extent of the Glen—The First
Improver—Troup and Glenlyon—The Macgregor's Leap—The Black
Wood of Chesthill—Protecting the Road with Trees—The Castle of
Carnban—A Bleak Spot—Arboreal Milestones—Culdares—Meggernie
Castle and Policies—The Plantations—What might yet be done in
Glenlyon,

LEAVING the property of Sir Robert Menzies at Coshieville, our journey naturally lies in the direction of Glenlyon, where, as we hope to be able to show, there is much that is interesting in connection with our present inquiry. Before entering the glen, there is also a good deal to command our attention. On leaving Coshieville, we turn to the left, cross Keltney Bridge, and enter the property of Garth, recently purchased from Mr Thomas Duff by Sir Donald Currie, the Member of Parliament for the county of Perth. Garth is intimately associated with our national history. Here may still be seen the fine remains of a very ancient stronghold, known as Garth Castle, which, for solidity of construction, and strength of position, is not to be surpassed. The keep or tower, of which only three sides remain, stands on a sloping triangular hill, about 100 yards long, facing the south, and on a bold promontory (of which the base is about 60 yards) formed by two branches of the Keltney Burn, deeply sunk in rocky channels, which thus constituted formidable moats ere joining each other some 60 yards below the Castle, where the noisy torrent proceeds, in a similarly rocky bed, for about 2 miles, when it flows into the River Lyon, a mile and a half from the junction of the latter with the Tay. The walls of the keep still remaining are from 60 to 70 feet high from the ground inside, but the earth has accumulated through the débris of the masonry falling in from time to time. A sloping passage of about 2 feet wide, which served as the staircase, is formed in the centre of the wall, and which evidently also went round the centre of the south and west walls, affording access to the different storeys of the building. The remains of the arches of the different storeys are still plainly visible, and the walls throughout are from 6 to 7 feet thick. The very fact of the staircase being in the thickness of the walls indicates, to the opinion of competent antiquaries, an extreme antiquity, for no such arrangement is met with in those Castles built at a comparatively later age. It is concluded therefore that "Garth Castle" was built at least 500 years ago, in the period when the Norman Barons exercised such independent sway far north of the Border; and it is believed that the Castle was erected by Alexander Stewart, Earl of Buchan and Lord Badenoch when, in 1379, he obtained the lands of Garth, and lands in Rannoch and elsewhere within the Athole district, by charter from his father, Robert II., for had the Castle been built at a later period by one of his descendants, there would certainly have been found evidence of the fact, considering the number of deeds still extant relating to each of them. It is also known to this day as "Caisteal-a'-Chuilein-Churta" or "The Castle of the Fierce Wolf," which quite accords with Lord Badenoch's historic appellation of "The Wolf of Badenoch." Mr Charles Poyntz Stewart in his handsome work, *Historic Memorials of the Stewarts of Forthergill*, recently printed for private circulation, and from which the above details have been gathered, considers, after a critical examination of the evidence, that he is fully justified in arriving at the conclusion that this stronghold owed its origin and name to "The Wolf of Badenoch." A tradition still exists in the district that Garth Castle, formed the prison house to which Lord Badenoch was consigned by his father, the king, as a punishment for burning Elgin Cathedral, and until he performed the penance enjoined by the Bishop, and made peace with the church. It would appear that he was not released from this imprisonment till his brother, Robert III., succeeded to the throne. Of the fact of his imprisonment there is no doubt, but as to Garth Castle being the actual prison house, there is only the tradition to support it. "The Wolf" appears to have sincerely repented of his lawlessness, for he was thought worthy of being interred in Dunkeld Cathedral, where a handsome monument to his memory is still to be seen. Major-General David Stewart, was a most distinguished member of the race of Stewarts descended from "The Wolf." Besides being the author of the familiar *Sketches of the Characters, Manners, and Present State of the Highlands, with Details of the Military Service of the Highland Regiments*, he is well known as a celebrated soldier. He was born in 1772 at Drumacharry House, Sir Donald Currie's residence, now called by him "Garth Castle." General Stewart was the head of the Drumacharry line of Stewarts claiming descent from "The Wolf." The direct male line of John Stewart of Garth and Fothergill, grandson of "The Wolf," and who died in 1475, through his eldest son, Neil Stewart of Garth, failed in 1577, and the succession devolved on the descendant of the above John's second son, Alexander Stewart of Bonskeid, whose male representative in the Athole district is now Mr Stewart Robertson of Edradynate, on whose property are the remains of an old castle having the same topographical characteristics as Garth Castle, and which is believed to have also been one of the strongholds of "The Wolf." The property was in the possession of the above-mentioned John Stewart in 1465. General Stewart entered the army in 1789 as an Ensign in the 42nd Regiment, and served in it with the highest distinction, rising step by step till he retired in 1814 with the rank of Colonel. On the death of his father and elder brother he succeeded to the family property in 1822, and three years afterwards he was promoted to the rank of Major-General. He was shortly afterwards appointed Governor of St Lucia, where he succumbed to fever in 1829. With the death of General Stewart, his branch of the Stewarts became extinct in the direct male line, and the property ultimately came into the hands of Sir Archibald Campbell, Bart., of Ava. He was also a distinguished soldier. After having seen a good deal of active service, he was, in 1824, appointed Commandant of the expedition against the Burmese, in which service his laurels were chiefly won. At the close of the war he received the thanks of both Houses of Parliament, of the Governor-General of India, and of the Court of Directors of the Honourable the East India Company, besides getting from the latter the handsome pension of £4000 a year, along with a gold medal. He became the laird of Garth, by purchase, and built the house of Drumacharry, which has recently been much enlarged by Sir Donald Currie, from plans prepared by Mr Heiton, Perth. In 1831, Major-General Sir Archibald Campbell was appointed Governor of New Brunswick, where he remained for about six years. He died in the year 1843, and the property was afterwards purchased by Colonel Macdonald Macdonald of St Martins, by whom it was sold to Mr Duff, who in turn sold it to Sir Donald Currie.

Although the property is not very extensive, there is some fine timber upon it. There are a number of very good larch and hardwood ornamental trees round Garth House. The Lombardy poplars, although not so numerous as they once were, are worthy of a passing notice. Several pretty large trees, principally larch, Scots fir, and ash, are to be found growing by the side of the burn, near the mansion-house. About six years ago, a very nice plantation was laid down by Mr Duff, a little to the north-west of the mansion-house, the wood being chiefly larch and Scots fir. A little further on, near the Free Church, there are a few large and aged sycamores. We next pass Glenlyon House property, belonging to Colonel Garden Campbell of Troup and Glenlyon House, and reach

> "Famed Fortingall, whose aged yew
> Still braves the tempest's shock."

Even apart from its "aged yew," Fortingall is a place of very great interest, as here we meet with the most northern known works of the Romans, and many valuable discoveries have been made bearing upon the Roman invasion of the country. It is even said, with some show of authority, that no less a personage than Pontius Pilate was born in this remote Highland district. The story told concerning it being the birthplace of the Roman Governor of Judea in the days of our Saviour is very circumstantial, and there is no reason to believe that it may not be absolutely true. We are told that a short time previous to the birth of Christ, Cæsar Augustus sent an embassy to Scotland, as well as other countries, with the view of endeavouring—what has been so often tried since—to effect a universal peace. The Roman ambassadors are said to have met Metellanus, the Scottish King, in this region, one of the ambassadors being the father of Pontius Pilate. As the story goes, a son was born to the ambassador at Fortingall while he was sojourning there on his laudable mission, and it is asserted that the son was the veritable Governor of Judea whose name is handed down to us in Holy Writ. It is, at all events, certain that such a mission was sent to Scotland by Cæsar Augustus about the time of the birth of Pontius Pilate, and that Metellanus received the ambassadors at Fortingall, where he was hunting and holding Court. The ambassadors brought rich presents with them, and the Scottish King, who was desirous of friendly relations with the Masters of the World, sent valuable gifts to the Emperor in return, and was successful in obtaining "an amitie with the Romans, which continued betwixt them and his kingdome for a long time after." The tradition may, therefore, be perfectly true. The remains of the Roman Camp are pointed out by the natives, with no small pride, although it requires some examination to trace its outline—

> " No towers are seen
> On the wild heath, but those that fancy builds,
> And, save a fosse that tracks the moor with green,
> It nought remains to tell of what may there have been.
>
> And yet grave authors, with no small waste
> Of their grave time, have dignified the spot
> By theories to prove the fortress placed
> By Roman hands to curb the invading Scot."

The camp is traditionally said to have been formed by Agricola, who fought a battle with the Caledonians in the neighbourhood. Many interesting Roman remains have been found from time to time in and about the site of the camp. Of these may be mentioned a Roman standard, the shaft of which encloses a five-fluted spear, and which is preserved at Troup House. In the prætorium of the camp was found a vase of curious mixed metal, and in shape resembling a coffee-pot. This was found about 1733, and is preserved in Taymouth Castle. Of late years a number of urns and flint arrow-heads have been picked up in and around the camp. The camp is situated about a quarter of a mile west of the village, the outline of the camp being about 1½ acre. The ramparts are almost entirely levelled with the ground, but can still be traced. The prætorium is remarkably complete, as also the marks of a deep fosse, which is supposed to have surrounded Agricola's headquarters. The ditch or outer trench is now in many places filled in, so that its course is not so easily followed.

The great object of interest to us in the meantime is the old yew in the churchyard. It is believed to be the oldest tree in Europe, and its age is estimated as high as 3000 years! The earliest accounts of this wonderful tree were given almost simultaneously, but independently, by the Hon. Daines Barrington, a Barrister, afterwards on the English Bench, and by Pennant. Barrington's account appears in a letter published in the *Transactions of the Royal Society* for 1769, in the course of which he says:—" I measured the circumference of this yew tree, and therefore cannot be mistaken when I inform you that it amounted to 52 feet. Nothing scarcely now remains but the outward bark, which hath been separated by the centre of the tree's decaying within these twenty years. What still appears, however, is 34 feet in circumference." Pennant also saw it in 1769, and in his description, published in 1771, he states that when he saw the tree its remains measured 56 feet in circumference. The middle part was decayed to the ground, but he states that, within memory, it was united to the height of 3 feet, Captain Campbell of Glenlyon having assured him that when a boy he had often climbed over or rode on the then connecting part. "Our ancestors," he proceeds, "seem to have had a classic reason for planting these dismal trees among the repositories of the dead; and a political one for placing them about their houses. In the first instance they were the substitute of the *Cupressus sempervirens;* in the other, they were the designed provision for the sturdy bows of our warlike ancestors,

> " Who drew
> And almost joined the horns of the tough yew."

Strutt, in his *Silva Britannica*, also refers to the Fortingall yew, and one of his etchings represents the tree with a much greater amount of foliage and branches than in the rough illustration given by Pennant ; so that the branches of the two divisions of the tree intertwine, and form one grand leafy head, resting on two hollow shells of trunk, which face one another. The gap in the trunk, however, is represented to be greatly larger than by Pennant, and with a funeral passing through it —the remark being made that this was the practice when funerals entered the churchyard. The late Mr Patrick Neill, who saw the tree in 1833, has given an account of its condition at that time. From this description, the tree appears to have undergone lamentable destruction during the few years which had elapsed since the date of Strutt's drawing. This observer says that large arms had been removed, and even masses of the trunk carried off, to make drinking-cups and other curiosities. In consequence, "the remains of the trunk present the appearance of a semi-circular wall, exclusive of traces of decayed wood, which scarcely rises above the ground. Great quantities of new spray have issued from the firmer parts of the trunk, and young branches spring up to the height perhaps of 20 feet. The side of the trunk now existing gives a diameter of more than 15 feet; so that it is easy to conceive that the circumference of the bole, when entire, should have exceeded 50 feet. Happily, further depredations have been prevented by means of an iron rail, whith now surrounds the sacred object." The Rev. Robert Macdonald, parish minister of Fortingall, in the New *Statistical Account of Scotland*, gives some additional information regarding the partial destruction of this venerable tree. Writing in 1838, he says :—" At the commencement of my incumbency, thirty-two years ago, there lived in the village of Kirkton a man of the name of Donald Robertson, then aged upwards of 80 years, who declared that, when a boy going to school, he could hardly enter between the two parts ; now a coach and four might pass between them ; and that the dilapidation was partly occasioned by the boys of the village kindling their fire at Baeltainn at its root." It is very sad to think that a tree which has survived the storms and vandalism of 3000 years, and concerning which so much interest is felt, should suffer at the hands of mischievous boys. The tree, we are glad to say, is now thoroughly protected, and is taken charge of by the parish minister, the Rev. David Campbell.

The latest scientific writer on the Fortingall yew is Sir Robert Christison, Bart, who, in the course of a paper on "The Exact Measurement of Trees," read at the Edinburgh Botanical Society on 1st July, 1879 and appearing in their *Transactions* gives a very minute description and measurement of the tree. The result of his examination is to convince him that the present apparently inconsistent condition of the ruins of the tree may be reconciled with the old descriptions. "Little

FORTINGALL YEW.

information as to its rate of growth is to be got from sections of the Fortingall yew itself. The only available parts remaining are the outermost portions of the old trunk,—representing its growth probably long after it had become a shell, and consequently impaired in vitality,—or the branches, in which in all trees the annual rings of wood, for the same periods of time, are much narrower than in the trunk. In all parts of the tree which I have examined the rings are very fine, so fine in general as not to be counted without the aid of a lens." In concluding his paper, Sir Robert remarks :—" It is better to use the general rules formerly arrived at, according to which the tree, in the first place, may be assumed to have attained a girth of 22 feet in a thousand years. After that age, no information yet got warrants a rate of more than an inch in 35 years Taking the lowest measurement of Barrington at 52 feet, the difference will thus add 2000 years to the age of the Fortingall yew, making it in all 3000 years old when measured in 1768–9. The result is startling, but not so improbable as may at first be thought, if it be considered that several English yews of scarcely half the girth are not without good reason held to surpass materially a thousand years of age, yet still appear to be in vigorous health, and steadily increasing; and that upwards of 3000 rings have been actually counted on the stump surface of a Californian *Sequoia*."

The wasted shell of this venerable tree still supports a wonderful amount of vigorous vegetation. The larger of the two shells of the ancient trunk forms about a third of a circle, measures 6 feet from one corner to the other, and bears crowded branches, which are covered with dense foliage, and form a head of 24 feet high. The other shell has a most fantastic form, and bears a vigorous head 16 feet in height. Outside the church wall, but almost touching it, is a fine young tree. It is about 28 feet from the nearest part of the old trunk, and is usually spoken of as the product of a shoot from the root of the old tree. Sir Robert Christison, however, says that he is not aware that the yew is prone to throw up suckers from its distant root-branches, and is disposed to believe that the younger tree sprung from a seed. It is a handsome, thriving tree, at least 150 years old, with a cylindrical trunk somewhat grooved, 53 inches in girth at 5 feet from the ground, branching at 8 feet, bearing a fine leafy head pointed like the pines in form, and 35 feet in height.

The history of the Fortingall yew can only be matter of conjecture. It is not unlikely that it may have been a venerable object of heathen worship before the introduction of Christianity, and that this led to the selection of the site of the building for

Christian worship. The yew has for very many ages been associated with funereal solitude ; and, for all we know to the contrary, the ground surrounding the Fortingall yew may have been a burying-place for a very great period. The yew is so associated in men's minds with ideas of death, that the poets hardly ever refer to it without giving expression to some ghostly sentiments. Blair, for instance, addressing himself to the grave, says—

> "Well do I know thee by thy trusty yew,
> Cheerless, unsocial plant, that loves to dwell
> 'Midst skulls and coffins, epitaphs and worms,
> Where light-heeled ghosts and visionary shades,
> Beneath the wan cold snow (so fame reports),
> Embodied thick, perform their mystic rounds :
> No other merriment, dull tree, is thine."

Gray's lines on the same sepulchral theme are more familiar :—

> "Beneath those rugged elms, that yew tree's shade,
> Where heaves the turf in many a mouldering heap,
> Each in his narrow cell securely laid,
> The rude forefathers of the hamlet sleep."

Sir Walter Scott, and almost every one of our best-known poets, speak of the yew in the same dismal tone ; and in the most ancient records we find it venerated as an emblem of the abode of the dead. As still further illustrative of the sacred character of the yew, a writer in *Notes and Queries* recently brought under public notice a superstitious belief which existed in some parts of the North of Scotland. "The yew," says this writer (Mr Edwin Lees, F.L.S., Worcester), "from its position in churchyards in proximity to the church, would seem to have been considered a consecrated tree, symbolical of eternal life. A curious Scottish superstition in reference to the yew as a consecrated tree is mentioned in the 'Recollections of O'Keefe,' published in *Ainsworth's Magazine*, vol. iii., as related by him. It is stated to be an idea in the north of Scotland that a person, when grasping a branch of churchyard yew in his left hand, may speak to any one he pleases, if he desire to do so, but, however loud he may call, the person spoken to will not be able to hear what is said, though the words will be audible to all around. O'Keefe mentioned that a man who wished to prejudice the clan against their chief without receiving punishment for his rashness, approached the chief when all his people were around him, and, bowing profoundly, as if to show his devotion, with the branch of yew in his hand, spoke in the most insulting and defiant manner for all around to hear. The result of this strange experiment may be easily conceived, but it is not stated." Whether the Fortingall yew was originally worshipped by the heathen natives of the district, and preserved to us from superstitious veneration, is now of little account. We have the fact before us, that so far as can be ascertained it is the most aged vegetable growth in Europe, and may probably have put forth its infant shoots as far back as the time of King Solomon. On this account it is one of the most interesting objects in Perthshire, and we trust that no thoughtless person will ever have it in his power to lay violent hands upon it, or disfigure it any further.

Glenlyon is not only the longest, but it is one of the grandest of those secluded valleys which abound in our Scottish Highlands. It extends in a westerly direction from the head of Fortingall to near Tyndrum, a distance of close upon 35 miles. The glen is indebted for its name to the River Lyon, which flows through it, and which rises on the south-east side of Ben-a-chastle, in a long south-westerly projection of the parish of Fortingall, close on the boundary with Glenorchy, in Argyleshire. The river runs for nearly 2 miles across the projection, and then describes the segment of a circle over a distance of $5\frac{3}{4}$ miles, between Fortingall on its left bank, and the most westerly section of Kenmore on its right. At midway of the $5\frac{3}{4}$ miles the river expands into Loch Lyon, which is $2\frac{1}{2}$ miles long by nearly 1 mile broad. On issuing from the loch, which is well stocked both with salmon and trout, the river continues its course through the glen till it joins the Tay, $2\frac{1}{4}$ miles from its cognominal lake, the entire length of the course of the Lyon being nearly 40 miles. As the whole of the glen is exceedingly narrow, the breadth of the river is at no place very great, and at some points the mountains encroach so much upon the struggling river, that there is only a space of 8 or 10 feet for its passage. Throughout its entire course, the river—now boisterously careering over rocky impediments, and now gliding joyously along its pebbly bed—flows amidst scenes of romantic grandeur and peaceful beauty, affording to the artist inexhaustible sketching ground, and to the poet an everlasting theme. History and tradition lend their valuable aid to increase the enchantment. The remains of numerous forts testify to the importance of the glen at a former period, and tradition speaks of many bloody encounters within its rugged defiles. The Lyon, which now sparkles so brightly, once ran red with the blood of the Macgregors, when, after a terrible struggle, the conquering Stewarts of Garth drove the kinsmen of Rob Roy out of the territory, wailing as they fled,

"Glenstrae and Glenlyon no longer are ours."

Up till a comparatively recent date, say less than 150 years ago, Glenlyon appears to have been almost destitute of timber, and, although it cannot yet be said to be as extensively planted as it might be, a very great improvement has been effected. As we shall afterwards endeavour to show, the credit of having first directed attention to the planting of Glenlyon is due to the same Mr Menzies of Culdares with whose name is associated the introduction of larch into this country. He was the proprietor of the greater part of the glen, and judging from the many noble trees which now embellish the property, and the system upon which they have been planted, he appears to have been a gentleman of superior taste, and possessed of a knowledge of arboriculture much beyond the age in which he lived, although very little information can now be gleaned regarding him. But he was not the only proprietor who, at this time, sought to utilise the waste land by planting, although there can be little doubt that the exertions of the other landowners in the glen were turned in this direction mainly through the influence and example of Mr Menzies. On the property of Colonel Garden Campbell of Troup and Glenlyon there is some very fine timber. On the hill-side above Fortingall there is a nice plantation consisting principally of hardwood of 50 or 60 years' standing. At Glenlyon House there are several fine ash and sycamore trees; and along the east bank of the river there is a fine strip of plantation, principally larch about 20 or 25 years old, extending from the Bridge of Lyon to the "Macgregor's Leap." "The Macgregor's Leap" is reached shortly after entering the glen, and claims attention, not so much on account of the distance which the haunted fugitive had to spring, as from the rugged footing on both sides of the river. At this point the glen opens up beautifully, and exposes, in front the Black Wood of Chesthill, which, with its timber of rich and varied tints, covering the slopes and heights, the dark recesses and the walls of rock, contributes in no small degree to the grand picturesqueness of this part of the glen. The wood extends to 127 acres, consisting principally of hardwood, Scots fir, spruce, and larch. The quality of the timber, especially the larch, is of the very best. Near Blackwood Cottage there is a remarkably good beech, girthing 17 feet 3 inches at 3 feet from the ground, and having a bole of 20 feet. There are several other large clumps of wood and single trees, principally beech, ash, and elm, scattered over the property. The larger single trees skirt the roadside, and the plantations rise from the river side almost up to the summit of mountains about 1700 feet above the level of the sea. The greater part of this wood was planted between the years 1730–40 by the Mr Menzies of Culdares of whom we have been speaking, and whose work of improvement in the glen here commences, Chesthill having then been part of his property.

At this place, and afterwards all along the road to Meggernie Castle, we were much struck with the systematic manner in which Mr Menzies appears to have gone about his work. The road is frequently cut on the face of precipitous hills, and we noticed that wherever the road was the least likely to be dangerous to travellers, he planted a row of trees for their protection. At that time, the land along the roadside was without fences, and but for the forethought of this keen arborist, there would not have been the slightest protection against accidents in those parts where they were likely to prove most serious. Many of these trees have now been cut down, but those that remain are sufficiently numerous to indicate the object of their being planted at such a spot. Although the greater part of the wood at Chesthill was planted by the gentleman to whom we have referred, some fine clumps of Scots fir and spruce were planted recently by the late Mr Stewart Menzies of Chesthill.

A little farther on we come to the ruins of the Castle of Carnban, formerly held by the Macgregors, whose stronghold is said to have been at Roro, farther up the glen. There is no authentic history connected with these ruins. It is said, however, that the Castle had a thatched roof, which was set on fire by a hot arrow during one of the many conflicts with neighbouring clans. The ruins are situated on a prominent spot at the top of the hill, on the side of which there is a round mound beautifully clothed with hardwood trees, with a few Scots firs intermixed. Continuing our journey, we soon reach Invervar, one of the most delightful spots in the whole glen, and which has been transferred to canvas by Mr Smart, R.S.A. The Lyon here assumes the character of a pretty broad stream, having a little island in the centre, covered principally with spruce and birch, while on either side of the river the bank is lined with tall, graceful birch trees. A steep heath-clad hill rises on the south side, and on the north there is a fine plateau. Looking up the glen, one ridge of dark, bleak mountains is seen projecting beyond another, until the view culminates in the rounded Craigellich, which seems to block the western passage. At this part of the glen a small plantation was laid down by the late Mr Stewart Menzies of Chesthill about twenty years ago, and contributes very much to the beauty of the scene.

We next come to Ruskich,—signifying "To Strip,"—where the wild Highlanders are said to have, at a critical moment in a fight, divested themselves of their habiliments, that they might continue the combat unencumbered. Here, at Duin Ruskich ("The Knoll of Ruskich"), we have a clump of spruce trees, but which have suffered severely from the great "Tay Bridge storm." There is also a very nice row of beeches along the roadside, planted for protection. To the west of Ruskich there is another small knoll in a very thriving condition, the wood consisting principally of larch between 40 and 50 years old. After this the country becomes comparatively barren for a few miles, so far as timber is concerned—the low-lying lands, however, being under cultivation. The absence of timber is here most noticeable, and although we occasionally meet with a solitary fir or birch

> "Moor'd in the rifted rock,
> Proof to the tempest's shock,"

it only serves to make the aspect more cold and melancholy. A few strips of plantations would be a very great improvement to the landscape, and might not be unprofitable. The bleakness of this part of the glen is somewhat relieved by the arboreal "milestones" planted by Mr Menzies of Culdares, who introduced the larch. These trees, which are very good beeches, were planted so as to indicate the distance from Meggernie Castle—one tree signifying one mile, two trees two miles, and so on until within two miles of Fortingall, where eleven trees were planted. At Cambusvrachan we enter what was, until 27th September 1883, the property of William George Steuart-Menzies of Culdares, by whose family it had been held since 1659. At the date mentioned, however, it was sold to Mr Bullough, cotton manufacturer, Accrington, Lancashire, for £103,000, exclusive of the timber, which has yet to be valued. It is the most interesting estate, from an arboricultural point of view, in the glen. There is no part of the glen where the improving hand of Mr Menzies of Culdares is more apparent, and there is no part of the glen where these improvements have been more enthusiastically continued throughout the past century. On entering the estate we immediately notice a beautiful little strip of Scots fir and larch planted alongside the Cambusvrachan Burn, and stretching well up the hill. Along the roadside there are many handsome beeches, and on a small island in the Lyon is a very large solitary ash tree girthing about 22 feet at about 3 feet from the ground.

Behind the shooting lodge at Innerwick there are several acres of a thriving plantation of Scots fir and larch planted ten years ago by the Trustees of the late R. Steuart-Menzies. At Bridge of Balgie there are several notable beech trees, for which we are indebted to Mr Menzies of Culdares. There are also many large natural weeping birches, one of which girths 8 feet $4\frac{1}{2}$ inches at 5 feet from the ground, with 20 feet of a bole. The wood in the plantations is very good, there being 50 acres of larch and 50 acres of Scots fir about 70 years of age. Immediately above the Bridge of Balgie, which connects the Killin Road with Glenlyon at a particularly rocky part of the river, the hillsides are thickly clothed with natural birch, many of the specimens being very large.

Leaving the public road at Bridge of Balgie, we enter the policies of Meggernie Castle, now the seat of Mr Bullough, and drive along a truly magnificent avenue of lime, beech, elm, and spruce trees of great size, and whose wide spreading branches form a complete arch over the approach which skirts the banks of the Lyon. This avenue was planted by Mr Menzies of Culdares, and is about 2 miles in length, circling beautifully round Meggernie Castle, which occupies a splendid situation in a park about 700 feet above the level of the sea. The measurements of several of the trees in this avenue are worthy of being recorded. One of the elms girths 19 feet 2 inches at 1 foot from the ground, and 16 feet 5 inches at 5 feet from the ground. Another girths 16 feet 3 inches at 1 foot from the ground, and 12 feet 3 inches at 5 feet from the ground. The lime trees in the avenue, with their knotty and leafy trunks, present a singularly beautiful and picturesque appearance. Several of them are of large size, one, which may be taken as a fair average of the whole, girthing 13 feet 3 inches at 2 feet from the ground, and 10 feet 5 inches at 5 feet from the ground. The beeches are also of respectable dimensions, one of them girthing 11 feet 3 inches at 5 feet from the ground. The Castle, round which this grand avenue circles, is about 300 years old, and is a fine specimen of the old Scottish baronial mansion. The older portion is a large square block of four storeys, with walls 5 feet thick, and the interior is fitted up with all the ancient requisites in the shape of dungeons, secret apartments, strongly barred gates, &c. One of the former places of horror has been converted into a commodious and well-filled wine-cellar. There is still, however, some evidence of the dreadful uses to which these dungeons have been put in the olden times; for there hangs the ugly hook to which many a Highland plunderer or poor prisoner of war has been suspended; and there is still the hideous cell, with the little feeding-hole in the door, where criminals of the deepest dye were wont to be confined. The entrance to the older part of the Castle is guarded by a ponderous iron gate, and an oaken door 3 inches thick. This older portion was slated as far back as 207 years ago, it having been previously thatched with heather. In this part of the building some beautiful pieces of tapestry, in splendid preservation, adorn the walls. In the entrance-hall there is a peculiar old bog-oak cabinet, elaborately carved with grotesque figures, and some fine old chairs of the same description. About thirty-five years ago a considerable addition was made to the Castle, much in keeping with the older portion.

The policies are ornamented with many remarkable trees. Not the least interesting of these are eight larches planted by Mr Menzies of Culdares at the same time as those at Dunkeld with which his name is associated. The Glenlyon trees have not had the advantage of such fine soil as those at Dunkeld, and they are consequently not nearly so large, but they are particularly fine trees for all that. The largest of the lot originally planted were cut down some seven or eight years ago. There are several very fine beeches and elms, evidently contemporaneous with those in the avenue, and about the same size. Some of the Scots firs in the grounds surrounding the Castle are also very large,—one at the river side, a little to the west of the Castle, girthing 10 feet 5 inches at 1 foot from the ground and 9 feet 8 inches at 5 feet from the ground; while another close beside it girths 9 feet 9 inches at 1 foot from the ground and 9 feet at 5 feet.

The arboreal features of this part of Glenlyon, as in the other parts, do not consist so much in the largeness of the individual trees as in the extent of planting which has been done in comparatively recent years. Previous to the days of Mr Menzies of Culdares,

> " The birch, the wild rose, and the broom
> Wasted around their rich perfume,"

throughout almost the entire glen, although many parts were utterly destitute of even this primitive covering. Now, however, much of the more unprofitable land, and the rugged hillsides often up to the summit, have been planted with larch or

Scots fir, giving a much more beautiful and warmer appearance to the landscape, and adding considerably to the value of the property.

Immediately on entering the approach to Meggernie Castle, we pass a very fine plantation of this character, the wood being between fifty and sixty years old, and of excellent quality. A little to the west of the flower garden, which is situated on nicely sloping ground at the back of the Castle, there is a magnificent plantation of Scots fir over 100 years of age. Great havoc was made in this wood by the storm of 28th December 1879. Adjoining this wood to the west, the Trustees of Mr R. Steuart-Menzies, under the direction of Mr P. M. Conacher, the factor, planted extensively ten years ago, so as to afford greater shelter, and provide a better background to the picturesque old Castle. These young trees, Scots fir and larch, are coming away beautifully, and already add very much to the beauty of the place. Above this young plantation there is a grand wood of about 1000 acres, consisting of larch and Scots fir of forty years' standing. Previous to this the ground was regular moorland, but with a good depth of soil in the lower parts. The wet portions were all thoroughly drained, and the ground was as well prepared as it could be under the circumstances. The trees were originally planted up to the skyline, but owing to the great exposure, they have not grown quite up to the summit, which is 1960 feet above the level of the sea. There are still, however, many trees quite close to the height of 1960 feet above the sea-level, being considerably higher that Turlum Hill near Crieff, hitherto reputed to be the highest wooded hill in Scotland. This is about the highest point in the glen proper, but at an elevation very little below this are several ridges,

"Where towering firs in conic forms arise,
And with a pointed spear divide the skies."

In front of the Castle, but on the opposite side of the river, there is a large extent of rich woodlands, the timber being chiefly Scots fir, intermixed with birch. Here there are several very large trees of great age, the measurements being much the same as the firs we have already noticed. A little to the south-west of the Castle there is a fine belt of Scots fir of about 70 years of age; and farther up the glen there is another small plantation of Scots fir, laid down about twelve years ago by Mr R. Steuart-Menzies, the father of the late laird. Along the south side of the river, near this spot, there is a row of magnificent beeches,—extending to about 100 yards in length,—evidently originally intended to be connected with the avenue leading from the approach. It is presumed that when Mr Menzies of Culdares planted those beeches, it was his intention to divert the river, by cutting off a sharp bend, so as to extend the avenue in this direction.

There is another very nice plantation, laid down by the late Mr R. Steuart-Menzies, about twenty-five years ago. It is entirely of Scots fir, and forms a delightful background to the Gallin Falls. These falls are within a few minutes' walk of the Castle, and are crossed by a small suspension bridge. The scene at this point is very grand. The Lyon rushes along a frightfully rugged bed, and forces a passage through adamantine rocks in many fantastic ways. In the deeper pools, the kingly salmon may be seen lurking, to recruit his strength for another of those desperate leaps which he has to take before gaining the quieter waters of the loch. Immediately above the wood, which forms a direct background to the falls, is another large area of old Scots fir, in a very flourishing condition, crowning the hill between the roads to Loch Lyon and the farm of Lochs. Immediately beyond the farm of Gallin there is a small plantation of Scots fir and larch, with a nice avenue of sycamore and elm, planted five years ago by the Trustees of the late Mr R. Steuart-Menzies. The following is a note of the extent of the woods on the Culdares property :—Cambusvrachan, about 4 acres; Innerwick, 8 do.; Kerromore, 7 do.; Bridge of Balgie, 15 do.; Barnpark, 15 do.; Dalreach, 25 do.; Castle Park, 30 do.; Gallin, 40 do.; plantation above the Castle, 1000 do.; natural Scots fir and birchwood opposite the Castle, 800 do.; and Cnocnocoirre (or resinous) Knoll, meaning the Knoll of Resinous Wood, 400 acres—in all, about 2350 acres.

The journey through the remaining portions of the glen is devoid of arboreal interest, the land being almost entirely pastoral. At Gallin Falls a small glen branches off to the right, and here the Connat flows through a lovely dell, and pours its torrents over a series of beautiful waterfalls. At the top of the dell are Lochs Damh and Girrie, two nice sheets of water, from which the Connat chiefly draws its supplies, and which afford good sport to the angler. After fording the Connat, we drive through a beautiful pastoral district, and pass Cashlie, where there is an old feudal tower, known as Caisteal M'Niall (Castle M'Neill). Along the whole of the route, we now find the remains of old feudal towers, namely, Caisteal Con a Bhacain, Caisteal an Duibhe, Caisteal an Deirg, and Sithean Camslai. At the farm of Innermean, a little beyond Pubil, we ascend Fingal's Seat (1070 feet), from which a most extensive view of the glen, east and west, is to be had. Loch Lyon is also in the midst of this farm. At the east end of the loch we have to quit our trap, and walk along a rough road skirting the side of Benvauich. On reaching the west end of the loch we turn a little to the north-west, and a mile and a half farther on we come to the march between the counties of Perth and Argyll, where the Culdares property terminates. After a short rest we descend the hill, and, about 6 miles from where we left our conveyance, we reach Auchbridge, and find ourselves within 3 miles of Tyndrum Station on the Callander and Oban Railway.

Much has been done to render Glenlyon one of the most beautiful, as it is the most extensive, of the Scottish glens, but a great deal yet remains to be done. Although there are some fine patches of arable land along the margin of the river, the glen, as a whole, is essentially pastoral, the blackfaced sheep and the sturdy West Highland cattle roaming over the hills free from all restraint. It deserves to be noticed, however, that the climate is so congenial in the low grounds that the less hardy Ayrshire cattle thrive amazingly; so much so, indeed, that Mr P. M. Conacher, the factor on the Culdares property, maintains one of the best Ayrshire herds in Scotland. The climate is generally wet and bleak, but the glen is not so much exposed to the severe storms of winter as one would naturally suppose; and the roads, notwithstanding the precipitous hills on either side, are seldom blocked with snow. The glen being chiefly pastoral, the absence of shelter over great tracts of the country cannot but be injurious to flockmasters. To the ordinary visitor, the absence of timber is specially noticeable along the sides of the numerous burns, and the many picturesque spots where, to quote the Poet Laureate,

> "The long waterfalls
> Pour in a thunderous plunge to the base of the mountain walls."

The exceptions which are met with only bring out in greater relief the many barren spots, and show how much beauty may yet be imparted to the glen. But the utility of more extensive planting will be chiefly recognised by those who can appreciate the advantage of shelter for the sheep, and the commercial value of the timber; and on these grounds alone the propriety of laying down plantations and clumps at the more exposed or leàst valuable portions of the glen cannot be doubted.

XXXVII.—OCHTERTYRE.

The Beauties of Ochtertyre—History of the Property—The Murrays of Ochtertyre—The Policies—The Mausoleum—A Fearful Tragedy—Craig-na-Cullach — The Principal Shrubbery — The Pinetum — Remarkable Trees—The Loch of Monzievaird—The Old Castle of Cluggy—Burns and the "Aiks" of Ochtertyre.

OCHTERTYRE, where "grows the aik," lies so close to one of the great tourist highways, the road between Crieff and Lochearn, that its more conspicuous charms are spread out before the most casual observer. Stretching along the slope of the lower terrace of the Grampians, with a fine southern exposure, its picturesque undulations, its tastefully-ornamented parks, and its abrupt, pine-clad heights, command the attention of all who have an eye for the lovely in Nature. The mansion-house, situated as it is high above the placid waters of the romantic little Loch of Monzievaird, and almost concealed by the surrounding woods, contributes very much to the grandeur of the landscape, which is acknowledged to be one of the most enchanting in Britain. But it is not until the policies are entered that the full beauties of the spot are disclosed; and as the proprietor, Sir Patrick Keith Murray, Bart., places few restrictions upon visitors, its inner attractions are not unknown. These attractions, wherever practicable, have been greatly enhanced by the hand of Art. The roads are skilfully led throughout the park, so that they are almost concealed from the mansion-house, and every obstruction to the expansive views around that could possibly be removed have been taken out of the way. The trees in the park, too, are so grouped as to be artistically-beautiful and harmonious, while the giants of the policies include some of the grandest specimens of their kind—

> "Is there a spot in Scotia fair
> So full of beauty rich and rare,
> Where Nature with a lavish hand,
> Has formed a perfect fairy-land?
> All gaze with wonder, and admire
> Thy beauties, lovely Ochtertyre!"

Ochtertyre has been in possession of the representatives of the present family, a branch of the house of Tullibardine, for upwards of four centuries. It is a notable circumstance that the father has been uniformly succeeded by his eldest son without any collateral relative intervening; and it is a no less notable circumstance, as we learn from two black marble tablets in the family mausoleum, that of the first four generations of the line, from 1424 to 1547, the Christian names of the heads of the family have been Patrick and David alternately, and the succeeding eleven generations have been Patrick and William alternately. Up till 1739 the family name was spelled "Moray," it having been altered to its present form by the third baronet. The Morays, as we have formerly had occasion to point out, acquired large possessions in Athole and Strathearn, and branched off in various distinguished families. The first of the Ochtertyre line owes his existence to the union of Sir David Murray of Gask and Tullibardine, who was knighted by James II. of Scotland in 1424, with Isabel, daughter of Sir John Stewart of Innermeath, Lord of Lorn. Patrick, the first laird of Ochtertyre, was the third son of this marriage, and he obtained the lands from his father, on his marriage with Isabel, daughter of Balfour of Montquhanie, ancestor of the Burley family. Patrick died in 1476, leaving a son, David, who became the second laird of Ochtertyre. David married Margaret, daughter of Henry Pitcairn of Pitcairn and Forthar and left two sons—Patrick, the third laird of Ochtertyre, and Anthony, who founded the Dollerie branch of the family. The eighth laird of Ochtertyre was a man of more than ordinary ability, and during his reign the family acquired a higher status than it had hitherto enjoyed. In 1669, he purchased the lands of Fowlis-Easter from Patrick, ninth Lord Gray, whose family had suffered severely by fines and oppression under the Cromwellian usurpation. In 1673, this laird, William, was created a baronet of Nova Scotia, with remainder to his heirs male. The succeeding laird, Sir Patrick, was a friend of the Revolution of 1688, and the Revolution Government entertained so good an opinion of his sagacity and integrity of character that he was one of those selected to distribute public money amongst the Highland chiefs to induce them to remain quiet. By his prudent and economical habits he amassed a fortune of about £18,000, which he expended upon the purchase of the barony of Monzievaird and other lands. His manorial seat was Fowlis Castle, where all his family were born. He sat in the last Parliament of Scotland as one of the four representatives for Perthshire. Although Sir Patrick was a friend of the Union and the Hanoverian succession, his heir, William, espoused Jacobite principles, and became a stout partisan of that party. He joined the Earl of Mar in 1715, and fought at Sheriffmuir, where he was made a prisoner, but soon after obtained his pardon. Sir William was highly educated and accomplished, having studied at Oxford, and made the tour of France and Italy, but the tenure of his inheritance was brief. It was he who altered the spelling of the family name, as already mentioned. When the Rebellion of 1745 broke out, the fourth baronet, Sir Patrick, held a commission as a Captain of the 42nd Regiment, and was taken prisoner by the victorious Rebels at Prestonpans, but was allowed to go on his parole to Ochtertyre. Sir William Murray, the fifth baronet, was also a soldier, but on the death of his father in 1746 he left the service for the purpose of devoting himself to the management of his property, to which he applied all the powers of an acute and cultivated mind. He was the promoter of everything tending to the improvement of the estate, and introduced the modern system of husbandry. He built the old mansion-house of Ochtertyre, which stood about 300 yards north-west of the present one. The sixth baronet, Sir Patrick, was born in the old house. He was also a great friend of rural improvement, and at the same time took a very active part in public affairs. In 1799, he was appointed King's Remembrancer for Scotland; in 1806 he was elected M.P. for the city of Edinburgh; in 1808, he was one of the Trustees for Manufactures and Fisheries; in 1810, he was appointed Secretary to the Commission on Indian Affairs; and in 1820, he became one of the Scottish Barons of Exchequer, a dignity which he held till his death in 1837. The younger brother of this laird, George, was the most notable member of the family, having obtained distinguished honours as a soldier and a statesman. After being educated at the High School and University of Edinburgh, he entered the army in 1789, as an Ensign in the 71st Regiment, exchanging subsequently into the 34th Regiment, and again to the Foot Guards. Beginning his career on the eve of the long war with Revolutionary France, he had soon the opportunity of distinguishing himself, and his promotion was rapid. In 1795, he became aide-de-camp to General Campbell on the staff of the expedition intended for Quiberon, but was sent to the West Indies, under Sir Ralph Abercromby. He shared in the expedition to Holland in 1799, and received a slight wound in battle. He afterwards received an appointment in the Quartermaster-General's department under Sir Ralph Abercromby, whom he accompanied to Egypt, taking part in the struggle which closed with the decisive battle at Alexandria. In the expedition to Copenhagen, he was Quartermaster-General to Sir John Moore, and fought with him in Portugal. He next served under Wellington throughout the Peninsular War. In 1812, he was promoted to the post of Major-General; and in 1813 he was created a Knight of the Bath. At the close of the war, he was appointed Governor of Canada, but as soon as he learned of the return of Bonaparte from Elba, he recrossed the Atlantic with troops for Europe, and, although arriving too late to be present at Waterloo, he was able to join Wellington before the entry of the British into Paris, and remained with the Army of Occupation until 1817. On returning home, he was repeatedly elected as M.P. for his native county of Perth, and held the posts of Secretary for the Colonies and Master-General of the Ordnance under the Duke of Wellington. In 1841, he was raised to the rank of General. He latterly devoted some attention to literature, and was the editor of the *Letters and Despatches of the Duke of Marlborough*. Returning to the lairds of Ochtertyre, Sir Patrick Murray was succeeded by his son William, who held a commission in the Black Watch, but ultimately retired from the army. By his marriage with Helen Margaret Oliphant, only child and heir of Sir Alexander Keith of Dunnottar and Ravelston, Knight-Marischal of Scotland, the family became intimately connected with one of the great historical houses of Scotland, Sir William at the same time assuming the surname of Keith. Sir William was universally recognised as the *beau ideal* of a country gentleman, taking a very great interest in the management of his property, and endeavouring to promote the interests of all around him. He was also distinguished for his devotion to literature and science, was an accomplished artist, and an excellent musician. He exerted himself in various ways to cultivate in others the talents which were so conspicuous in himself, personally undertaking the instruction of pupils who showed an aptitude for special studies and accomplishments. In 1851, he commenced the erection of a fine observatory at Ochtertyre, which was completed in the following year, and to which the public were freely admitted. He died on 16th October 1861, deeply regretted by all. The observatory was taken down shortly after his death. He was succeeded by his eldest son, the present Baronet, who was born on 27th January 1835. In early life, he entered the army, bearing a Captain's commission in the Grenadier Guards, but he has since retired. On the death of his mother, he succeeded to the Dunnottar and Ravelston estates, which have since been sold. He is the fifteenth in descent from the founder of the family.

Entering the policies by the eastern gate, we pass along a beautiful avenue leading directly to the mansion-house. Midway between the entrance gate and the mansion, we reach the family mausoleum, an appropriate and substantial erection occupying the site of the old Parish Church of Monzievaird—the scene, in 1511, of the fearful tragedy which Sir Walter Scott has narrated in his introduction to *The Legend of Montrose*. There was a long-standing feud between the Murrays and the Drummonds, and the result of a sanguinary battle fought at Knock-Mary, an eminence on the south side of the Earn, near Crieff, was that the Drummonds were defeated. While the Murrays were returning home with their booty, the Drummonds were unexpectedly reinforced by a body of Campbells who had come to revenge another quarrel with the Murrays. Observing their danger, the Murrays took refuge, unperceived, in the church, with their wives and children, but as their enemies were about to disperse, one of the Murrays, unluckily, shot an arrow from a window of the church, killing a Campbell, and revealing the presence of the fugitives. The building was surrounded and set on fire, and 160 men, with their wives and families, perished. Only one of the Murrays escaped, and that through the connivance of a Drummond, who was so persecuted by his clan for this act of humanity that he had to take refuge in Ireland. The church was afterwards rebuilt, and continued the Parish Church until the beginning of the present century. Before proceeding further through the policies, we ascend the finely-wooded hill of Craig-na-Culloch ("The Cock's Hill"), which commands a prospect that could not be excelled anywhere. Southwards the view is only limited by the Ochil range, which is seen stretching from Dunblane to Dunning, with the Fife Lomonds standing prominently behind. Between the spectator and the Ochils there is spread out as fine an expanse of country as the eye could wish to rest upon, embracing the whole of the town of Crieff, and its gorgeous surroundings, with Muthill, Auchterarder, and the woods of Tullibardine in the distance. To the west, the view extends over the Forest of Glenartney to Ben Voirlich, Ben More, Ben Ledi, and other mountains scarcely less elevated. At the foot of the slope lies the Loch of Monzievaird, covering some 30 acres, reposing in dreamy quietude, and reflecting on its surface the foliage of the graceful trees which skirt its shores. At every point the eye catches some new beauty as it wanders over this extensive panorama, and it is almost with a pang that the back is turned upon so ravishing a spectacle. The hill is chiefly covered with young timber of the commoner varieties.

The principal shrubbery, laid out by the present proprietor's grandfather, over eighty years ago, is one of the most interesting spots within the policies. It extends from the mansion to the garden, exactly a quarter of a mile. Grass walks, 12 feet broad, divide it almost to the middle, where it breaks off into three large divisions, each separated by grass walks of the same width, joining together at the upper end. Of late years portions which had become overgrown by the commoner varieties of deciduous and evergreen shrubs, have been rooted out and levelled, with the view of having a complete pinetum. One division has been planted with nearly all the varieties of *Abies* and *Picea*, while the entire half of the top or upper division is being planted with the true pines. In one of the centre divisions upwards of thirty distinct varieties of *Taxus* and their allies have been planted, and are making most satisfactory growth, a great attraction being a line on one side of Golden Irish Yew, which has a charming effect. On the lower or narrow part are two broad borders on each side of the walk above referred to, in which are planted many of the older and more interesting shrubs, and the greater number of the fine specimens of hollies for which this place is so famed, the collection including nearly 100 varieties. Amongst the most notable trees in the portion set off as a pinetum is an *Abies Morinda* or *Smithiana*, supposed to be one of the original trees raised by the late Mr Smith, gardener, Hopetoun. It is in a most healthy state, and girths 6 feet 6 inches at 1 foot from the ground, and 5 feet 5 inches at 5 feet, the height of the tree being 39 feet 6 inches. There is a fine *Abies Menziesii*, 57 feet high, with a girth of 6 feet 10 inches at 1 foot, and 4 feet 10 inches at 5 feet. There is also a very nice *Wellingtonia gigantea*, with a girth of 8 feet 7 inches at the ground. In the same neighbourhood, on the side of the terrace walk, there are three very good cedars of Lebanon, measuring as follows:—No. 1, 9 feet 7 inches at 1 foot from the ground, and 8 feet 5 inches at 5 feet, with a bole of 11 feet; No. 2, 9 feet at 1 foot, and 8 feet 1 inch at 5 feet, with a bole of 19 feet; and No. 3, 9 feet 2 inches at 1 foot, and 8 feet 5 inches at 5 feet, from the ground, with a bole of 14 feet. There are also in the principal shrubbery, several fine hardwood trees. The most notable of all is a grand old ash, which has braved the storms of about 400 winters. It has a girth at 1 foot from the ground of 34 feet 10 inches, and at 5 feet the girth is 20 feet 8 inches, with a spread of branches of about 70 feet. It stands at an altitude of 380 feet on a loamy soil, with a subsoil of trap rock, and has a southern exposure. A slit about 5 feet from the ground reveals that the trunk is hollow, but the opening is too narrow to do more than admit a man's arm. The curious are sometimes tempted to push through a stick or umbrella to ascertain the extent of the cavity, and sometime ago, a lady was unfortunate enough to let go her hold of an umbrella, which was heard to fall, with a thud, within. Here it must remain until the tree succumbs to the axe or the ravages of Time, the skeleton perhaps giving rise to some curious speculations in the distant future. Proceeding a little farther, we were attracted by a fine elm bush growing out of a 1-inch jumper-hole in a large boulder from which curling-stones are made. Immediately outside the hole, the bush has a girth of 7 inches, and it shows a yearly growth of 3 inches. The marvel is where the bush draws its sustenance. Besides the old ash above referred to, there is another old tree of the same variety of considerable size, but so covered with ivy that its girth cannot be accurately ascertained, but it may be roughly set down as 15 feet 6 inches at about 3 feet from the ground. There is a fine larch, also well covered with ivy, with a girth of 14 feet 3 inches at 1 foot from the ground, and 13 feet at 5 feet, and a bole of 24 feet. A very nice Spanish chestnut girths 11 feet 6 inches at 1 foot from the ground, and 9 feet 9 inches at 5 feet. There is also an unusually handsome purple beech girthing 8 feet 3 inches at 1 foot from the ground, and 6 feet 9 inches at 5 feet, with a beautiful straight bole of 18 feet. The shrubbery is further ornamented with a number of very good Portugal laurels. One has a girth of 7 feet 1 inch at 1 foot from the ground, and 4 feet 6 inches at 5 feet, with a clean bole of 12 feet; and another girths 6 feet 3 inches at 1 foot, and 5 feet 2 inches at 5 feet.

In proceeding from the shrubbery to the mansion-house, the way lies along a splendid terraced walk, 10 feet broad, extending east and west for 330 yards, and terminating at the west in an octagon built of strong masonry, and coped with granite boulders. A very fine effect is produced by the masonry being covered with ivy, some of the plants of which have been propagated from those growing at the study window at Abbotsford. In the park between the terrace walk and the lake, we noted the following excellent beeches:—No. 1, 15 feet 9 inches at 1 foot from the ground, and 13 feet 2 inches at 5 feet, with a bole of 5 feet; No. 2, 15 feet 11 inches both at 1 foot and 5 feet, with a bole of 6 feet; No. 3, 14 feet 11 inches at 1 foot, and 10 feet 7 inches at 5 feet, with a bole of 12 feet; No. 4, 13 feet 11 inches at 1 foot, and 11 feet 1 inch at 5 feet, with a bole of 7 feet; No. 5, 14 feet 3 inches at 1 foot, and 12 feet 2 inches at 5 feet, with a bole of 9 feet; and No. 6, 15 feet 8 inches at 1 foot, and 11 feet 7 inches at 5 feet from the ground, with a bole of 6 feet. There is also an exceedingly pretty fern-leaf beech between the house and the lake. Two Scots firs make the following measurements:—No. 1, 10 feet 5 inches at 1 foot from the ground, and 8 feet 8 inches at 5 feet, with a bole of 12 feet; No. 2, 11 feet 3 inches at 1 foot, and 9 feet 6 inches at 5 feet, with a bole of 15 feet.

Our way next lies in the direction of the Loch of Monzie-vaird, with its 30 acres of bright water studded with beautiful islets, and indented with charming little bays. Sir Patrick Keith Murray very generously grants permission to a number of the leading inhabitants of Crieff, and those who keep summer lodgings, to fish upon this fine sheet of water, providing them with keys that they may use the boats in the house when desired. The surroundings of the lake, as has already been pointed out, are exceedingly pretty, the ruins of the old Castle of Cluggy rendering the scene peculiarly picturesque. The peninsula upon which the castle stands anciently bore the name of "The Dry Isle," probably in contradistinction to "The Cairn," or artificial islet opposite to it in the lake, and which tradition alleges to have been the prison of the castle. In the year 1467, the castle was termed an old place of strength—*antiquum fortalicium*—and there is a tradition that it was once held by the Red Comyn, whom Bruce slew in the church at Dumfries. For a considerable number of years after the death of the first Laird of Ochtertyre, the Dry Isle appears to have been in other hands than those of his family. Some legal documents, dated 1488, show that David Drummond, son of the Lord Drummond, alleged that the isle and castle pertained to him by reason of tack, and he held possession accordingly, the Lords of Council sustaining his allegations. Later on, in 1525, the Tullibardine family are found in possession of the isle and the loch; and in 1542 they were included within the barony of Trewin, which was erected in favour of that house. About a hundred years afterwards, they are found in full possession of the family of Ochtertyre. At that time the castle was still intact with its fosse and drawbridge, and during the war with Cromwell the ancient fortalice was the abode of the laird, who found it a place of considerable security in a troubled country. On the south of the lake is the scene of the Battle of Monzievaird, where Kenneth IV. was killed, in 1003 ; and at the west end there is a large mound marking the spot where the victims of the plague which ravaged the district during the reign of Charles I. are buried. The lake is girt with a varied and most suitable selection of trees, the richly contrasting foliage producing a delightful effect. A broad bank bounds the north-west portion of the lake, thickly studded with large oaks, believed to be those which excited the admiration of Burns when he wrote :—

> "By Ochertyre there grows the aik,
> On Yarrow Braes the birken shaw ;
> But Phemie was a bonnier lass
> Than Braes of Yarrow ever saw."

This bank was formerly loaded with shrubbery and the common varieties of trees, but about ten years ago, Sir Patrick, very judiciously, cleared these all out, so as to bring the "aiks" into greater prominence. Many of these oaks are grand trees, ranging from 3 to 4 feet through at the surface of the ground. The largest one girths 18 feet 1 inch at 1 foot from the ground, and 16 feet 2 inches at 5 feet, with a bole of 12 feet. It was in June 1787 that Robert Burns visited Ochtertyre, where he was hospitably entertained by Sir William Murray, the fifth baronet, and Lady Augusta. Sir William's fair relative, Euphemia Murray of Lintrose, who was afterwards married to Lord Methven, one of the Judges of the Supreme Court of Scotland, was then on a visit to Ochtertyre, and her charms powerfully inspired the muse of the poet. In addition to "Phemie, the blythest lass," the national poet has left another souvenir of his brief sojourn at Ochtertyre in the lines on scaring the waterfowl on Loch Turret :—

> "Why, ye tenants of the lake,
> For me your watr'y haunt forsake ?
> Tell me, fellow creatures, why
> At my presence thus you fly ?
> Why disturb your social joys,

> Parent, filial, kindred ties?—
> Common friend to you and me,
> Nature's gifts to all are free:
> Peaceful keep your dimpling wave,
> Busy feed, or wanton lave:
> Or beneath the sheltering rock,
> Bide the surging billow's shock."

When Burns walked the shores of Loch Turret, the scene was that of a Highland desert, "welcome from its loneliness to the heart of the poet," but now it is richly clad with thriving pine, and is a regular resort of the tourist. The same service has been rendered to other parts of the property where the necessity or expediency of planting was apparent, and the work of extending these plantations is one which is constantly in progress.

XXXVIII.—LAWERS.

The Beauties of Lawers—History of the Property—A Thrilling Incident—
The Old Chapel—Early Planting—Notable Trees—"The Queen"—
The Largest Scots Fir in the World—Newer Coniferous Trees—
Plantations.

No traveller along the public road between Crieff and Comrie can escape noticing and admiring the magnificent situation of the mansion-house of Lawers, the seat of Colonel David Robertson Williamson. Nor can he fail to observe that the imposing building, with its tasteful turret and pillars of the Doric and Ionic orders, is in every way worthy of the situation. While the mansion is the first object that will strike the eye of a stranger, the mind is immediately diverted from its artistic design and elegant proportions to its gorgeous surroundings,—gorgeous in arborous beauties of the rarest kind, and in a combination of all those natural features which have made our Scottish scenery famous. Standing on the lower ridge of Fordie Hill, at an altitude of about 250 feet above sea-level, its windows command a prospect difficult to excel for extent and variety. Far away to the west is seen "lone Glenartney's hazel shade," with the cleft and cairn-crowned top of Ben Voirlich peeping over the rugged hills. With the aid of a good field-glass, the red-deer may occasionally be seen from the drawing-room windows as they roam in majesty over the ancient and Royal Forest of Glenartney. Across the plain on which the village of Comrie nestles amidst the richest foliage, the dark hills of Aberuchill cast their shadows, but only to heighten the effect of the scene. Southwards, a great part of the Vale of Strathearn, with its serpentine river, fertile fields, and umbrageous slopes, lies before us, while Turlum raises its pine-covered shoulders towards the east. The house itself, with its fine colonnade, is, as we have said, a large and imposing building. It was originally a plain old Scottish mansion, the main portion being one of the oldest buildings in Strathearn. The house was remodelled in 1738, from a design by William Adam. At that time the possessor was Colonel Campbell who fell at the battle of Fontenoy, after having only had the privilege of sleeping in his new house one night. Various alterations have since been made upon the building. When in possession of Colonel Williamson's grand-uncle, Lord Balgray, a Senator of the College of Justice, a new face was put upon it, transforming it into the present picturesque structure. Lawers was not the original name of the property, this name being given by a branch of the Campbells of Glenurchy, who left the foot of Ben Lawers several centuries ago, and settled in this district.

The estate at present covers about 50 square miles of country, —including fruitful arable land, rich and beautiful timber, lonely moorland, and bold rocky mountains,—but, at the time of the Reformation, it was known to have been much larger, extending along the north shore of Loch Earn, and reaching the waters of Loch Tay. The property was then in the hands of a branch of the Campbells, and the laird of that day, Colonel Campbell of Lawers, was distinguished as a warm friend of the Protestant cause, and held the command of a regiment of the Reformers. His successor, Sir John Campbell of Lawers, was knighted by James VI. in 1620, and in 1633 was created Earl of Louden, through marrying Margaret Campbell, Baroness of Louden. He afterwards became Chancellor of Scotland, and took a prominent part in resisting the arbitrary measures of Charles I. He was also so firmly opposed to Cromwell that he was excepted by him out of his Act of Grace, but pardoned in 1654. At the Restoration he was deprived of the Chancellorship, and fined in the sum of £12,000 Scots as well. His brother, who succeeded him at Lawers, was equally hostile to the Commonwealth. He raised a regiment to oppose the "Protector," by whom, however, he was defeated, with heavy loss, at Inverkeithing. His son and successor, Colonel Campbell, was also a man of note, and regarding him a very thrilling story is told. While in the discharge of some duty, he gave grievous offence to the powerful Clan M'Gregor. Determined to be avenged, the M'Gregors collected a strong force, and attacked Lawers House at midnight, with the intention of murdering the unfortunate Colonel. Regardless of the heartrending entreaties of his wife, they were about to deal a death-blow to their victim, when he begged of them to be conducted to the little chapel on the property in order to perform his devotions before dying. The request was complied with, but on the way to the chapel he succeeded in inducing his captors to accept a ransom of 10,000 merks, to be paid at a tavern in Balquhidder on the following Monday. Punctual to his engagement, he appeared with the money, but he had arranged that while the money was being paid, a troop of cavalry should surround the house, and capture the freebooters. This they succeeded in doing. His enemies offered to procure a ransom to recover their liberty, but the Laird of Lawers had learned from experience to put no trust in their tender mercies, and they were all conveyed to Edinburgh, and publicly executed. The chapel referred to is one of the oldest ecclesiastical buildings in Strathearn, and was, till the close of last century, the burial-place of the family of Lawers. This interesting ruin stands to the south-west of the mansion, and is credited with being the scene of many a tragedy—

> "When chiefs from far Glenlyon,
> And clansmen from Loch Tay,
> Brought to the fane their offerings,
> And fought the holy day."

The chapel is surrounded by a fine grove of yews, and the design and origin of two parellel and magnificent avenues of oak trees, which stretch from the Earn until they seem to be lost on the wooded heights of Lennoch and Drummond-Ernoch, indicate that they were originally planted as an ornament to the chapel, as they are much older than the mansion. The girth of two of these oaks is 20 feet 8 inches at 1 foot, and 13 feet 5 inches at 5 feet from the ground; and 20 feet 7 inches at 1 foot, and 12 feet 6 inches at 5 feet. Two of the yew trees girth 11 feet 3 inches at 1 foot, and 18 feet 6 inches at 5 feet; and 10 feet 9 inches at 1 foot, and 18 feet 6 inches at 5 feet from the ground.

The estate was acquired in 1784 by General Archibald Robertson of the Royal Engineers. At his death, in 1813, the property devolved on his niece, Miss Boyd Robertson, who, by her mother, was the lineal descendant of the famous Zechariah Boyd, and Bishop Boyd of Glasgow. In 1814, she married her cousin, David Robertson Williamson, to whom we have already referred as a Judge of the Court of Session, under the title of Lord Balgray, and who was the son of Alexander Williamson of Balgray. Lord Balgray studied for the Bar, and passed as an advocate in 1783; Sheriff of Stirling in 1807; and was raised to the Bench in 1811, his title being taken from his property in Dumfriesshire. Lord Balgray was not only a highly-talented Judge, but he was an energetic country gentleman, and did much to embellish the estate and forward agricultural improvements. The first of a large collection of cups and medals in the possession of Colonel Williamson—who succeeded in 1852, on the death of his grand-aunt, Mrs Robertson-Williamson, widow of Lord Balgray—was won in the arena of the Highland and Agricultural Society in 1829 by Lord Balgray, with an ox, which, judging from its representation on canvas, was of no ordinary merit. The efforts of the present laird in the same direction are too well known to require recapitulation. It is not in the showyard alone that he has earned well-deserved distinction. The workers' cottages erected by him are perfect models, and gain the admiration of all who see them; while throughout the entire estate, improvements more or less extensive, have been going on for the last thirty years. Apart from the large sums expended on improvements by Lord Balgray, thirty miles of fencing have been erected on one farm alone,— that of Innergeldie,—and about £50,000 has been spent by Colonel Williamson on buildings, drainage, road-making, fencing, and planting. The object of this extensive fencing is to keep sheep from wandering from one farm to another, thus preventing them from being unduly disturbed by the shepherds.

A good deal of the planting was done by Lord Balgray, but from the monster trees to be met with in great numbers throughout the property, it may safely be concluded that Lawers has been well wooded for centuries, the trees in the neighbourhood of the old chapel, or those originally planted for its embellishment, being very old, and many of them of great size. Some of the grandest of these trees have succumbed to gales in recent years. The grandest of the whole, a sycamore, popularly known in the district as "The Queen," and which stood in the field known as the Pond Park, was blown down by the gale of December 1881. An estate paper describes this tree as having a circumference of 24 feet in the year 1781, when the kitchen garden was sheltered by its branches. When it was blown down, the girth at 1 foot from the ground was 38 feet, and at 5 feet the girth was 17 feet 6 inches. An elm blown down by the same gale girthed 16 feet at 1 foot, and 12 feet at 5 feet from the ground. An ash tree, still standing, near the Pond, has a girth of 25 feet at 1 foot, and 17 feet 6 inches at 5 feet. A fine lime, near the remains of the chapel, girths 24 feet 6 inches at 1 foot, and 18 feet 4 inches at 5 feet. There ere also some grand beech and Spanish chestnut trees in the Long Avenue, one of the former having a girth of 21 feet at 1 foot, and 15 feet 9 inches at 5 feet; and one of the latter a girth of 19 feet 3 inches at 1 foot, and 14 feet 8 inches at 5 feet from the ground. Near this avenue, there is an irregularly-shaped field rejoicing in the name of the "Bellman's Acre," so called because it was occupied for ages as compensation for the services of that functionary. It is believed that "the acre" was sold about the middle of last century, the price received being the loan of a horse for a journey to Perth and back. There are also several magnificent specimens of the Scots fir, Lawers having the reputation of having grown the largest Scots fir in the world. This tree was, unfortunately, blown down in 1850, and it was of such a marvellous size, that in the following year a section of it was shown at the International Exhibition at the Crystal Palace, where it attracted much attention. The girth of the tree when it was blown down was 17 feet 7 inches at 1 foot from the ground, and 13 feet 9 inches at 5 feet. There are still two trees standing which nearly approach to this size—one in the Long Avenue, and the other near the waterfall. The former has a girth of 17 feet 5 inches at 1 foot, and 13 feet 3 inches at 5 feet from the ground; and the latter a girth of 17 feet 3 inches at 1 foot, and 12 feet 5 inches at 5 feet. A few years ago an enormous silver fir was blown down, the measurement as it lay on the ground being as follows:—length of tree, 102 feet; girth at 1 foot up, 15 feet 3 inches; do. at 5 feet, 12 feet; do. at 20 feet, 10 feet 10 inches; and do. at 27 feet up, 10 feet 6 inches. The bole was 54 feet in length, and the cubic contents amounted to 404 feet. From the estate paper to which reference has already been made, we learn that about the year 1731 an enclosure extending to about 60 acres was planted with a variety

SCOTS FIR AT LAWERS.

of trees, which, in 1781, were reported as being in a thriving condition. During the years 1764, 1767, and 1768, a quantity of natural or hagg wood is mentioned as having been cut, and the produce sold for £1500—a circumstance which is of some interest as illustrating the difference in the value of this wood at the end of a century, the amount being probably double what would now be realised.

Many of the newer coniferous plants found an early home at

Lawers. In 1786, an *Abies Canadensis* was planted by General Robertson, and when it was blown down by the Tay Bridge gale, it was regarded as the largest tree of the kind in Great Britain. An *Abies Douglasii* was planted in 1840, and now girths 7 feet 6 inches at 1 foot, and 6 feet 4 inches at 5 feet from the ground. A *Cedrus Deodara* planted at the same time now girths 5 feet 3 inches at 1 foot, and 4 feet 9 inches at 5 feet. Between 1849 and 1866 a considerable number of *Abies Douglasii*, *Cryptomeria Japonica*, and *Araucaria imbricata*, were planted. During 1868, about 200 *Picea nobilis*, and over 300 fancy pines of different varieties, were planted. All the varieties planted are growing well, the *Abies Menziesii* and the *Picea Pinsapo* luxuriating in low damp situations. The *Picea nobilis*, however, is growing with more vigour than any of the other varieties, especially where planted in alluvial soil, and on a declivity with a southern exposure, some of those planted in 1868 having attained a height of 25 feet. The *Picea nobilis* is not eaten by hares or rabbits, although these destructive animals eat the bark of the *Wellingtonia gigantea*, *Picea Nordmanniana*, and *Cupressus Lawsoniana*, to such an extent that it has been found necessary to protect them with wire-netting.

Planting upon an extensive scale was commenced by Lord Balgray in the year 1815, when 600 additional acres of larch, Scots fir, oak coppice, and spruce, were laid down; but this work was evidently undertaken to replace timber which had been cut, as several trees of an earlier date are still standing. About £30,000 worth of timber has already been taken out of this plantation alone. The Scots fir is of especially fine quality, as is indicated by the extreme redness of the bark, and the deep blue colouring of the leaves. Col. Williamson acts upon the principle of clearing the ground in preference to thinning, and replanting the ground,—larch following oak coppice, Scots fir following larch, and *vice versa*. After procuring the plants from the nurseryman, they are kept for three years in the home nursery before being planted out. In planting, care is always taken to dig a proper pit for the roots, the slitting system being carefully avoided. For ten or twelve years the Colonel has discarded the cruelty of traps for destroying the rabbits, as he finds that the ground can be as effectually cleared by ferrets—any rabbits remaining in the holes being dug out. This was a laborious work to begin with, but now the ground has been so honeycombed with these repeated diggings, that very little labour is required to clear out the rabbits at any time. A mixture of cow-dung and hot lime is also applied to the bark of the young plants at intervals until they grow beyond the reach of danger from vermin; and this plan is not only found to be effectual for the purpose originally intended, but is also most beneficial for the plants. The ground throughout the plantations, which extend in all to about 1000 acres, is thoroughly drained by deep surface cuttings, and everything possible is done to promote the free growth of the timber, the Colonel taking a lively personal interest in this as in all other departments of his beautiful property.

XXXIX.—STROWAN.

Situation—History of the Property—Improvements—Tom-a-Chastile—Ancient Marketplace of Strowan—Origin of Name—Remarkable Trees—Plantations, etc.

Strowan, like the other estates in Upper Strathearn, enjoys natural advantages of a very high order. Situated in the centre of the strath, it commands, from various points, a full view of all the beauties of this delightful district, and has within its own grounds attractions that are second to none. The present proprietor, J. T. Graham-Stirling, is a cadet of the house of Airth, being the second son of the late Thomas Graham of Airth, who, till the birth of his second son, inherited the property of Strowan from his maternal granduncle, General Sir Thomas Stirling, Bart., and took his name. General Sir Thomas Stirling was a brother of Sir William Stirling of Ardoch, and purchased the estate from Lord John Murray, afterwards Duke of Athole. The property was left by the General for the benefit of the second son of the present proprietor's father, before Mr Graham-Stirling was born,—his father being in possession from 1808 to 1811, when the present laird was born. Mr Graham-Stirling thus became laird of Strowan the moment he was born, his father occupying the estate as trustee until his son had attained his majority.

During the minority of the present laird, large additions were made to the property, and many valuable improvements were carried out, about 700 acres being planted all round the property, and a great deal of draining and fencing executed. It was his father's desire to extend the boundaries of Strowan, if possible, but, as sufficient contiguous land could not be had at the time, the money to be devoted to this purpose was ultimately invested in the purchase of the estate of Coulgask, part of the Gask estate. The father of Mr Graham-Stirling died in 1836, and in the following year the present laird took possession, having previously been an officer in the Black Watch since November 1827. Immediately on taking up his residence at Strowan, Mr Graham-Stirling entered energetically into the work of further improvements, commencing with an extensive drainage scheme. He has since built almost every farmsteading on the property, and erected almost all the existing fences, most of the hedges at that time having been destroyed by the wintering of hoggs. One of the first, and one of the most important of the improvements to which he devoted himself was the erection of a strong embankment on the north side of the Earn. At this period, the whole of that portion of the estate was liable to be frequently and suddenly submerged by the waters of the Earn, which sometimes flooded the farmhouses to a depth of 2 or 3 feet. The erection of this embankment, which was commenced in 1842, may justly be described as a stupendous undertaking, it being about three miles in extent, and most substantially built. The spirit of the river did all it could to frustrate the work of confining its waters, by rising in revolt two or three times unexpectedly, and washing away the works that had been laboriously constructed. When the work was at last successfully accomplished, it proved to be the greatest improvement that could have been effected, as a magnificent stretch of the strath could then be ploughed and reaped, without fear of the produce being swept away by the overwhelming waters of the Earn. Altogether, Mr Graham-Stirling has spent from £20,000 to £30,000 upon improvements since he entered into possession of the property. More fortunate than his father, Mr Graham-Stirling has been successful in extending the property very considerably, having at different periods acquired the adjoining lands of Mill of Fortune, Auchingarrick, and Drummond-Ernoch (signifying "Drummond the Irishman") — the latter being a small property which was gifted by the Murrays to the Drummond, who, after the burning of Monzievaird Church, had to take refuge in Ireland until the Murrays were strong enough to recal him from exile, and reward and protect him.

While these operations for the general improvement of the estate were in progress, Mr Graham-Stirling was not unmindful of the amenity of his own dwelling. It is, unfortunately, situated quite close to the public highway,—no wall bounding the road, or fence protecting the lawn,—but so tastefully are the grounds laid out, so judiciously are they planted, and so well kept is the road, that it is with difficulty one can realise that the road is anything but a private one. Originally, nothing could be seen from the front windows of the mansion but a thicket of coarse bushes, while the lawn was no more attractive—Mr Graham-Stirling remembering of corn and lint being grown upon it. In 1838, the thicket of wild shrubs was removed, and a very fine view opened up to Turlum Hill. The mansion-house was built by Sir Thomas Stirling, but the present proprietor has greatly extended it. In 1866, a large addition was made to it, a new front, and several additional rooms and offices being erected. The drawing-room window now looks directly across to the fine hill of Turlum on the south, and to the east on Tom-a-Chastile, crowned by the lightning-shattered obelisk to the memory of Sir David Baird, the hero of Seringapatam. The monument, which was erected in 1832 by his widow, Lady Baird Preston, is built of huge blocks of Aberdeen granite, some of the stones weighing 5 tons, and it rises to a height of 82 feet, being an exact copy of Cleopatra's Needle, but without the hieroglyphics. It was struck by lightning on the 28th May 1878, when the top was very much damaged, and a portion thrown down. No attempt has been made to restore it. A writer in *The Beauties of Upper Strathearn* says,—" Not long ago there were some picturesque Scots firs there that added immensely to the effect of the view, and gave a rare beauty to the monument; but it is too little merely to say that they were cut down, for it was a wanton outrage on good taste, and a disfigurement of one of the most beautiful features in the valley of the Earn." The hill of Tom-a-Chastile is believed to have been the site of the ancient Royal Castle of the Earn, a proud fortress of the Earls of Strathearn. It was destroyed by fire, and tradition asserts that many noble persons fell victims to the conflagration, including Joanna,

Countess of Strathearn, daughter of Earl Malise, who married the Countess of the Orkneys. The story of her death is a melancholy and a tragic one, and is thus succinctly told by the author of the work which we have just quoted,—" Joanna had been married to the Earl de Warenne, who seems to have induced her to abandon Scottish loyalty, and enter into a conspiracy against Robert Bruce. She was tried for conspiracy and treason before the Scottish Parliament, at Scone, in August 1320, and, along with Lord Soulis, a fellow-conspirator, was sentenced to perpetual imprisonment. The earldom of Strathearn then reverted to the Crown, and it is believed that Joanna, the unfortunate Countess, was committed a prisoner to the dungeons of her father's castle. A few years ago, when the obelisk was erected, the dust and rubbish and burned *debris* of 500 years were cleared away. The workman removed with difficulty what seemed to them a part of the solid rock, but in doing so they discovered a chamber beneath them. Amid the ashes and dust with which it was thickly strewn, there were a few household articles, some gold and silver ornaments,— betokening the rank of her who wore them,—and the remains of ancient bones. And thus, as is often the case, the monument to honour and bravery rests on the tomb of misfortune."

The policies of Strowan are ornamented with many fine trees, most of the varieties adapted for this country being represented. One of the first trees to attract attention is a very beautiful lime, a little to the south-west of the mansion, and which owes its interest to the fact of it shading the cross which marked the ancient market-place of Strowan before it was transferred to Crieff. This market is a memorial of the day of St Rowan, in the Scottish Roman Calendar, to whom Strowan owes its name. St Rowan lived about the middle of the seventh century, and is described as a learned and brave ecclesiastic. He bequeathed a piece of land for the support of a bellman ; and this land, together with the bell, has come into the possession of Mr Graham-Stirling, who may now be said to be the hereditary bellman. The bell is about 9 inches in height and 5 in diameter, and has no tongue ; so that it must have been struck from without. There is no inscription, and the metal is coarse in quality. The old church of Strowan is a beautiful ivy-covered ruin, a little to the south-east of the house. Amongst those buried in the churchyard is Margaret Murray, daughter of James Murray, fiar of Strowan, to whom we alluded, in describing Ardoch, as having thirty-one children. The remains of Lieutenant Graham-Stirling, eldest son of the present laird, and who fell gallantly at Tel-el-Kebir, while leading his company in the Black Watch, also lie here. The churchyard is surrounded by some very noble trees. Amongst these are several splendid ash trees, the largest of which girths 13 feet 7 inches at 3 feet, the bole being without a branch until a height of 20 feet is reached. There is another ash girthing 12 feet 3 inches at 3 feet, and several others nearly the same size. There are two very large elms—No. 1 girthing 12 feet 1 inch at 3 feet, and No. 2 girthing 10 feet 9 inches at 3 feet. There is also an oak girthing 10 feet 9 inches at 3 feet, with a fine bole of 22 feet. The largest of the beech trees girths 10 feet 8 inches at 3 feet, and the largest lime tree girths 11 feet 8 inches at 3 feet. There are several very fine silver firs, with an average girth of about 10 feet, and a height of 80 feet. Numerous larch and spruce trees also approach the same size. The largest of the sycamores girths 10 feet 3 inches at 3 feet; but there is one at Glasscorry girthing 13 feet 5 inches at 3 feet from the ground. There are several very fine oaks on the farm below Lawers House, the average girth being about 11 feet.

Passing the churchyard, and proceeding for several hundred yards eastward, we get a grand view from the Bridge of Strowan, a gap in the wood laying the whole of the country up the river fairly before us. Some of the finest trees on the property have been blown down. Amongst these was an elm containing nearly 320 cubic feet, and believed to have been one of the finest in Scotland. A sum of £50 was once offered for this tree and an ash which grew beside it. Altogether, about 6000 trees have been blown down by recent gales. Amongst the most recent of the improvements in the neighbourhood of the house is the construction of an artificial lake—5 acres in extent—at Upper Strowan, and which is as useful as it is ornamental, it being utilised for the purpose of driving a sawmill, and for trout-fishing.

One side of the famous Turlum Hill, the finest wooded hill in Scotland (1260 feet), is on the Strowan property. A small portion of the hill was first planted by General Sir Thomas Stirling, and the remainder was planted between the years 1818 and 1827. Originally, the hill was covered with tufty heather and stones, and was of very little value. Now it not only carries a load of magnificent timber, but under the umbrageous shelter there is a rich crop of grass, affording sustenance to a large number of sheep and cattle. At first Scots firs were planted on the top, but they suffered severely from the weight of snow which rested on their branches, and they were gradually thinned out to make way for larch,—a few of the better firs, however, being left.

As is the case with most plantations, the larch here is to a certain extent infested with the bug, the woods with a northern aspect being least affected. Mr Graham-Stirling's opinion regarding the disease is that the larch being suited to an earlier climate than this, the disease is caused by the exudation of the resin being checked by the early frosts, and sending the sap back into the tree. He has experimented on trees, principally silver firs, which had died 5 feet at the top, by slicing them down with a hedge-bill as far as could be reached, with the result that the trees at once commenced to shoot up new tops. There are altogether 700 acres of wood on the Strowan estate proper, and rather less than 200 acres on the other properties connected with it.

XL.—DUNIRA.

The Dundasses of Dunira—Dunmore Hill—Dunira House—The Policies
—The Steading—The Plantations.

SITUATED between Comrie and St Fillans, the estate of Dunira
lies in the centre of one of the grandest spots in Perthshire, and
in itself constitutes one of the chief attractions of a district rich
in everything that is majestic and lovely. Within its boundaries
there is a wide diversity of scenery, Highland and Lowland, and
from the more elevated portions we can see spread out before us
all the unrivalled beauties of Upper Strathearn. The estate is
the seat of an important branch of one of our great historic
families, the present proprietor being Sir Sidney James Dundas,
who succeeded his father as third baronet in 1877. "The
Dundasses," says Lord Woodhouselee, in the *Transactions of
the Royal Society*, "are descended of a family to which the
historian and the genealogist have assigned an origin of high
antiquity and splendour, but which has been still more
remarkable for producing a series of men eminently distinguished for their public services in the highest offices
of Scotland." They are generally believed to have sprung
from the Dunbars, Earls of March, who derived descent
from the Saxon Princes of England. Amongst the first of the
name we read of is Searle de Dundas, who is frequently mentioned in the time of King William the Lion, and who died
early in the following reign. The Dunira branch of the family
trace their lineage to Sir James Dundas of Arniston, whose only
son, Robert, created a baronet on 24th August 1821, was the
ancester of Henry, Viscount Melville. Robert, second Viscount
Melville, sold the property of Dunira, in 1824, to Sir Robert
Dundas, Bart. of Beechwood, whose line is still in possession.
Sir Robert Dundas was one of the principal Clerks of the Court
of Session, and Deputy to the Lord Privy Seal in Scotland.

His only son, David, an advocate, became the second baronet—the present proprietor, as already stated, being the third. The father of the first Lord Melville was Lord President of the Court of Session. His son, Henry Dundas, was, therefore, naturally trained for the bar, to which he was called in 1763, and soon attained a large and lucrative practice. His success in his profession made his promotion rapid. In 1773, he became Solicitor-General; in 1775, Lord-Advocate; and in 1777, Joint-Keeper of the Signet for Scotland. At the same time, he sought Parliamentary honours, being elected Member for the County of Edinburgh in 1774, retaining this seat till 1787, when he was chosen Member for the City of Edinburgh, which he represented till 1802. In 1782, he was made a Privy Councillor, and Treasurer of the Navy; and when Pitt became Premier, he was appointed President of the Board of Control under the new East Indian system. He became Home-Secretary in 1791, and Secretary of War in 1794—a post which he held till the commencement of 1801, when he retired with Pitt. In the following year, he was raised to the Peerage by the titles of Viscount Melville and Baron Dunira, the first title being taken from Melville Castle, in Mid-Lothian, the seat of the father of his first wife. When Pitt again became Premier, Lord Melville was appointed First Lord of the Admiralty. He was afterwards impeached for his management of the funds of the navy, but was acquitted on all the charges by large majorities, and immediately restored to the Privy Council, although he did not afterwards hold any public office. It has been said of Lord Melville that he possessed the greatest share of power ever intrusted to a Scotchman since the Union, except for a short time to Lord Bute. He had a great capacity for hard work, and was a "clear, acute, and argumentative speaker." He died in 1811, and his friends in Perthshire paid a marked tribute to his memory by erecting a magnificent obelisk on the top of Dunmore Hill at Dunira, a little to the north of the village of Comrie. It stands 72 feet high, and is built of a granite-like stone found at Innergeldie, in the neighbourhood. A statue of him was placed in the Parliament House, Edinburgh; and a column, surmounted by his statue, also stands in the centre of

St Andrew's Square, Edinburgh, all testifying to the high estimation in which his services were held.

Although in the very midst of a rugged, mountainous country, which might be supposed to dwarf the proportions of such a monument, no grander spot than Dunmore Hill could have been selected for perpetuating the memory of the distinguished laird of Dunira. From almost every point of view it is a conspicuous object in the landscape, and the giant heights behind only serve to provide it with a nobler background. Dunmore Hill rises somewhat abruptly to a height of about 800 feet, and its sides are beautifully clothed with tall and graceful pines. At its feet, slightly to the south, lies the pretty little village of Comrie, reposing as peacefully at the confluence of the Ruchill, the Lednock, and the Earn, as if it had never experienced the dreaded shock of the earthquake, or was haunted by the memory of the bloody battle of Mons Grampius, which several hold was fought in this neighbourhood—a supposition which receives some corroboration from the presence of traces of a Roman Camp, called Victoria, by certain antiquarian authorities. Immediately at the foot of the hill is the deep ravine through which the Lednock forces its way and plunges into the "Devil's Cauldron," reminding one of the famous lines of Byron—

> "The hell of waters, where they howl and hiss,
> And boil in endless torture—while the sweat
> Of their great agony, wrung out from this,
> Their Phlegethon, curls round the rocks of jet,
> That gird the gulph around in pitiless horror set."

This splendid waterfall, one of the most remarkable in Scotland, is grandly shrouded with wood,—oak coppice, larch, Scots fir, and weeping birches—which conceals the waters so completely, that until they are closely approached their presence is only revealed by the deafening noise of their impetuous rush into the basin, where they hiss, and foam, and shriek, and writhe, "like a demon newly plunged into Tartarus." Terribly wild as the scene actually is, the genial Gilfillan beautifully points out a "little bit of blue," when, in a lecture on his native village, he tells us that above the northern edge of the falls, "where the first plunge begins, you see the clear, blue sky shining through the scattered trees with an ineffably fine effect of relief, and suggesting the contrast between the eternal calm of the heavens, and the turbulence of the earthly life of man." Taking our stand on the huge bare rock on the top, known as the "Drawing-Room," and extending our view beyond the immediate surroundings of Dunmore Hill, we are greatly surprised at the vastness of the extent of country before us. Stretching far over the vale of Upper Strathearn, we decern the whole range of the Ochils, and can even trace one of the Lomonds of Fife peeping over the top. Looking towards Dundee, again, the Law is distinctly visible. On the north, rises the great peaks of Ben Chonzie, and towards the south we look into the mouth of Glenartney Forest. To the south-west, the Aberuchill Hills lift up their massive forms; and in the rear, between the ridges, we can see the conical head of Ben Voirlich towering above its compeers. Westwards, the eye is delighted with the sparkling waters of Loch Earn, as they glisten in the sunshine, and impart a gleam of brightness to the dark mountains around. The wood on Dunmore Hill is principally larch, and some of the trees have attained a great size. The finest larch which we measured girthed 11 feet at 1 foot from the ground, and 9 feet 4 inches at 5 feet. Amongst the hardwood trees, the most notable we met with is a beech girthing 11 feet 10 inches at 1 foot, and 11 feet at 5 feet.

Not the least interesting feature in the landscape as seen from Dunmore Hill, is the most beautiful mansion-house of Dunira, standing conspicuously out at the foot of the larch-clad Crappisch Hills. The old house of Dunira stood a little to the west of the present mansion, and was built by Lord Melville. The new house was built by the late Sir David Dundas in 1852, from designs by Mr Bryce, Edinburgh. It is of the mixed baronial style, and is a very beautiful structure, thoroughly in keeping with its magnificent surroundings. It is built entirely of Bannockburn freestone, which had to be carted all the way from Greenloaning, the nearest railway station in those days. It is three storeys in height, with an underground flat, and is surrounded by ample offices harmonising with the character of the building. The lawn surrounding the mansion is laid out in four fine terraces, the making of which involved a great deal of excavating and carting, the ground having originally been bare rock and heather. The main entrance, beautifully ornamented with heather and shrubbery, was excavated out of solid rock for about 15 or 20 feet, with a breadth of 25 feet. A wash-house, of a very unique description, and concealed by over-hanging trees, was built at the same time as the mansion. It is in the Swiss style, and is roughly built of quartz and hill stones, with hot lime run in. Many beantiful white quartz stones are stuck in throughout the building.—the whole having an exceedingly pretty effect, with the additional advantage of being exceptionally strong.

None of the trees in the policies or the woods are very aged, but in both there are many grand specimens of different varieties. Attention is first attracted to a very beautiful weeping birch, growing on a little mound south of the flower garden. It formerly stood too close to the south-west corner of the house, and was removed to its present site about twenty years ago, it being regarded as far too fine a tree to be destroyed. Even at that time it was an immense tree to be transplanted, its removal actually costing about £50. A large part of this outlay was caused by the tree sitting on a rock, which had to be blasted in order to free the roots. It was transplanted without sustaining any injury, and is now one of the most striking objects in the policies. The girth at 5 feet from the ground is 6 feet 7 inches. Several very beautiful purple beech trees stud the park, and present a nice contrast. A thick shrubbery has been planted on an embankment in front of the house; and on the opposite side of the shrubbery walk is a strip planted as a pinetum, and containing some excellent specimens. There is a nice *Abies Menziesii* girthing 4 feet 10 inches at 1 foot from the ground. *Cedrus Deodara* has attained a height of about 25 feet; and there are several very good araucarias over 30 feet high, as well as some fine specimens of *Cupressus Lawsoniana*. There are three thriving specimens of *Thuja Craigiana*, sent to the late Sir David Dundas from British Columba. A *Picea nobilis* has a height of 51 feet 6 inches. Three very promising specimens of *Wellingtonia gigantea* have also to be noted. The largest reaches a height of 36 feet 6 inches, and girths 9 feet at the ground; the second largest is 34 feet high, with a girth of 8 feet; and the third is 30 feet high, also with a girth of 8 feet. This strip is very suitably situated for the finer varieties of conifers, being admirably sheltered by the embankment and surrounding hills. A really splendid golden yew grows close to the main entrance to the house, the circumference of the plant being 54 feet. Amongst the hardwood trees in the policies the most notable is an elm growing by the side of the approach, and girthing 14 feet 2 inches at 5 feet from the ground, and a number of beech trees at the side of the avenue leading to the the farm, girthing from 10 feet to 13 feet 3 inches at 5 feet from the ground.

At the farm itself there is much that is interesting. In 1879 a splendid new steading was built, and other improvements have been carried out since Mr Charles Dundas, brother of Sir James, became Commissioner on the property. The plans for the steading were prepared by Mr Ewing, Glasgow, and everything is admirably fitted up for the rearing of high-class stock, a fine shorthorn herd having been formed upon a foundation drawn from Keir, Lawers, and other celebrated herds. About twenty years ago, a large hay barn was erected,—amongst the first in the country,—and it has been found so serviceable that in 1881 another was built, double the length, or about 150 feet. Near the steading a beautiful row of workmen's houses were erected about 15 years ago, each house containing three bedrooms, kitchen, scullery, front and back entrances, garden plots, water inside, and other conveniences to be found in houses of a good class in large towns. The landsteward's house, built at the same time, is very conveniently situated between the steading and the workmen's houses. The gardens extend to $4\frac{1}{2}$ acres, have a fine southern exposure, and are sheltered from all the cold winds. There are about 360 feet of glass-houses, including vineries, peach-houses, stove-house, green-house, &c. The family burying-ground is situated in the very heart of the woods, to the west of Dunira House. The situation is rendered very impressive by the majestic trees which surround it on every side; and the cemetery, with its fine avenue of Irish yews about 15 feet high, is very tastefully laid out.

DUNIRA.

The plantations extend to about 1500 acres, and lie principally on the hills from Comrie to within a mile and a-half from St Fillans. A walk through the heart of these extensive plantations proved most interesting. The timber is chiefly larch and spruce, with a mixture of oak coppice. The larch and spruce are almost all full grown, being from 60 to 100 years old, but transit being difficult, and the times far from favourable, the axe is not being brought into use. No further planting has been done, with the exception of small patches, within the past twenty years, but a commencement has just been made with an additional 60 or 70 acres, it being intended to plant about 10 acres each year until the work is completed. The new plantations will all be laid down with larch—poplars, however, being planted on damp sites. The trees being mostly matured, there are, consequently, many grand specimens to be found throughout the plantations. In the wood behind the garden, a common spruce, blown down, girthed 12 feet above the swell of the roots, with scarcely any difference at 5 feet up, while the length of the tree was 108 feet. This was one of very many that looked equally large. Some very fine larch trees also grow here. One of those measured girths 10 feet 8 inches at 1 foot from the ground, and 8 feet 5 inches at 5 feet. The trunk rises to a height of 40 or 50 feet before a branch is reached, and the tree is altogether 120 feet high. A great many of the larches range from 80 to 120 feet in height. There are several extensive and beautiful glens on the property, all of which are richly loaded with valuable timber. It was in one of these glens that Hogg, in his *Queen's Wake*, describes his beautiful creation, " Kilmeny," as wandering up alone :—

> " Bonny Kilmeny gaed up the glen ;
> But it wasna to meet Dunira's men,
> Nor the rosy Monk of the Isle to see,
> For Kilmeny was pure as pure could be.
> It was only to hear the yorlin sing,
> And pu' the blue cress-flower round this spring ;
> To pu' the scarlet hypp, and the hind berrye,
> And the nut that hang frae the hazel tree."

In Glen Boltachan, to the north-west of Dunira House, there are numerous grand waterfalls, the source of the stream being

Loch Boltachan, a nice little sheet of water, only three miles from the mansion. The Boltachan comes down at times with such tremendous rapidity that it has been found necessary, at great expense, to causeway the bed of the stream for half-a-mile at the most dangerous part of its course to prevent sudden flooding. The glen is lined with ash, oak, larch, and spruce, the combination producing a charming effect. A walk has been made on each side of the glen, and numerous bridges have been thrown across the stream at the most picturesque points. In the other glens, the same care has been taken to bring out their beauties, and the plantations are very accessible, being intersected by no less than 35 miles of roads.

XLI.—CARSE OF GOWRIE.

The Carse a late Acquisition from the Sea—Erection of Embankments—Extent of the Carse.—Kinfauns: History of the Property—The Castle—Kinnoull Hill—Kinfauns Long Famous for its Trees—An Old Chestnut—Other Notable Trees.—Gray: Enormous Ash Trees and Cedars of Lebanon—Other Notable Trees.—Seggieden: Collection of Hard-Wooded Plants — Notable Trees. — Fingask: The Castle—Trees and other Objects of Interest—Kinnaird Castle.—Gourdiehill: The Finest Holly Tree in Great Britain—Other Notable Trees.—Megginch: Origin of the Name—History of the Property—Notable Trees.—Rossie-Priory: The Collection of Newer Coniferous Plants.—Castle Huntly: History of the Property—Enormous Scots Fir—Other Notable Trees.—Glencarse.—Pitfour.—Errol Park.—Murie.—Balruddery.—Orchards of the Carse of Gowrie.

THE CARSE OF GOWRIE is more noted for the fertility of its agricultural land than for the extent of its woods and forests. Still, it is not without interest as illustrating the improvement which has been effected upon the face of the country during the past century by judicious planting, and it is not without a respectable catalogue of notable trees. The whole of the carse, indeed, may be said to bear testimony to the improving spirit of the age, as it is within quite modern times that it has earned for itself the distinction of being "The Garden of Scotland." Whether we look at the rich agricultural land, or the great extent of its orchards, it is no misnomer to characterise the Carse of Gowrie as "The Garden of Scotland," although, like the rest of the country, it has not in recent years been equal to its fruitful reputation. This, however, we trust, will only be temporary, and whenever the seasons are favourable for the growth and ingathering of the fruits of the earth, the Carse of Gowrie is likely to hold its own. It is pretty generally believed that the whole of the Carse of Gowrie is but a late acquisition from the sea—the flat face of the country and the names of the places supplying tolerably conclusive evidence upon this point. The names of many of the places begin or end with "inch,"—that is, island,—such as Inchture, Megginch, &c., which, in all likelihood, were the names, with a variation in the spelling—"Inse-tower," "Meg-inse," &c.—they went under when these lands were either islands or sandbanks. The *Annual Register* of 1760 supplies some reliable information as to what the Carse actually was at a not very remote period. "Some old written instruments," we are told, "mention Errol as a place standing to the south of the Tay, though it stands a long mile to the north of the river at present. The inhabitants of the country have a tradition that the course of the Tay in former ages was by the foot of the hills to the north of Errol, and to this day show the very holes in rocks to which the ships' cables were fastened. But if the Tay ran so far to the north, as there is great reason to believe, all the lower ground to the south of Errol would be drowned, and that Firth would be twice, if not thrice, as broad as it is in our times. The inhabitants of Perth remember to have heard their fathers say that in the high Hill of Kinnoull they have seen the remains of staples and rings, with other conveniences for shipping, as in a harbour. At a village two miles above Perth, and far from the Tay, some workmen, draining a peat marsh, found the ring, stock, and shaft of an anchor, with a great log of wood standing erect in the earth, to which it is conjectured the ships' cables were fixed. The children of the workmen are still alive to attest this fact." We do not, however, require to go so far back for satisfactory evidence as to the reclamation of the Carse from the sea. Until about 1825, a great portion of the lands of Pitfour was under water; but in that year the proprietor commenced the erection of strong embankments, which have been the means of adding considerably to the family property. Other lairds carried out similar work. Up till about the year 1780, the sea continued to make serious encroachments upon the lands in the neighbourhood of Errol. So serious were these encroachments that the house and offices on the farm of Daleally had to be removed at two different periods to a more inland situation. About 1836, Mr John Lee Allen, then the principal proprietor in the parish of Errol, set to work to reclaim the land by the erection of extensive embankments, and he succeeded by this means in recovering from the Tay, and adding to his estate, about 100 imperial acres of deep, rich soil. At the same time, a broad belt of reeds was planted along the shores, by " dippling," at an expense of £12 per Scotch acre. Indeed, up till the year 1760, as we stated in the chapter on " The Past and the Present," the Carse of Gowrie was disfigured with large pools of water, and it is from that period that the major part of the work of reclamation may be dated.

The Carse of Gowrie extends the whole way from Perth to Dundee, a distance of 22 miles, and within these limits we shall find much that is of interest. The first estate which falls to be considered is Kinfauns, the property of Mr E. A. Stuart Gray. This estate was anciently a possession of the family of Charteris, to whom it belonged for several centuries. It is understood that the first proprietor of the estate of the name of Charteris was Sir Thomas de Longueville, or Charteris, a native of France, and of a considerable family in that kingdom. He received a grant of the lands of Kinfauns from King Robert about the year 1314. The traditional story regarding him is that, when at the Court of Philip IV., in the end of the thirteenth century, he killed a French nobleman, with whom he had a dispute, in the presence of the King, and that, being refused pardon, he made his escape. For several years he infested the seas as a pirate, under the name of the " Red Reaver," from the colour of his flag, and was at length captured by Sir William Wallace, who interceded for him with such good effect that he procured his pardon about 1301. He accompanied Wallace on his return to Scotland, aided in his principal exploits, and continued ever after his faithful adherent. He appears to have proved equally faithful to Bruce, as a grant of lands was the reward of his bravery and his services. Amongst the relics carefully preserved at Kinfauns Castle is the sword of Sir Thomas Charteris, a ponderous double-edged weapon 5 feet 9 inches long, $2\frac{1}{2}$ inches broad at the hilt, and proportionally thick, with a round knob at the upper end nearly 8 inches in circumference. The property continued in the possession of the Charterises until the seventeenth century, and during that period they took a prominent part in national affairs, and supplied Perth with not a few Lord-Provosts. The property afterwards passed to the Blairs of Balthayock, who had long been the bitter enemies of the Charterises. About the middle of the seventeenth century, the Hon. Alex. Carnegie, a younger son of David, Earl of Northesk, married the daughter of Sir William Blair of Kinfauns, and their descendant, Miss Blair, married, in 1741, John Lord Gray, by which marriage the estate came into the possession of the noble family of Gray. This family is of Norman extraction, the first of them landing in England with William the Conqueror, who was a near relative. Several of the noblest families in England are of this stock, including the lovely but unfortunate Lady Jane Gray. The first of the Grays in Scotland was a younger son of Lord Gray of Chillingham, who came to this country in the reign of William the Lion, about the year 1200. A descendant of this Sir Andrew Gray, espoused the cause of Bruce, who rewarded him with several grants of land, including the barony of Longforgan, in the Carse of Gowrie. The Grays were, accordingly, as conspicuous in the history of their country as the Charteris family. The old Castle of Kinfauns, which was removed at the beginning of the century to make way for the present noble edifice, was a building of very great age. The building of the present Castle was commenced in 1820 by Francis, fifteenth Lord Gray. The Castle was designed by Smirke, and the style of architecture is the castellated, of a simple and imposing character, exceedingly well suited to its elevated site, and to the grandeur of the surrounding scenery. The building stands on a raised terrace, 40 feet in width, with circular bastions projecting at the corners, and a flag tower of 84 feet. The principal rooms occupy the east and south fronts of the Castle, and command a most delightful prospect of the river, and a great extent of the fertile country surrounding it. The greater part of the Hill of Kinnoull is on the Kinfauns estate; we have already had the opportunity, in the course of the chapter on Dupplin, of pointing out the share which a former laird of Kinfauns had in the planting of this beautiful hill, and giving to it that sylvan aspect which is now so much admired. On the Kinfauns side of the hill, the ascent is exceedingly steep, and clothed with a profusion of thriving trees. Large pieces of the rock above occasionally break away, and carry down numbers of the trees along with them. A little to the east of the Castle, the ground begins to rise again, on a steep but smooth ascent, forming the west side of another beautiful hill, clothed all round with splendid timber. This hill is sometimes called the Binn Hill, on account of its conical shape, and sometimes the Tower Hill, from its being surmounted by a picturesque tower upwards of 80 feet high, and built about 1813 by Lord Gray as an observatory. From the top of this hill (which is a place of public resort) a magnificent bird's-eye view is got of the Castle and grounds. A little further to the east there is another hill, bearing a strong resemblance to Kinnoull Hill, and the top of which commands a view extending over the whole of the Carse of Gowrie to the German Ocean. Towards the south and west, the view extends across the fertile lands of Lower Strathearn, as far as Auchterarder Moor. Kinfauns Castle has been famous for its trees from an early period. As far back as 1789, we find Dr Walker making special mention of some very large Spanish chestnut trees. He does not, however, give any details regarding them beyond the statement that the largest of them had been cut down in October 1760, and measured 22 feet 8 inches in circumference. He remarks further, that this tree was supposed by the proprietor to be above 200 years old. When cut, the trunk was found to be entirely decayed, but all the branches had leaves and fruit upon them the year in which the tree was removed. One of these old chestnuts still grows on a bank in front of the Castle, and appears to be perfectly healthy, although it has suffered considerable damage of late, having lost two of its limbs from high winds. It grows at an altitude of about 100 feet above sea-level, on a light loamy soil, with a subsoil of gravelly clay. It is about 75 feet high, and girths 17 feet 3 inches at 1 foot from the ground, and 14 feet 6 inches at 5 feet. The tree has a southerly exposure, and is sheltered from other quarters by rising ground and adjacent trees. Amongst the other noteworthy trees in the immediate neighbourhood of the Castle is a fine larch, a little to the west, rising to a height of about 90 feet, with a spread of branches of from 50 to 60 feet. The girth at 1 foot from the ground is 11 feet, and at 5 feet, 8 feet 3 inches. This tree, however, is evidently giving way, and may endanger the Castle should it yield suddenly to a westerly wind. A remarkably good acacia grows directly under the Library windows, being between 50 and 60 feet high, with a bole of 12 feet, and a girth of 7 feet 8 inches at 1 foot from the ground. To the west of the house there is also a nice specimen of the deciduous cypress, between 30 and 40 feet high. In the shrubbery there are many trees of more than average size. One of the many fine sycamores has a bole of 45 feet, and an entire height of 80 feet, with a girth of 11 feet 7 inches at 1 foot, and 9 feet 3 inches at 5 feet. A very handsome beech girths 11 feet 8 inches at the narrowest part of the bole. A spruce, with 20 feet of a bole before a single branch is reached, and a height of nearly 90 feet, girths 10 feet 9 inches at 1 foot, and 8 feet 6 inches at 5 feet, being one of the heaviest spruces on the property. There are numerous oak trees girthing about 12 feet at 1 foot, and 10 feet at 5 feet from the ground. The other notable trees in the shrubbery include a splendid weeping ash, a large and handsome purple beech, a nice specimen of the variegated maple, two very good cedars of Lebanon, a grand compact Portugal laurel, over 150 feet in circumference; silver stripped oak, &c. There are also several notable trees, particularly elms, in other parts of the grounds. The largest of these elms grows on a bank below the West Approach, and not far from the Castle. It is about 200 years old and has a height of 70 feet, and contains about 460 cubic feet of timber. The girth at 1 foot from the ground, exclusive of all excrescences, is 23 feet, and the girth at the narrowest part of the bole, about 7 feet from the ground, is 17 feet. At about 9 feet from the ground, the tree breaks into two great limbs, the larger of which girths 13 feet, the other being very nearly the same size. The spread of branches is about 100 feet. The soil here is black loam, with a subsoil of sandy loam. The exposure is southerly, and the tree is sheltered from other quarters by rising ground and adjacent trees. It is altogether a splendid elm, and is so much prized that £50 was offered for it about forty years ago. Other two very good elms grow on either side of the garden gate. The larger one girths 11 feet 3 inches at 1 foot, and 9 feet 9 inches at 5 feet; and the other girths 10 feet at 1 foot, and 8 feet 8 inches at 5 feet,—the bole of both trees being about 20 feet. Within the garden there is a beautiful weeping elm of the Camperdown variety. When the grounds were originally laid out, the greater part of the expense was incurred in the planting of the finer varieties of the deciduous trees, and there is, consequently, not a very large collection of the newer conifers. These, however, are well represented by several magnificent specimens in the garden. Amongst them is a splendidly-furnished araucaria, planted 30 years ago, which has now reached a height of 28 feet. Beside it is a most promising *Wellingtonia gigantea*, planted in 1860. Mr Macdonald, the gardener, received a present of this plant in a thumb-pot, when it was no larger than a man's finger, and he kept it in a cool house for two years previous to its being planted out. Four years ago it lost about 8 feet of its leader, but it still has a height of 28 feet. The girth at the ground is 9 feet 3 inches. A fine *Picea Lowii*, planted at the same time as the *Wellingtonia*, also lost about four years' growth six or seven years ago, although it is at present nearly 40 feet high. An attempt was unsuccessfully made to form a leader, but the plant has since formed a very good leader for itself. The tree is growing at the rate of about 2 feet per annum, and is altogether an exceptionally fine plant. The girth at 1 foot from the ground is 6 feet, and at 5 feet the girth is 4 feet. A *Cedrus Deodara*, also planted at the same time, has gained a height of 36 feet, with a girth of 5 feet at 1 foot, and 3 feet 9 inches at 5 feet. A *Cupressus Lawsoniana*, planted seventeen years ago, is also a very symmetrical and handsome tree, with a height of 28 feet, and a diameter of 9 feet. An *Abies Douglasii*, planted 22 years ago, has a height of 70 feet, although it lost its leader about 12 years ago. The stem is perfectly straight, and the growth is from 2 to 3 feet per annum. The present girth is 6 feet 9 inches at 1 foot, and 4 feet 6 inches at 5 feet. There is also a beautifully-spreading specimen of *Abies Canadensis* 35 feet high. At about 2 feet from the ground the trunk, which girths 10 feet 3 inches at 1 foot from the ground, breaks into three stems, two of which girth 5 feet, and one 4 feet 7 inches at the point of division. The plantations are very valuable, and extend to fully 700 acres.

Large and interesting as many of the trees at Kinfauns really are, they are all completely eclipsed by the magnificent specimens to be seen at Gray, another property belonging to Mr Stuart Gray, but situated at the opposite extremity of the Carse. The soil in this portion of the Carse is unusually rich, which may account for the extraordinary growth of most of the trees. Looking, first of all, into the garden in front of the farmhouse of Benvie, we meet with two enormous ash trees. The larger one girths no less than 32 feet at the base, 25 feet at 1 foot from the ground, 19 feet at 3 feet, and 17 feet at 5 feet, the bole being about 27 feet. This tree, however, unlike its companion, is not particularly sound, although its proportions are truly noble. The other ash, which is quite close, girths 15 feet at 1 foot, and 11 feet 6 inches at 5 feet. There are fully 40 feet of a straight clean bole, and the timber is very good. Several very fine trees of different varieties grow in the East Den of Gray. An elm, with a girth of 15 feet at 1 foot and 12 feet at 5 feet, is a particularly good park tree, but it is not in a position to be seen to proper effect. There is a larch 100 feet in height, with a girth of 13 feet 10 inches at 1 foot, and 10 feet 6 inches at 5 feet; and a silver fir almost the same size, the girth being the same as the above at 1 foot, and 10 feet at 5 feet. There is a large silver fir in the policies, the girth at 1 foot being 14 feet 6 inches, and at 5 feet the girth is 13 feet. The most notable trees in the policies, however, are two great cedars of Lebanon. The larger one girths 27 feet at the ground. At 3 feet from the ground, the tree breaks into four limbs, the principal one being 14 feet 6 inches in girth, and the others about 8 feet. The tree is about 60 feet high, and it has a wonderfully grand head. It has at different times given indications of yielding to the furious gales of recent years, but Mr. Stuart Gray is so solicitous about its safety that he has given instructions that

CARSE OF GOWRIE.

no expense is to be spared in endeavouring to maintain it. Recently four 1½-inch wire cables were erected, at a cost of £20, to support it; and for the present the tree seems to be thoroughly secure. The lesser one, which is only a few yards distant, girths 20 feet at 1 foot from the ground, and also breaks into a number of branches at 3 feet up, the principal branch being 17 feet 6 inches in girth. Both of these are very remarkable trees, and are a source of great attraction to the grounds. The finest sycamore in the policies girths 12 feet 6 inches at 1 foot, and 10 feet at 5 feet. A horse chestnut girths 14 feet at 1 foot, and 12 feet at 5 feet. There is a very fine park oak girthing 17 feet at the narrowest part of the bole, being 4 feet up. The first branch is reached at 10 feet from the ground, and the girth of the largest branch is 12 or 13 feet—the second largest being about 8 feet. There are altogether about 600 or 700 cubic feet of timber in the tree, for which as high a sum as £40 was offered a number of years ago. Another very nice oak with a bole of 30 feet, girths 14 feet at 1 foot, and 10 feet 6 inches at 5 feet. The kirk and village of Liff, which are in the immediate neighbourhood, are also surrounded by some very fine oaks.

Leaving the property of Mr. Stuart Gray at Kinfauns, the most convenient place to visit is Seggieden, the seat of Colonel Drummond Hay, whose family is a branch of the Hays of Leys. The Colonel is a well-known naturalist, and amongst the rarities to be met with at Seggieden is a very complete collection of hardwooded plants, flowering shrubs, and fine foliaged evergreens. The collection embraces the most choice of the hardy exotic shrubs, and they are so artistically grouped as to display their fine foliage and flower to the best advantage. The remarkable trees upon the property are not very numerous, but several of them are deserving of notice. There is a fine specimen of the *Haguenensis* variety of *Pinus sylvestris*, reared from seed brought home by Mr Louden from one of the forests of Germany about the year 1818 or 1820, the seed having been given to the gardener at Seggieden. A few very good beech trees are scattered over the policies. The largest one girths 17 feet 4 inches above the swell of the roots, and 11 feet 10 inches at 5 feet from the ground. There is another girthing 14 feet above the swell, and 11 feet 6 inches at 5 feet. The latter has a beautiful spread of branches, several of the limbs being so ponderous that they have to be supported by strong props. A horse chestnut girths 13 feet 1 inch at the narrowest part of a bole of 12 feet 6 inches, and 14 feet at 5 feet from the ground, —there being a great many heavy limbs and a magnificent top. A grand poplar tree grows quite close to the railway. The girth at 1 foot from the ground is 17 feet, and at 5 feet the girth is 16 feet. There is a bole of 12 feet, and many of the branches are equal in girth to what would be esteemed very large trees. There is another poplar—old, and completely covered with ivy —which girths 14 feet at the narrowest part of a splendid bole. An old and gnarled, but very beautiful Spanish chestnut tree has a girth of 11 feet 6 inches at 1 foot, and 9 feet 1 inch at 5 feet. The best of the oak trees grows in the East Approach, and is in every respect a grand tree, with a bole of 32 feet, and a girth of 14 feet 6 inches at 1 foot, and 12 feet 2 inches at 5 feet. A lime tree has a girth of 15 feet 1 inch at 1 foot, and 11 feet at 5 feet. On the south side of the garden, on the road leading to the river-side, there is a row of lime trees, little inferior to the one just mentioned. Two fine elm trees, each 80 feet high, grow together. The girth of the larger one is 14 feet 5 inches at 1 foot, and 11 feet 10 inches at 5 feet; and the girth of the smaller one is 13 feet 3 inches at 1 foot, and 10 feet 6 inches at 5 feet. A splendid branchy Scotch yew, with a bole of 10 feet, girths 10 feet 3 inches at 1 foot, and 9 feet 6 inches at 5 feet. There are a number of walnut trees of great size, the best of them girthing 14 feet 1 inch at 1 foot, and 11 feet 1 inch at 5 feet from the ground. A fine ivy-covered ash in front of the house girths 11 feet round the narrowest part of the bole. There are only a few acres of plantation.

In Fingask we have a place of great natural beauty and historic interest. As the home of the Threiplands, it has always been closely indentified with the efforts of the Stuarts to regain the Crown. It was originally built as a place of strength, and in 1642, during the Civil War in the time of Cromwell, it stood a siege. Within its hoary walls the Chevalier St George, son of James II., spent the night of 7th January 1716, while on his way from Glamis to Scone, where he was proclaimed King. In 1746 the Government troops completely ransacked the Castle, demolished the fortalices, and razed a great part of the building to the ground, in consequence of the attachment of the Threipland family to the cause of the unfortunate House of Stuart. A portion of the ancient Castle, however, seems to have been preserved, as one part bears the venerable date of 1194. Since its partial destruction in 1746, the building has been greatly enlarged and modernised, but the castellated form is still preserved. The Castle stands on the brink of a deep glen, amidst richly-wooded eminences, and is surrounded by a large acreage of gently-undulating grounds, tastefully laid out in flower gardens, shrubberies, and enticing walks. From its situation, the Castle commands an extensive prospect over the great valley below, which here opens out in one vast amphitheatre, through which the Tay rolls for many miles, until it is lost in the German Ocean. There are very few notable trees at Fingask. Its charm lies in the exquisite taste with which the grounds are laid out in the quaint Dutch style. Its fantastically-cut Irish yews,—which are here to be seen in forms we have never met elsewhere,—and its beautifully-winding walks, with their numerous surprise views, are sights which can never fade from the memory. The present proprietor, William Murray Threipland, has only recently succeeded, but the late laird, Sir Patrick Murray Threipland, lived for a long period on the property; and did not a little to embellish it. Numerous sculptured groups are placed in the grounds, illustrative of incidents chiefly drawn from the works of Burns; and here and there plates have been placed bearing favourite pieces of poetry, suggested by the scene in which they are placed,—the following from the scene in the forest of Arden, in *As You Like It*, being a fair example :—

> " Are not these woods
> More free from peril than the envious court ?
> Here feel we but the penalty of Adam,
> The season's difference ; as the icy fang
> And churlish chiding of the winter's wind ;
> Which when it bites and blows upon my body
> Even till I shrink with cold, I smile, and say,—

> This is no flattery : these are counsellors
> That feelingly persuade me what I am !
> Sweet are the uses of adversity ;
> Which, like the toad, ugly and venomous,
> Wears yet a precious jewel in his head;
> And this our life, exempt from public haunt,
> Finds tongues in trees, books in the running brooks,
> Sermons in stones, and good in everything.
> I would not change it."

In one of the walks we met with a tree bearing this inscription, "Lady Effie's Gean, 1742," which may be regarded as a favourite of other days. The most notable tree of all is an *Araucaria imbricata*, planted in 1842. It is one of the best-furnished trees in the country, and is close upon 40 feet high. It is planted on a rock, with very little soil. There is a characteristic-looking Scots fir, bearing an inscription to the effect that Sir William Allan had sketched the tree, and embodied it in his famous painting, "The Lost Child." There is another inscription testifying that the board upon which the inscription is printed is of elder wood cut from a tree which marked the spot where the Earl of Mar raised the standard of revolt in 1715. The policies and plantations contain many fine timber trees, but there are no others calling for special remark. At Kinnaird Castle, in the immediate neighbourhood, there are a few very good sycamores. Although now in the possession of the Threiplands of Fingask, this castle—which presents an excellent specimen of the sort of dwelling in use when safety received more consideration than comfort—was formerly the property of the noble family of Kinnaird, whose present seat is at Rossie Priory. It was afterwards held by the Threiplands, but forfeited in 1716, and was re-acquired by the late Sir Patrick Murray Threipland as lately as 1853. Sir Patrick restored the walls of the old castle, and roofed it in harmony with its original character, and it is now a favourite resort with pleasure parties from Perth and neighbourhood. It is picturesquely surrounded with wood, and from its battlements a magnificent view can be had. One of the chief attractions at Fingask and Kinnaird is an interesting and valuable collection of Jacobite relics.

Glencarse, the property of Mr Thomas Greig, is also a very interesting place. Along the principal entrance a very pretty effect is produced by a line of Portugal laurels, about 12 feet high, alternating with Irish yews, about 15 feet high. One of the most notable features in the policies is the fine collection of *Wellingtonia gigantea,* there being ten of them ranging from 30 feet high to 42 feet high, and some of them bear cones. The largest one was planted about 1854, having been presented to Mr Greig by the late Mr Gorrie, Rait. The exact height is 42 feet 2 inches, with a girth of 10 feet 8 inches at the ground. Two of the same variety planted on the Prince of Wales' marriage day, 10th March 1863, have also attained a considerable size—the height of the one being 38 feet 2 inches, with a girth of 8 feet 3 inches at the ground; and the height of the other 34 feet 8 inches, with a girth of 7 feet 7 inches at the ground. The pinetum also contains many fine specimens of other varieties, including *Crytomeria Japonica,* the best specimen of which girths 5 feet at the ground; an *Araucaria imbricata,* over 30 feet, presented to Mr Greig by the late Mr Turnbull of Bellwood; *Abies Canadensis, Abies Albertiana, Abies excelsa, Abies Menziesii, Abies orientalis, Biota orientalis, Biota tartarica, Cedrus Deodara, Juniperus Chinensis, Juniperus Virginiana glauca, Libocedrus Chilensis, Picea Cephalonica, Picea lasiocarpa, Picea nobilis, Picea Pinsapo, Picea Benthamiana* (bearing cones), *Taxodium sempervirens, Thujopsis borealis,* &c. The hardwood trees include several very fine specimens. Near the house there is a really magnificent sycamore girthing 18 feet at 1 foot, and 13 feet 7 inches at 5 feet from the ground, with a grand bole of 30 feet, and an entire height of 80 feet. There is another fine sycamore, and a very curious one, growing at the back of the garden. Four stems rise from one stool, and two of these stems are connected, about 12 feet up, by an arm of the tree in such a way as to form a strong horizontal bar, without there being the slightest indication of the point of junction. Considering the height of the tree, the connection must have been formed naturally. There are many fine oaks, one of them—measured at random as a fair specimen—girthing 12 feet 8 inches at 1 foot, and 9 feet 3 inches at 5 feet. The elm trees have also several splendid representatives. One of them girths 16 feet 8 inches at 1 foot and 11 feet at 5 feet, with a bole of 12 feet; and another girths 16 feet at 1 foot, and 14 feet at 5 feet. A beautiful beech, on the side of the principal entrance, has a girth of 14 feet 10 inches at 1 foot, and 11 feet 6 inches at 5 feet, with a bole of 20 feet. An ash quite close to it girths 15 feet at 1 foot, and 10 feet at 5 feet. The plantations on Glencarse Hill extend to about 100 acres, and contain several grand specimens of larch and silver fir. One of the larches, which claims attention from its great height, is about 120 or 130 feet high. The stem is a really magnificent one, the whole of the branches being mere twigs. The girth of this tree is 8 feet 4 inches at 1 foot, and 6 feet 11 inches at 5 feet. Several of the silver firs reach a height of fully 100 feet. One of them girths 14 feet 7 inches at 1 foot, and 12 feet at 5 feet, with a bole of about 60 feet. Another one near it is almost equally good, the girth at 1 foot being 13 feet 4 inches, and at 5 feet, 11 feet 11 inches. Since Mr Greig came into possession of the property improvements upon an extensive scale have been carried out, including drainage, fencing, covering of cattle courts, and improvements of steadings, while a large number of neat and commodious workmen's houses have been erected.

At Pitfour, the property of Sir James Stewart Richardson, Bart., there are several trees deserving of notice. The largest of the hardwood trees is an elm girthing 18 feet 8 inches at 1 foot, and 12 feet at 5 feet, with a bole of 12 feet. A sycamore girths 17 feet at 1 foot, and 12 feet 6 inches at the narrowest part of the bole, being 5 feet from the ground. A silver fir also girths 17 feet at 1 foot, and 12 feet 4 inches at 5 feet. Another silver fir has a girth of 13 feet at 1 foot, and 12 feet 4 inches at 5 feet. Both of these are very fine trees, reaching to a height of about 90 feet. There are a number of good cedars of Lebanon, with an average girth of about 8 feet. The best of the larches girths 15 feet at 1 foot, and 9 feet 8 inches at 5 feet. The newer conifers are largely represented, and include a number of fine specimens of *Abies Douglasii*, one of which has a circumference of 30 yards, and has a noble appearance; *Cedrus Atlantica, Picea Lowii, Picea Nordmanniana, Picea Webbiana, Abies nigra, Picea Cephalonica, Picea Pinsapo, &c.*

Proceeding further down the Carse, we have our attention directed to Gourdiehill, a beautiful little estate chiefly under fruit trees. Until recently, it was the property of Mr Robert Matthew, but a few months ago it was purchased by Armand Lacaille, Pollockshields, Glasgow. Here there is a superbly grand holly tree, said to be the best-grown tree of the kind in Great Britain. The girth at the ground is 7 feet 3 inches, and at 5 feet up the girth is 6 feet 6 inches, there being scarcely any variation in this girth until a height of 28 feet is reached, when the first branch is met with. The whole of the stem is as smooth as a hewn pillar, and it has a very symmetrical ball-shaped top of about 18 feet in diameter. The tree, although it must be of a great age, is sound and healthy, and is a most striking and beautiful object in the lawn fronting the house, where it is situated. There are also some very good oaks and sycamores at Gourdiehill. An oak in front of the house girths 13 feet at 1 foot, and 10 feet at 5 feet, with a beautifully-straight stem of 40 feet. Another oak close beside it is a little less in girth, but has an almost equally good bole. The oaks also include one girthing 11 feet 6 inches at one 1 foot, and 9 feet at 5 feet, with a straight bole of 45 feet; and another having a girth of about 11 feet at 5 feet, and a bole of 15 feet. There are two very good sycamores, one having a girth of about 12 feet at 5 feet, and a bole of 26 feet; and the other a girth of 10 feet, and a straight, clean bole of 33 feet. An elm has a girth of 13 feet 6 inches at 5 feet, with a bole of 12 feet, and a very large top of wide-spreading branches. There are a number of very large fruit trees, but we shall refer to these more particularly when we come to speak of the orchards for which the Carse has long been famous.

The next property of interest in the Carse is that of Megginch, the seat of John Murray Drummond. The original name of the place was Melginch, and the change to the modern form gave the late Dean Stanley the opportunity of immortalising this beautiful estate in graceful verse. A few years before his death, Dean Stanley, accompanied by Mrs Drummond of Megginch, Canon Rearson, the Rev. Dr Alex. Laing, Newburgh, and others, paid a visit to Lindores Abbey, and in the course of conversation Dr Laing asked Mrs Drummond why her property was not called by its old name, Melginch. The Dean immediately inquired what the old name meant, to which Dr Laing replied that it meant "the bare island"—the true Gaelic being *maol*, signifying bald or bare, without foliage, and *innis*, an island or place of pasture. The Dean made no further remark at the time, but on his return to Megginch he recited the following lines:—

> "Bleak, bare, and bald an island stood
> Above old Tay's wide weltering flood;
> No tree, no shrub, no floweret crown'd
> The precinct of this barren ground.
>
> But now how changed that dismal scene—
> All deck'd with ever-varying green:
> The darksome yew, the gladsome rose,
> The alley's shade, the bower's repose.
>
> So still may Melginch rude and bare
> Give way to Megginch sweet and fair;
> The softened word its story tell,
> And break its own transforming spell."

The property of Megginch which originally included in it the three Earldoms of Errol, Tweeddale, and Kinnoull, was formerly held by a branch of the family of "Hay of Errol," which owes its distinction to the gallantry of that Hay and his sons who turned the tide of battle at Luncarty about the year 990. The land, according to the historical narratives, was granted as a reward for the gallantry displayed at that battle. After a triumphal entry into Perth, Kenneth III. convoked an Assembly of States at Scone, where it was decreed that the valiant Hay should have his choice of the greyhound's course or the falcon's flight, as his reward. He preferred the falcon's flight, and the noble bird, on being unhooded on a hill in the vicinity of Perth, never paused in his flight till he reached the confines of the parish of Errol, where he alighted on a large stone, still bearing the name of the "Hawk's Stane." All the intervening land was gifted, and was afterwards either apportioned to different members of the family, or sold. The oldest charter belonging to what was originally the estate of Errol, and constituting it a barony, was granted in the time of William the Lion (1166 to 1214). The eldest branch of the Hay family was created Earl of Errol in 1452, in the time of James II. The estate and the Earldom continued united till 1634, when the property was sold to John Drummond, eighth Baron of Lennoch, in Strathearn. The eldest branch of the Hay family ended in a female, who, in 1720, was married to the Earl of Kilmarnock. Several of the Drummonds of Megginch occupy an honoured place in the annals of our country. John Drummond of Megginch sat as Member of Parliament for Perthshire from 1727 to 1734; Adam Drummond, ninth of Lennox and fourth of Megginch, was a captain in the first American War, in the reign of George II., and represented the Forfar Burghs, and afterwards Shaftesbury, in Parliament; Admiral Sir Adam Drummond, the sixth laird of Megginch, was a highly-distinguished naval officer; and his brother, Sir Gordon Drummond, was a celebrated General. The eldest son of Sir Gordon Drummond commanded the Brigade of Guards at the Fall of Sebastopol, and died in 1856, in command of the 1st Battalion of the Coldstreams. William Russel, second son of Sir Gordon Drummond, was Lieutenant on Board H.M.S. *Satellite*, and was shot at Callao, South America, while protecting British subjects during an insurrection. The Admiral became the husband of Lady Charlotte Murray, eldest daughter of John, Fourth Duke of Athole, and relict of Sir John Menzies, Bart., the present proprietor being their eldest son. Malcolm, the son of the present proprietor, is a Lieutenant in the Grenadier Guards, to which his father formerly belonged. The Admiral took a great interest in the estate, and effected several improvements, including the erection of a new front to the picturesque Castle of Megginch, the oldest entire building in the parish, the inscription upon it being,— "Petrus Hay, ædificium extruxet, A.D. 1575." Amongst the attractions of the interior of the Castle is one of the largest private ornithological collections in the country, formed by Captain H. M. Drummond, 42nd Royal Highlanders, one of Sir Adam Drummond's son's, now Colonel Drummond Hay, well known as the President of the Perthshire Society of Natural Science. The plantations at Megginch are not very extensive, but there are several magnificent single trees, principally oaks, for which the strong soils of the Carse are best adapted. In the Kingdom field there are two really splendid oaks. The larger one girths 16 feet 3 inches at 1 foot, and 12 feet 9 inches at 5 feet, with a bole of 10 feet; and the other girths 14 feet 7 inches at 1 foot, and 11 feet 7 inches at 5 feet, with a bole of 12 feet. Both of these trees have grandly spreading heads and present a noble appearance. At Inchcoonans Farm there is even a finer specimen, the girth at 1 foot being 17 feet, and at 5 feet 12 feet 3 inches—the bole being no less than 25 feet. There is another very nice oak in front of the Castle, girthing 12 feet 4 inches at 1 foot, and 11 feet 9 inches at 5 feet, with a nice round bole of 15 feet. Amongst the first objects to attract attention in the policies are the fine clumps of aged yews, originally planted to adorn the four corners of the old kitchen garden, which is now devoted to the cultivation of flowers. The clumps consist of about ten trees each, growing close together, and girthing from about 6 to 10 feet at 5 feet from the ground. Two very fine gean trees arrest attention, the girth of the larger one being 11 feet 9 inches at 3 feet with a bole of 17 feet. There is also a twin walnut of remarkable size—the girth at the place where the trees break off from the parent stem being 13 feet 2 inches and 10 feet 8 inches, the boles being 18 and 20 feet. The other hardwood trees include an elm 15 feet in girth at 1 foot from the ground, and 10 feet at 5 feet; a fine old lime tree in the flower garden, girthing 16 feet at 1 foot, and 14 feet at 5 feet; a very clean-growing beech, girthing 16 feet 9 inches at 1 foot, and 11 feet 2 inches at 5 feet, with a grand bole of 25 feet; a Spanish chestnut, girthing 16 feet 4 inches at 1 foot, and 13 feet 10 inches at 5 feet, with a bole of about 35 feet. There are also a great many trees in the policies closely approaching the measurements which have just been given. The best of the larches girths 12 feet 7 inches at 1 foot, and 10 feet 3 inches at 5 feet, with a bole of 20 feet. A very beautiful and promising *Wellingtonia gigantea* grows behind the gardener's house, the plant having been grown here from seed sent from California to Mr Matthew, late of Gourdiehill. The girth at the ground is 9 feet, and the height of the tree 35 feet. There are also several very fine araucarias. One of the most delightful features in the grounds is the extensive and well-kept avenues, the principal one, leading from the castle to the curling pond, extending to fully three-quarters of a mile in length, and forming a charming walk.

One of the most interesting places in the Carse of Gowrie, alike to the casual visitor and to the ardent arborist, is Rossie-Priory, the princely seat of the Right Hon. Lord Kinnaird. It occupies a magnificent site on the upward slope of Rossie Hill, on the Braes of the Carse, and is situated about 130 feet above the level of the sea. The Priory, which is built of a very fine stone to be had in great abundance upon the estate, was founded in 1807, from designs by Mr Atkinson, and was completed in the year 1817. It is, from every point of view, a splendid building, and a decided attraction to the whole of the eastern end of the Carse. If the appearance of the Priory from the level of the Carse be beautiful, no less lovely is the charming landscape as seen from the Priory. In front, the rich alluvial Carse spreads out east and west, every foot of it cultivated to the highest degree, and dotted here and there with stately mansions and comfortable homesteads. Between the Carse and the County of Fife, the Tay pours its ample waters—here broadening out into the estuary, and giving a fine tone to the whole scene. Beyond the Tay, the eye wanders over a rich expanse of cultivated land, till the view is closed by the swelling Lomonds and the uplands of Fife. Away to the right, the prospect is grander and more rugged,—closed in by the mountains of Perthshire, which, far on the horizon, are seen tossing their peaks to heaven. To the east, the Tay—widening into sea—stretches to the horizon; while the smoke of Dundee tells of the restless activity which has made the Tay there as famous for commercial enterprise as its higher reaches are for sylvan beauty and glorious landscape. The pleasure grounds, gardens, and policies of Rossie-Priory are extensive and beautiful, stately trees, extensive lawns, and far-spreading parks, adorning the stately edifice in every direction. West of the pleasure grounds is the hamlet of Baledgarno, or town of Edgar, so called from a castle erected by King Edgar on a hill behind; and close to the hamlet is Baledgarno Den, a deep and beautifully-wooded ravine, through which a rapid hill torrent tears its impetuous way. South of the pleasure-grounds of Rossie-Priory are the ruins of Moncur Castle, hoary with age, and round which there cluster legendary stories of Pictish wars. The noble family of Kinnaird takes its rise as far back as 1176, Randolph Rufus having obtained a grant of the lands of Kinnaird from King William the Lion; and George Kinnaird having been knighted by Charles II. in 1661, and afterwards created Lord Kinnaird of Inchture in 1682. The late Lord Kinnaird and his predecessor were also men of distinction, both as politicians and as improving landlords. The present Lord Kinnaird sat as Member for the burgh of Perth from 1832 to 1839, and continuously from his re-election in 1859 till 1878, when he was raised to the Peerage on the death of his brother. The parks at Rossie-Priory are studded with many beautiful hardwood trees, but the chief attraction is the splendid collection of newer coniferous plants. There is a grand specimen of *Abies Douglasii*, planted in 1826, with a girth of 11 feet at 1 foot, and 9 feet 6 inches at 5 feet from the ground. The branches, which curve gracefully along the ground, have a circumference of 150 feet. A *Cedrus Deodara* rises to a height of fully 40 feet, and an *Abies Atlantica* has a height of 50 feet, the whole of it being bole. *Taxodium sempervirens* grows to the height of 45 feet, with a girth of 12 feet 6 inches at 1 foot, and 9 feet at 5 feet from the ground. The best specimen of *Wellingtonia gigantea* girths 9 feet at the ground, two of them girthing exactly the same. *Abies Menziesii* girths 11 feet at 1 foot, and 10 feet at 5 feet. Amongst the most flourishing of the newer coniferous trees mention may also be made of *Cryptomeria Japonica, Picea Lowii, Araucaria imbricata, Picea grandis, Picea Nordmanniana, Abies Albertiana, Picea magnifica, Cupressus macrocarpa,* and *Cupressus Lawsoniana.* Experience has shown that *Pinus excelsa* does not thrive very well in any situation, it being seemingly too tender for the climate. A very fine poplar tree reaches a height of 110 feet, with a girth of 14 feet at 1 foot, and 12 feet at 5 feet from the ground.

At Castle Huntly, the property of George Frederick

Paterson, there are also a few trees of exceptional interest. The Castle itself is a very conspicuous object in the landscape. Occupying the summit of a precipitous rock, it is almost inaccessible on every side, except the north-east. On the west side, the Castle rises to the great height of 130 feet, and, as the walls are fully 10 feet thick, it must have been a place of considerable strength when warriors were the regular occupants of its lofty turrets. The Castle, which was built about the middle of the fifteenth century, is in first rate preservation, and, although it has been converted into a modern residence, there can still be seen all the "conveniences" of an ancient stronghold,

> "The battled-towers, the donjon keep,
> The loophole gates, where captives weep."

The dungeon is a very gruesome-looking place, constructed out of the solid rock, and capable of accommodating a large number of unfortunate captives. Several instruments of torture are still to be seen, and who can tell how many poor victims they have held in their cruel grasp? The Castle is altogether one of the best specimens of an old baronial residence in Scotland, and it is so substantially built, and is so carefully preserved, that there is every probability of its continuing to adorn the landscape of this part of the Carse for many centuries yet to come. From its prominent position, and its great height, it commands one of the grandest views that can be imagined. The prospect embraces the entire Carse of Gowrie, with its numerous gentlemen's seats, the whole stretch of the river from above Newburgh to the German Ocean, a large portion of the coast of Fife, including the Lomonds, a considerable extent of Lower Strathearn, with the Ochils in the distance, and right over the elevated Braes of the Carse to the higher range of the Sidlaws. The Castle was built by the second Lord Gray of Fowlis, and the tradition is that he named it after a daughter of the Earl of Huntly. The property was sold to the Earl of Strathmore in 1615, and in 1672 the fortalice became known as Castle Lyon, and bore his name until 1777, when it was purchased by an ancestor of the present proprietor, who, having married a daughter of Lord Gray, the descendant of the founder, naturally restored the old name. Perhaps the most notable of all the trees at Castle Huntly is a grand old Scots fir, about 250 years of age, and which is in some respects as remarkable a tree of the kind as is to be seen in the country. The girth of the tree at the ground is 24 feet. At 1 foot up the girth is 16 feet, and at 5 feet it is 15 feet. The trunk then swells out until it has a girth of about 30 feet, and carries a magnificent head. The tree has a noble appearance, but there are indications of its breaking up. Another Scots fir, much younger, and in good condition, girths 11 feet 6 inches at 1 foot, and 11 feet at 5 feet, with a bole of 15 feet. A very fine yew girths 10 feet at 1 foot, and 9 feet 2 inches at 5 feet from the ground, with a bole of 9 feet, while the circumference is 68 yards, the branches sweeping the ground in beautiful style. There is another specimen of the same variety closely approaching this one in size. There is a splendid sycamore girthing 16 feet at 1 foot from the ground, and 12 feet at 5 feet, with a bole of 30 feet. A great ash, girthing about 30 feet at the ground, stood at the stable-door until it succumbed to the fury of a gale, but an excellent idea may be formed of its size from the trunk, which now forms a summer-house at Longforgan manse. There are still several very good ash trees throughout the property, as well as oak, sycamore, elm, beech, ash, and horse-chestnut trees. There is also a nice little mixed plantation of about 60 acres.

There are still a few places in the Carse of Gowrie to which reference ought to be made. At Errol Park, the property of Mrs Mollison, there are a lot of very good trees. In front of the house there is an avenue of excellent horse chestnuts, girthing from 9 to 14 feet at 5 feet up. There is also a row of yews, girthing from 7 to 10 feet. An oak girths 13 feet at 5 feet, with a bole of 12 feet. One of the most noteworthy of the trees is a beautiful specimen of *Taxodium sempervirens*, about 30 feet high, and 7 feet 6 inches in girth at about 3 feet from the ground. The other trees include very good specimens of cedar of Lebanon and *Cedrus Deodara*. At Murie, the property of Mr J. B. Brown-Morison, there are a lot of very fair trees. At Balruddery, the present proprietor, Mr White, has effected many valuable improvements, and is still engaged in enhancing the value and improving the amenity of the estate.

The house has been enlarged until it has become one of the finest in the Carse ; and the policies, which were already ornamented with many fine young park trees, have been of late vastly improved.

The Carse of Gowrie, as we said at the outset, has long been famous for its orchards. Many of these are of considerable extent. The Gourdiehill orchard, for instance, is regarded as the largest, as it is one of the best in Scotland. This orchard extends to 35 acres, and the trees, which number about 7000, are beautifully planted in rows. Most of the trees are from 20 to 60 years of age, but there are a number of very old trees, some of them being from 200 to 300 years of age. A few of these are of considerable size. The largest is a pear tree girthing 10 feet, and having a bole of 12 feet, with an entire height of 65 feet. The trees are all in a good-bearing state, and in some years upwards of 200 tons of apples and pears have been gathered. Great attention is bestowed on the selection of the best varieties, both as regards quality and bearing. Some years ago, medals were gained both in Edinburgh and London for a selection of six unknown varieties, and the fruit is so highly esteemed that many prizes have been secured at international and other shows. The next largest orchard in the Carse is that of Monorgan, near Longforgan, on the Castle Huntly Estate. It extends to about 25 acres, and, although it was probably planted about 150 years ago, it is still in a fair state of bearing. At Seggieden there are about 22 acres of orchard. Most of the trees are between 50 and 60 years old, and are in a good state of bearing, decaying trees being regularly replaced with new ones. This orchard has been let on a 19 years' lease, now nearly expired, at an annual rent of £210, without the undercrop. Orchards let on lease generally bring from £8 to £10 per acre. At Glencarse, there are about 8 acres of orchard, including a good many fine English damson trees—Mr Greig being the first to plant these trees in the Carse upon a large scale, about twenty years ago. There are now as many damson trees as there are apples and pears, as they are found to be a more profitable crop. At Seaside and Fingask, there are each about 16 acres under orchard ; at Grange Orchards there are about 12 acres ; at Castle

Huntly, 10 acres; Port Allen, 9 acres; and at Kinfauns, Waterybutts, Glendoick, Pitfour, Flatfield, Megginch, Horn, Newbigging, Muirhouses, Bogmiln, Powgavie, Ballindean, Balruddery, Raws, &c., there are orchards from 2 to 10 acres in extent. A good number of fruit trees are also planted in the neighbourhood of Errol and other places. Most of the orchards are let each season, and the prices, consequently, vary considerably. Since the commencement of the manufacture of jelly upon an extensive scale, the price of the inferior kinds has increased very much. Formerly, when there was a prolific crop, apples could hardly find a market, but now, even when there is an exceptionally large crop, the price seldom falls below 4s per cwt. for inferior sorts, while it has sometimes risen as high as £18 per ton. Early table apples generally bring a good price, but the late and better sorts do not realise the prices they once did, on account of the large importations from America. Scotch apples are considered the best for preserving, being more tartish than either English or Irish; while the American, Dutch, French, and German apples are too soft and sweet for jelly, and generally run into syrup. About one-third of the orchards consist of pear trees. This fruit realises quite as good a price as in former years, but for some time past the seasons have been unfavourable for the late kinds. Gooseberries, strawberries, and other small fruits are also very extensively grown in the Carse.

XLII.—MISCELLANEOUS.

Remarkable Trees round Perth—Wallace's Yew at Elcho—Historical
  Yew at Culfargie—Kinmonth—Killour Pinetum—Stanley—Bonhard
  —Murrayshall — St Martins — Cardean — Clathick — Cultoquhey—
  Monzie Castle—Delvine—Snaigow—Stenton—Birnam and Pitcastle—
  Gigantic Ash at Logierait—Bonskeid—Lude—Kilbryde Castle—
  Argaty—Lanrick Castle—Cambusmore and Strathyre—Invertrossachs
  —Leny—Edinchip—Conclusion.

FOR picturesqueness of situation, it may safely be said that
Perth has no equal in Scotland. Flanked on the north and
south by its magnificent Inches—the finest public parks
possessed by any town of the same size—it has, on the east
and west, an extensive range of delightful slopes, while at a
little distance it is surrounded by bold and fir-clad hills, which
impart to it a truly romantic grandeur, as seen from a slight
elevation. We have already alluded to the more prominent of
the wooded eminences surrounding the ancient and historic city,
but there yet remain to be noticed several arboreal features of
interest in closer proximity to the town. The Inches, as a
matter of course, provide us with the richest material. The
South Inch, although the smaller of the two, is much better
wooded, and the trees are of greater size and age. It is believed
that trees were first planted here in the early part of last century, and, generally speaking, the original plan has been well
adhered to. The trees have been laid out in avenues and groups,
and these are so well arranged, and present such an imposing
appearance, as to excel everything of the kind in any other
town in Scotland. However much the grandeur of the Inch
may be enhanced by the surrounding scenery, encircled as it is
by the happiest combination of the beauties of nature and
of art, there can be little doubt that the beauty and arrangement of the trees add materially to the completeness of the

picture. There are, altogether, four avenues. The principal one lines the Edinburgh Road, which intersects the Inch, separating it into eastern and western divisions of unequal proportions; a second skirts the side of the river; the third leads from the Edinburgh Road to Craigie Bridge, and the fourth leads from the entrance at Marshall Place to Craigie Bridge. Both the Edinburgh Road Avenue, and the one next the river contain several exceptionally good trees. Looking, first of all, at the trees on the west side of the Edinburgh Road, we have a number of large specimens of their kind. An elm girths 16 feet 6 inches at 1 foot, and 11 feet 6 inches at 5 feet; an ash girths 16 feet 7 inches at 1 foot, and 11 feet 10 inches at 5 feet; and a beech, the finest of all the beeches on the Inch, has a girth of 11 feet 9 inches at 1 foot, and 10 feet 9 inches at 5 feet, with a grand bole of 20 feet. There is a larger beech on the eastern side, but it is not such a good specimen. The girth of this tree is 15 feet at 1 foot, and 12 feet 5 inches at 5 feet, with a clean bole of 18 feet to the first limb. An ash, on the eastern side, girths 18 feet 5 inches at 1 foot, and 15 feet at 5 feet, with a grand bole of 20 feet; an elm next to it girths 11 feet at 5 feet; and there are beeches girthing 9 feet 8 inches, and 11 feet respectively at 5 feet from the ground. The one side of the avenue next the river consists of elms, and the other side almost entirely of sycamores. The sycamores are especially good trees, the average value being about £5 each, but one or two are worth nearly double that sum. The best of them girths 10 feet 3 inches at 5 feet up, with a bole of about 12 feet. There is a fine ash in this avenue, the girth being 12 feet 7 inches at 1 foot, and 10 feet 6 inches at 5 feet, with a bole of 14 feet. The avenue from the Edinburgh Road to Craigie Bridge, has a roundel about the centre, and contains several very good hardwood trees, many of them aged. The eastern division of the Inch, is further ornamented by two finely-balanced horse-chestnuts, with seats around them. The one next the river girths 14 feet 11 inches at 1 foot, and 11 feet at 5 feet, and the other girths 9 feet 10 inches at five feet. Many of the trees in the avenues have been blown down, but the Town Council has judiciously filled up the vacancies, chiefly·

with sycamores, so that all the avenues are quite complete. There are no trees on the North Inch specially noteworthy for their size, although there are several fair specimens, and a very good avenue. About the centre of the Inch there are two specimens of *Wellingtonia gigantea* planted, with a great deal of ceremony, on 10th March 1863, as memorials of the marriage of the Prince and Princess of Wales. The exposure and the soil of the North Inch, however, are not favourable for this variety, and the trees were so seriously affected by the severe frost of 1880–81 that they are now almost dead. The larger of the two has attained a height of about 15 feet, and the smaller one is about 12 feet high. Considerable care is exercised over the Inches by the Town Council and the inhabitants, decayed trees being removed from time to time, and others planted in their place. In the spring of 1860, the Council requested Mr William M'Corquodale, Lord Mansfield's forester, to draw up a report on the state of the trees on both Inches, with suggestions for their future management. Mr M'Corquodale presented a very interesting and exhaustive report, and his suggestions were generally adopted. One portion of his report, however, aroused violent opposition. Struck by the naked appearance of the North Inch, he suggested that the line of trees at the top of the Inch should be continued along the river side to the bridge, and that an avenue should be formed. The Town Council cordially approved of this suggestion, but an agitation against it was created in the public mind, and the proposal was abandoned. Besides the trees on the Inches, there are a few others worthy of notice. At the side of Craigie Burn, near the railway bridge, there are two very aged and gigantic Huntington willows, the larger one girthing 13 feet 7 inches at 5 feet, and the other 10 feet 10 inches at 5 feet. At Balhousie Castle, the residence of Mr John Shields, there is a remarkably large and well grown beech tree, close by the side of the lade which runs through the policies. There is a girth of 18 feet at 1 foot, and 15 feet 2 inches at 5 feet from the ground. About 8 feet up a small branch breaks out, but, apart from this, there is a grand bole of 30 feet, the entire height of the tree being 95 feet. Within the policies there are some beautiful coniferous trees, the most notable being a pair of shapely and most promising *Wellingtonias*. Notice ought also to be taken of the nice row of lime trees planted in Tay Street. Ever since the formation of this fine promenade, the propriety of planting a row of trees was pressed upon the authorities, but various obstacles stood in the way of having the proposal carried out. Lord Provost Hewat, however, took the proposal in hand, and, with the aid of a number of the citizens, had the trees planted in February 1882. In regard to the selection of the trees, the Lord Provost acted on the advice of Colonel Drummond Hay of Seggieden, in every respect a competent authority. The trees were selected from the nursery of Mr Francis Waterer, Knapp Hill, Surrey, and they were planted under the superintendence of Mr David Black, landscape gardener, Dundee. The trees were unusually well-trained, and, when planted, ranged from 20 to 24 feet in height. Due care was taken to promote their growth by planting them in black soil and old turf, and they are thriving as well as could have been wished, and are a decided ornament to the street.

Starting from the county town, we do not require to travel very far until we find something which demands even more than a passing notice. We refer, first of all, to WALLACE'S YEW tree at Elcho, in the parish of Rhynd. Whether regarded in the light of its great age, or the historical associations connected with it, the yew at Elcho is the most interesting tree in this district. In girth, it measures 11 feet 6 inches at 1 foot from the ground. At the height of 4 feet 6 inches it throws out three large horizontal limbs, the spread of the branches being fully 50 feet in diameter. It is said to have been planted by the great Scottish Patriot himself, and the tradition is by no means improbable. During a very severe storm in the spring of 1881 several of the branches broke under the weight of the superincumbent snow, and had to be lopped off.

There is another interesting yew on the Earl of Wemyss' property at CULFARGIE, near Abernethy. It is neither so old nor so large as Wallace's tree, but it is also closely linked to the history of the past. It was under the shade and shelter of this tree that the Rev. Alexander Moncrieff, one of the four fathers of the Original Seceders, and the laird of Culfargie, preached to his flock, when, in 1740, he was ejected by the General Assembly from his charge. The tree consists of six distinct stems arranged in the form of a circle, leaving a recess in the centre, which, at the height of 5 feet, measures, internally, 22 feet in circumference. At the ground, outside, the girth is about 20 feet, and the spread of the branches is over 40 feet.

Before leaving this district, we have to notice the beautiful little property of KINMONTH, extending to about 1000 acres, and which was purchased in 1864 by the trustees of the late Sir George Simpson, Governor of Hudson Bay Territories, for his son, J. H. Pelly Simpson, the present proprietor. It consisted originally of two small estates—the larger one, Fingask, belonging to Mr M'Gill of Kinbuck; and the other to Mr P. Small Keir of Kindrogan. On neither property was there a house fit for the proprietor's residence, so that the new proprietor had to erect a mansion for himself, make roads, lay out policies, and plant the trees necessary for ornament or use. By his labours in this direction, he has wrought a complete transformation in the place. After occupying two or three years in erection, the mansion was completed in 1876. It is a most substantial structure in the old baronial style, and, situated as it is on the south-eastern shoulder of the Hill of Moncreiffe, it commands a fine view of Lower Strathearn and the Carse of Gowrie. Advantage is taken of its elevated position, and the steep and shelving nature of the ground to the south of the house, to form, with much artistic skill, a spacious lawn and terraced walks, which are ornamented, in fine taste, with patches of flowers and shrubs, interspersed with numerous fine ornamental trees, reflecting great credit on the gardener, Mr Methven, under whose supervision the grounds were laid out. The gardens are also laid out most artistically. When the estate was purchased, there were about 30 acres of fine thriving plantation, consisting of Scots fir, larch, and oak. Since Mr Simpson came into possession, about 25 acres additional have been planted in positions selected for effect in the vicinity of the mansion. The finer varieties of the newer coniferous plants are well represented along the terraces and borders. Indeed, the skill and labour bestowed in forming this, one of the latest of Perthshire's noble mansions, have produced a noticeable improvement upon this part of the county, and time will no doubt impart additional charms, as the many fine ornamental trees develop their elegant forms and graceful foliage.

KEILLOUR PINETUM, now the property of James Scott Black, has long been known to arboriculturists, and although it has not in recent years received the attention which might have been bestowed upon it, it still possesses a good deal of interest for those who note the progress of the newer conifers. It now forms part of the Balgowan property, but it was at one time attached to the Methven estate, and to Mr Smythe of Methven we are indebted for particulars regarding its early history. From the memoranda in his possession, we learn that a commencement was made with the planting of that part of the moor of Keillor in which the pinetum is situated in the year 1823, when 30,000 trees were planted. It was not till the year 1831, however, that the planting of the pinetum proper was commenced. It was in the year 1831, and the four following years, that the greater part of the newer conifers were planted, including *Pinus Cembra, Pinus inops, Pinus Nepalensis, Pinus Strobus, Pinus ponderosa, Pinus Banksiana, Pinus rigida, Pinus Austriaca, Pinus monticola, Picea Fraserii, Picea Webbiana, Picea Pinsapo, Picea Cephalonica, Abies microcarpa, Pinus exelsa, Abies Douglasii, Abies Menziesii*, &c. A number of seedling larches from the Tyrol, from Blair-Drummond, and other places, were also introduced. Later on, *Araucaria imbricata* was planted, and several other rare pines, all of which were planted by the late Mr Robert Smythe of Methven and his landsteward, the late Mr Thomas Bishop. Some idea of the extent of the work accomplished is conveyed by the fact that during the three years following 1831 no less than from 60,000 to 70,000 of the newer conifers were planted by these gentlemen. The following are the present girths of a few of those planted in 1835 :—

KEILLOUR PINETUM—STANLEY.

	1 FOOT FROM THE GROUND.	5 FEET FROM THE GROUND.
Abies Douglasii,	15 ft. 6 in.	14 ft.
Picea nobilis,	5 ft. 6 in.	5 ft.
Abies Menziesii,	6 ft.	4 ft. 10 in.
Picea Cephalonica,	5 ft. 2 in.	4 ft.
Pinus monticola,	6 ft. 7 in.	5 ft. 9 in.
Pinus ponderosa,	5 ft. 4 in.	3 ft. 8 in.
Pinus Austriaca,	5 ft.	4 ft. 5 in.
Pinus Cembra,	3 ft. 10 in.	3 ft. 3 in.
Weeping Larch,	3 ft. 9 in.	3 ft. 5 in.
Common Larch,	5 ft. 8 in.	4 ft. 2 in.
Scots Fir,	4 ft. 1 in.	3 ft. 8 in.
Araucaria imbricata,	3 ft.	3 ft.
Silver Fir, ...	5 ft. 11 in.	5 ft.

The *Abies Douglasii, Pinus monticola, Picea nobilis*, and a few others, average about 65 feet in height; the cembras, larches, and Scots firs, from 55 to 60 feet; and the araucarias about 35 feet. There are several younger plants thriving very well, particularly *Abies Albertiana, Cedrus Deodara*, and some of the *Thujii*.

STANLEY, as an old historic residence, possesses an interest that is much wider and deeper than the extent of its acres would seem to imply. Originally the property of the Nairnes, it is intimately associated with the ill-fated cause of the Stuarts, and it is not a little remarkable that, after passing through a series of vicissitudes it has again come into the possession of a representative of the old family. The present proprietor is Colonel Frank Stewart Sandeman, son of Glas and Margaret Stewart Sandeman, of Bonskeid, one of the oldest strains of the Stewarts, being in direct line from the Wolf of Badenoch, and through him to King Robert II. Colonel Sandeman is also grand-nephew of Scotland's most distinguished poetess, Lady Nairne. What is known as "Lady Nairne's Tea House" still stands upon the apex of the hill behind Stanley House, where it commands a most extensive view, encircled by an unbroken chain of distant mountains. Colonel Sandeman is a J. P. for the counties of Perth and Forfar, and has long been prominently identified with the volunteer movement, being the commander of the Forfarshire Artillery. Stanley House, built about 450 years ago, could not have been placed in a more desirable situation. Standing on the right bank of the Tay, on a beautiful haugh about 30 feet above the level of the river, and surrounded by many grand old trees, it is finely sheltered on the north and east by a crescent-shaped hill about 150 feet high, thus leaving it open to the warm, genial breezes of the south, yet protecting it from the biting winds of the north-east. The room from which Lord Nairne made his sudden flight from the home of his ancestors is still used as the dining room. During the '45, his lordship was sitting in this room when his outposts gave warning that the Royal troops were crossing the hill to capture him. Seizing the more important of Prince Charlie's papers, he forded the river in the neighbourhood, and made his escape to Pitlochry. The property was twice confiscated, and on the last occasion it was purchased by the Duke of Athole of that day, who was a cousin of the attainted laird. The son of Lord Nairne, Major-General Nairne, married Miss Oliphant of Gask, "The Flower of Strathearn," whose poetic genius was the means of the family title being restored. It having come to the knowledge of George IV., when in Scotland in 1822, that Mrs Nairne was the authoress of *The Attainted Scottish Nobles*, he was led, two years afterwards, to restore the forfeited title, making Major William Murray Nairne, fourth Lord Nairne, and his wife Baroness Nairne. The title, as explained in our account of Meikleour, is now merged in the Marquisate of Lansdowne. The mansion was first called Stanley House, either about 1683, when on the death of Robert, first Lord Nairne, his daughter, the Lady Margaret Nairne, wife of Lord William Murray, fourth son of the Marquis of Athole, by Lady Amelia Sophia Stanley, daughter of the Earl of Derby, succeeded to the estate, the supposition being that it was so called in honour of her mother-in-law; or in 1703, when, on the death of the Duke, the Duchess Dowager of Athole came to reside at Stanley House, the residence of her son. The place has been greatly altered since it was the home of the Jacobite Nairnes, but many of the grand old trees still remain. The original approach to the house has been altered to suit the exigencies created by the erection of the famous Stanley Mills, but the fine broad avenue, with its imposing beeches, is almost intact. The best of the beeches here has a girth of 16 feet 10 inches at 1 foot from the ground, and 12 feet at 5 feet, with a splendid bole of 16 feet. Another girths 16 feet 6 inches at 1 foot, and 11 feet 6 inches at 5 feet, with a bole of 12 feet. There are several girthing about 11 feet at 5 feet, and one, although not very thick, has a splendid bole of 30 feet. Along the bank of the river, and directly in front of the house, there is a fine row of large beeches, the best of them girthing 15 feet 6 inches at 1 foot, and 12 feet 6 inches at 5 feet, with a bole of 20 feet. Many of the other varieties of hardwood trees are represented by noteworthy specimens. One of the finest oaks in the county grows on the hillside behind the house. It girths 16 feet 8 inches at one foot, and 12 feet 10 inches at 5 feet, with a magnificent bole of 25 feet, and a splendid top. The tree is in perfect health, and still growing. There are two black Italian poplars of large size, the one having a girth of 22 feet at 1 foot, and 12 feet 6 inches at 5 feet, with a bole of 20 feet; and the other with an equally good bole, girths 20 feet 6 inches at 1 foot, and 12 feet 4 inches at 5 feet. There are numerous elms and sycamores about 90 feet high, although not very thick. Close to the house are a number of aged yews of large dimensions, and in a most flourishing state. The girth of one at 5 feet is 12 feet 8 inches, and the girth of another is 9 feet. Within the policies there is a very remarkable round castle of great age. The construction of it is altogether unique, as may be seen from the engraving which we give of both the exterior and the interior. In the olden times it must have been a place of great strength, and a dungeon, now almost filled up, is supposed to have had a connection with Stobhall on the opposite side of the Tay. The castle, which is called Inchbervis or Inverbervie, is believed to have been regularly resorted to by Wallace. There is a tradition that it was at one time a religious house in connection with the Abbey of Dunfermline, and the charters show that the proprietors were held liable in the yearly tribute of a grilse. From this it would appear that the Stanley stretch of the Tay was as famous in the olden times as it is at the present day. There are not two miles of

(Interior.) WALLACE'S CASTLE. (Exterior.)

the river better known than the Stanley waters, or that afford better sport with salmon during the whole of the fishing season. All anglers are familiar with such pools as the "Cat Hole," "Hell's Hole," and "Pitlochry Head." It was at the latter pool where the historical John Briggs was depicted by Leach in *Punch* landing his first fish in his arms after his tackle was broken, the sketch being drawn at Stanley House. Although it is scarcely within our province to describe Stanley Mills, still these are so closely bound up with the history of the property that they cannot be altogether passed over. Up till 1775, the "Yett House," or gate house, a very peculiar old building still standing, was the only erection on the site of these great cotton mills. Far-seeing men selected the site on account of the magnificent water power, which, coming from the Tay with a fall of 25 feet, is led to the mills by tunnelling the hill behind. These mills were the second to be erected in Scotland, if not in Britain, and were built under the auspices of the famous Arkwright. The date of the erection of the mills was 1785, when Dempster & Co. feued the land from the Duke of Athole. In 1800 the mills and lands were sold to Craig & Co., who carried them on till 1814. They were stopped from that date till 1823, when Dennistoun, Buchanan, & Co. got possession of them, spending £160,000 on their improvement, and employing about 1200 workers. On the death of Mr Buchanan, the mills and estate passed into the hands of Mr Howard, Burnley, Lancashire, who carried them on till the great cotton famine consequent upon the American war, when he was obliged to stop the works, with the result that the place became a living picture of Goldsmith's "Deserted Village." The works were afterwards re-opened to a limited extent, and carried on till about seven or eight years ago, when the property was on the brink of passing into the hands of a Limited Liability Company in England, with the view of making the most out of the mills and estate. The principles upon which this Company proposed to work the property, may be inferred from the fact that one of the methods for obtaining a dividend was to have all the trees cut down and sold. Happily, Sir Douglas Stewart of Murtly and Grantully came to the rescue and, through his kindness, the property was secured by his relative, the present owner. Colonel Stewart Sandeman has brought a vast amount of energy and technical skill to bear upon the working of the mills, which are now being successfully conducted. The mills are carried on in conjunction with Manhattan Works, Dundee, and Colonel Stewart Sandeman is associated in the management with his brother, Mr Charles Stewart Sandeman of Springlands. Until the works passed into the hands of the present management, the trade was entirely with India, the "Stanley 30s." being a household word for nearly a century. Many changes, however, have taken place in recent years. Entirely new machinery has been introduced, and the market is now altogether local, the jute and linen factories of the three neighbouring counties taking the whole production for salvages and cotton warps. The specialties include cotton belts, ropes for transmitting power, &c., which promise ere long to make the Stanley Mills as well known at home as they formerly were abroad. What is, however, more in accordance with our present purpose, the Colonel venerates the old trees and their associations, so that there is no danger of their being removed by him, as would certainly have been done had the property been acquired by the Limited Liability Company.

BONHARD, the seat of Mr Alexander Macduff, has a very pleasing aspect on the south of the Sidlaws. The property has been in possession of the family since 1742, and the present proprietor is the fifth laird in succession, each of whom has borne the name of Alexander. A curious and interesting tombstone, the gravestone of the family during last century, stands in a sheltered nook in the policies, and bears 1747 as its earliest date. The stone, which is very richly carved, is surmounted by the family arms, while at the foot there are a skull, cross-bones, and an hour-glass. The motto, "Memento Mori," surrounds the lower part. The most prominent part of the property, as seen from a distance, is Springfield Hill, with its picturesque tower, in imitation of an ancient fort. The best view to be had from it is towards the west, where the city of Perth, and the whole of Strathmore as far as Coupar-Angus and Blairgowrie, are spread before us, with the Logiealmond and Crieff hills in the background. The mansion is a fine modern structure, completed in 1849, and stands on the site of the old Castle of Springfield. A different site was at first proposed, but the presence of the aged trees round the ancient castle was the determinating influence in the final selection. Close to the mansion-house is the beautiful Den of Bonhard, through which flows the delightful little stream of Annaty Burn. The Den is only about a quarter of a mile in length, but within this short stretch are concealed many sylvan beauties. A pleasant winding walk runs through the Den, and leads to a celebrated "Wishing Well" called Dowie's Well, frequently resorted to as a refreshing and curative spring. The Den is chiefly under young oak, the older standards having been recently removed, as being fully matured. Many of these were over 150 years old, and splendid specimens for their age. A few good beeches, about 10 feet in girth have been allowed to stand. One of the features of the Den is a group of good Scots firs of the Braemar variety, growing upon an artificial mound in a bog. The trees have evidently been planted for some special purpose, as there are traces of a walk having led to them through the bog. Before entering the policies, attention cannot fail to be attracted to a fine avenue of old limes bordering the public road. In summer the road is perfectly arched over with the foliage of these fine old trees, which afford a grateful shade. The west approach to the house is lined with an avenue of young oaks, transplanted over 30 years ago. They have been excellently trained, have fine clean stems, and are doing very well. Amongst the most notable of the trees in the policies, is, to adopt the language of Clare, the rural poet, a

> "Huge elm, with rifted trunk all notched and scar'd."

The girth at 1 foot from the ground is 15 feet 2 inches, and at 5 feet the girth is 12 feet. The bole is fully 40 feet. It has a grand head, splendidly furnished, making it altogether a most noble tree. There are several other elms of good size. A beech, with 15 feet of a bole, has a girth of 14 feet 2 inches at 1 foot, and 11 feet 6 inches at 5 feet. The best of the oaks girths 13 feet 3 inches at 1 foot, and 9 feet 3 inches at 5 feet; and the best of the ash trees girths 12 feet 4 inches at 1 foot, and 9 feet 9 inches at 5 feet. Within the last twelve years a large number of the newer coniferous trees have been planted throughout the policies, and they are all thriving very well. An *Araucaria imbricata*, planted about 33 years ago, is fully 35 feet high. It bore a cone last season, but the interesting fruit was unfortunately blown off by a high wind, and no trace of it could afterwards be found. About six years ago, a small pinetum was formed in a convenient space connecting the garden and the pleasure grounds, and has been made most ornamental. On the way to the garden a little pond has been extended and beautified. A number of spruce trees encircle the pond, the water of which has been diverted at a higher level so as to form a waterfall. The effect has been considerably heightened by the formation of a rockery under and around the trees, and the erection of a neat rustic bridge across the narrowest part of the pond. The specimens in the pinetum include *Cedrus Deodara, Wellingtonia gigantea, Picea Nordmanniana, Picea nobilis, Picea Lowii, Abies orientalis, Picea magnifica, Picea grandis, Abies Parryana glauca*, as well as a number of the commoner varieties. The kitchen garden is a little out of the common. On the north, east, and west sides it is enclosed by a brick wall; and on the south, where there is a splendid exposure, a neat iron railing, upon a coping of about 18 inches, has been erected so as to secure the full advantage of the sun. The garden has thus a much lighter appearance than if there were a south wall, and the trees in the pleasure grounds are so arranged that they provide ample shelter from the wind. In a park near the garden there is an extensive dovecot, bearing the date 1709, and containing 1492 nests. Amongst the more important relics of antiquity on the property are two Druidical circles on the eastern portion of the estate. There are nine immense stones standing upright in each, and surrounded by beech trees, which appear to have been originally planted as a hedge. Balcraig, the residence of the Misses Mercer, is on the Bonhard property. The house was originally called Springfield Cottage, but the Misses Mercer have added to it very materially, and greatly enhanced the amenity of the place by acquiring new ground, and laying out flower and kitchen gardens. The place is nicely sheltered on the east, and by some fine beech trees on the north, and commands a magnificent view of the Grampians. Springfield Hill, to which reference was made at the outset, is chiefly clothed with spruce and fir, planted after larch, which was cut down on attaining maturity.

MURRAYSHALL, in the parish of Kinnoull, and about three miles from Perth is an old mansion built in 1664 by Sir Andrew Murray (a younger son of the Balvaird, now the Mansfield, family), and having been modernised according to the taste of last century, was again restored in 1864 to somewhat of its old aspect. A daughter of Sir Andrew Murray, who inherited the estate, married a younger son of the then Graeme of Balgowan, and through this marriage the late John Murray of Murrayshall, on the death of Lord Lynedoch, and afterwards of Mr Robert Graham (both without issue), became the representative of the Graemes of Balgowan. He took the name of Graham, as required by the entail of Bertha. This small property on the banks of the Almond did not share in the disentail of Lynedoch and Balgowan, which Mr Robert Graham effected, in consequence of a flaw discovered in those entails. Mr Murray Graham having died without issue, he was succeeded by his nephew, Captain Murray Graham, R.A. The estate contains some trees that are worthy of notice. The most important of these are four fine old elms. They are short boled, but make splendid park trees. The largest one girths 15 feet 10 inches at 1 foot, and some of the limbs girth 7 feet. Another, with a girth of 14½ feet at 1 foot, has a spread of branches 26 yards in diameter. The others girth 14 and 12 feet respectively at 1 foot, with very little difference at 5 feet up. There is a peculiar circle of old hornbeams (being an old arbour grown up), the largest of which girths 10 feet 10 inches at the narrowest part of the bole; a Spanish chestnut girths 10 feet 6 inches at 1 foot, and 9 feet 1 inch at 5 feet; an ivy-covered ash, girths 9 feet at 5 feet; a tulip tree, 4 feet 4 inches at 5 feet; a cedar of Lebanon, with a short bole, but a fine branchy, ornamental head, girths 9 feet 2 inches at 3 feet. There is also a very good specimen of the silver holly. The newer coniferous trees are well represented, and there are several splendid silver firs in the Den. Murrayshall Hill is one of the most prominent points of the Sidlaws, and considerably higher than the hill of Springfield. It is surmounted by an obelisk erected in 1850 to the memory of Thomas Graham, the celebrated Lord Lynedoch. Besides comprehending all that can be seen from Springfield Hill, we have here views of the whole of the Carse of Gowrie, a portion of Dundee, and the county of Fife as far as St Andrews. This hill comprises spruce and fir, and has a bare top, with the remains of an old wood.

ST MARTINS has long been favourably known to arborists and others as a place of more than ordinary interest, and under the present proprietor, Colonel W. Macdonald Macdonald, it has been considerably improved. The place was originally known as the Kirklands, but was changed to St Martins, on the property being acquired by the Macdonalds. The purchaser of the property was William Macdonald, W.S., of Ranathan, and the circumstances under which he acquired it are rather interesting. While on his way north he happened to stay at the old inn, now used as the landsteward's house, and he was so strongly impressed with the beauties of the place, that he there and then expressed his resolution to purchase it should it ever come into the market. He was a man of considerable prominence in his day. Along with the then Sir John Sinclair, Bart. of Ulbster, his most intimate and lifelong friend, he was the originator o the Highland and Agricultural Society, as well as its first Principal or Hon. Secretary, and that body has on several occasions placed on record its high sense of the ability with which the duties of that office were discharged. At the general meeting on 29th June 1792, it was unanimously resolved, as a mark of the grateful sense which was entertained of the Secretary's unremitted and spirited attention to the objects and prosperity of the institution, that a piece of plate should be presented to him, with a suitable device and inscription. Mr Macdonald, while expressing his warm acknowledgements to the Society for the compliment, asked that the motion be deferred till the funds of the Society were in a better position, as, in the meantime, he considered the repeated approbations of his conduct a sufficient return for his exertions. The proposed presentation was, accordingly, delayed, but at the general meeting on 30th June 1800, a Committee suggested the propriety of the Society's voting a piece of plate, valued fifty guineas, to Mr Macdonald, and that he should be requested to sit for his portrait to Sir Henry Raeburn. This suggestion was cordially carried out, and the painting, bearing the following inscription, was placed in the Society's hall:—" William Macdonald, Esq. of St Martins, Secretary of the Highland Society of Scotland, as a mark of their regard and esteem, and of the high opinion they entertain of his services to the institution. 1803." A duplicate, considered to be the better of the two, was, at the same time, presented by the Society to Mr Macdonald, and is now in the dining-room at St Martins Abbey. Several pieces of plate were on various occasions presented to Mr Macdonald, and are highly valued as heirlooms at St Martins. Next year, Mr Macdonald resigned the Secretaryship, and was elected Honorary Treasurer. Still further to show its respect for Mr Macdonald of St Martins, upon his death, in 1814, a marble monument was erected inside St Martins Church, bearing the following inscription :—" To the memory of William Macdonald, Esq. of St Martins. Born 2nd April 1732; died 17th May 1814. Secretary to the Highland Society of Scotland from the time of its original institution till the year 1804, when he exchanged that appointment for the office of its Treasurer, which he filled till his death. This monument was erected by the Society in testimony of its deep and grateful sense of his eminent and important services." He was succeeded by his son, William Macdonald, advocate, who married Grisel, daughter of Sir William Miller, Bart. of Glenlee, a Lord of Session by the title of Lord Glenlee. Mr William Macdonald died without issue, when the estate came into the possession of the present proprietor, his cousin. Colonel Macdonald Macdonald is the representative of the branch of the great Clan of Macdonald which settled on Deeside after the Lord of the Isles engaged the King's forces at the Battle of Harlaw in 1411. There is no authentic account of this branch of the clan till 1630, when mention is made of Alexander Macdonald, known by the name of Marcus. His great-great-grandson was James Macdonald of Ranathan, father of the purchaser of St Martins. He fought for the Stuarts in 1715, joining with a younger brother, the standard of the Earl of Mar. Colonel Macdonald still owns two splendid claymores, Andrea Ferrara's, worn by his ancestors in that famous campaign, and which are made of metal so pure that the weapons can easily be bent double. This family did not go out in the '45. The Clanranald did so, but the ancestors of the Colonel always held to the Lord of the Isles, from whom they descend. Colonel Macdonald still retains, near Balmoral, two acres on the Hillside, being the burial place of the Macdonalds of Ranathan. The two last generations, however, lie under the Church of St Martins. Colonel Macdonald Macdonald is the only son of Major-General James Alexander Farquharson of Oakley, Governor of the Windward Isles, who died in 1834, by Rebecca, eldest daughter and co-heir of Sir George Colquhoun, Bart., of Tilly-Colquhoun; and a grandson of John Farquharson of Booughdarg in Glenshee, a military officer, and lineal descendant of Findlamore, by Christina, eldest daughter of James Macdonald of Ranathan. Upon the death of General Farquharson in 1834, from whom the present laird of St Martins inherited Glenshee, a magnificent monument was erected to his memory by the Government at the scene of his official labours, in connection with which he earned considerable distinction. Colonel Macdonald shortly afterwards obtained a photograph of this monument, and has erected a *fac simile* of it at the Dove Craig Park, St Martins, overlooking a beautiful artificial lake, with an island in the centre. A silver epergne of great beauty and massive proportions was presented to the Governor by the planters of St Lucia for the prompt suppression of a servile rebellion, very similar to the disastrous one which broke out more recently under Governor Eyre. William Farquharson Macdonald, on succeeding his uncle, the last male representative of the house of Colquhoun, in 1838, took the name of Colquhoun, but did not assume the baronetcy of Nova Scotia. On the death of William Macdonald of St Martins, Colonel Macdonald was obliged by his cousin's will to adopt the name and arms of Macdonald solely, and he, accordingly, became William Macdonald Macdonald, but his children bear both the names of Farquharson and Colquhoun. Colonel Macdonald claims the chieftainship of the Colquhouns, as nephew of the last heir male Sir Robert Barclay Colquhoun, of Colquhoun and Luss. He has claimed to be called out, through his mother, as co-heiress of his father, Sir George Colquhoun of Colquhoun and Luss, in the Nova Scotia baronetcy, created by Charles I. in 1625, in the same way as is often done with female baronies of England after they have been long periods in abeyance, but the claim has not yet been recognised. The honour of belonging to this distinguished family is one of which Colonel Macdonald may well be proud, as there are few periods of Scotland's history in which the heads of this ancient race did not perform their duty to their country, and adhere to the noble motto, *noblesse oblige.* Their intermarriages were with the first families of Scotland, including twice with the royal line of Stuarts and twice with the Grahams of Montrose. Among many distinguished marriages of the daughters of the House of Colquhoun that of Margaret, daughter of Sir John Colquhoun, Lord High Commissioner of Scotland, one of the greatest men in Scotland in his time, to Sir William Murray of Tullibardine, in 1459, connects the House of St Martins with all the Perthshire Murrays. This lady is said to have borne her husband 17 sons. Two marriages with the House of Montrose connect this branch of the Macdonalds with all the Perthshire Grahams, while other marriages less direct connect them with the Drummonds of Drummond Castle. The mansion-house, a massive and commodious building, was erected towards the close of last century, but has been considerably enlarged and adorned by the present proprietor, who has styled it St Martins Abbey, in consequence of monastic remains in the vicinity, and also of his intention gradually to change the architecture into an ecclesiastical baronial pile. The Colonel it may be remarked, holds his lands in the vicinity of the Abbey from the monks of Dunkeld, now represented by the Dukes of Athole. The policies and grounds were planned and laid out 25 years ago by Mr Craiggie-Halket of Cramond, a famous landscape-gardener. The trees were so thinned as to present handsome groups, avenues, and vistas, and shortly afterwards a splendid grass drive, extending from the mansion to the Coupar-Angus Road, was laid out by Mr Roos, London, who also planned some of the groups of trees along the route of this grass drive. The whole of this work was carried out under the personal superintendence of Colonel Macdonald, who is well known for his knowledge and taste in matters connected with landscape gardening. Many of the avenues are of exceptional interest. Stretching from the gardens to the Upper Walk leading to the house, is a magnificent double avenue of newer conifers, the inside consisting of about 40 *Wellingtonia giganteas*, and the outside of the same number of specimens of *Picea Nordmanniana*, the effect being at once unique and pretty. This avenue was planted by electors of the Angus Burghs after the rapid attack made by Colonel Macdonald on the seat of the Right Hon. W. E. Baxter in 1868, in which the Colonel defeated his opponent in the old constituency of 1200, but was terribly beaten by the new constituency of 7500, just created by his leader, the late Lord Beaconsfield. This attack is known to history as "The Raid of Montrose." The avenue bears this name, and an inscription on a handsome stone, records the circumstances under which it was planted. Perhaps the most remarkable of the newer coniferous trees at St Martins is a splendid *Wellingtonia*, girthing 9 feet 3 inches at the ground, and reaching a height of 25 feet. This tree was presented to the Colonel by his guardian, the late Sir P. Murray Threipland, Bart., of Fingask, as an acknowledgment of an annual supply of birches sent from St Martin's woods, not, as the Colonel facetiously remarked, for whipping the school-boys, but for clearing the way for the curling stones of their elders. The Colonel has appropriately named the tree after its donor. Another walk leading from the house to the garden is lined with very fine elms and sycamores, many of which are of great size. The best of all these is an elm in a park on the south side of the garden, a really splendid specimen. The girth at 1 foot from the ground is 22 feet, and 14 feet at 5 feet, with 15 feet of a bole, and a magnificent top. There are about half-a-dozen of elms closely approaching this size. One near the stable has a grand bole of about 40 feet, with a girth at 1 foot up of 12 feet, and 10 feet at 5 feet. The beeches are numerous and large. The best of them girths 14 feet at the narrowest part of the trunk, and although the bole is only 7 feet, it supports an enormous number of gigantic branches, which give the tree a most majestic appearance. Amongst the other trees deserving of special notice is a beautiful sycamore with a bole of 30 feet, and a girth of 13 feet at one foot, and 9 feet 9 inches at 5 feet up. At Friarton, and near one of the policy parks, there is a large ash marking the site of an ancient monastery. Although there is not now anything to indicate the presence of an ancient ecclesiastical edifice, there are a number of old people still living who remember of the last remnants of it being removed. The girth of this ash at 1 foot up is 19 feet, and at 5 feet the girth is $11\frac{1}{2}$ feet. A pinetum was formed about twenty years ago, and is stocked with many thriving specimens of the best-known varieties. At Cairnbeddie, which was at one time a separate property, there is a relic of the reign of Macbeth. Here he had one of his principal strongholds, and it was not until he fancied that he was insufficiently secure at Cairnbeddie that he built the great Castle of Dunsinane. The remains at Cairnbeddie were examined by the late John Murray Graham of Murrayshall, during the minority of Colonel Macdonald, his ward, and a considerable number of objects, but of no great interest, were discovered. Most of them are still at Murrayshall. Tradition has handed down the story, contrary to the version of Shakespeare, that it was here the tyrant met the weird sisters, and a remarkable conglomerate boulder is still pointed out as "The Witches' Stone." At Bandirran, a beautiful property in the neighbourhood, acquired sometime ago by Colonel Macdonald, there is a commodious and substantial mansion-house, surrounded by a number of aged trees of good size. Some of the finest of these are hedgerow trees on the road between St Martins and Bandirran. These include two large beech trees, the one having a girth of $15\frac{1}{2}$ feet at 1 foot, and $11\frac{1}{2}$ feet at 5 feet, with a bole of 24 feet; and the other has a girth of 14 feet at 1 foot, and 11 feet at 5 feet, with a bole of

14 feet. Two silver firs near the mansion girth 15 feet and 11½ feet respectively at 5 feet up, with boles of about 20 feet. The celebrated hill known as the King's Seat, a little to the east of Dunsinane, is on St Martins estate, in the parish of Cargill, where the Colonel is, after the Drummonds of Drummond Castle, the largest heritor. The greater part of the hill, which rises to a height of about 1200 feet above the level of the sea, is moorland, but Colonel Macdonald, with excellent taste, intends planting groups of trees, and forming lakes. From the laird of Dunsinane, he has obtained permission to make a green drive through that property, so as to connect St Martins Abbey by a ten miles ride with this beautiful hill. The plantations upon the property, including some valuable timber at Glenshee, are very extensive. A large proportion of this wood has been planted since the present laird succeeded to the property. During the forty years he has been in possession, Colonel Macdonald has planted at least thirty miles of hedges, and otherwise done much to improve the general aspect of this part of the country, which, up till about 1745, was a bleak moor, with the exception of the valley of St Martins, extending about 10 miles from the Sidlaws to the Tay, but not above half-a-mile broad.

At CARDEAN, in the extreme north-east corner of Perthshire, and partly in Forfarshire, we have an excellent example of what can be done within a very few years to improve the appearance of a property, when taste, energy and capital are brought into play. Situated in one of the most delightful parts of the fertile valley of Strathmore, the natural surroundings easily lend themselves to artificial adornment, and these have been taken full advantage of by the present proprietor, James Cox of Cardean and Clement Park, Lochee, the senior partner of the eminent firm of Cox Brothers, Dundee. Although it is only recently that Mr Cox became a landed proprietor in Perthshire, he has all his life occupied a prominent position in the adjoining County of Forfar, where he is a Deputy-Lieutenant, a J.P., and an ex-Provost of Dundee, and can trace his descent from two notable families in Perthshire, both hailing from the Carse of Gowrie. On the one side he is connected with the family which was instrumental in saving the life of Wallace, while making his escape from Dundee, after killing the Governor's son; and, on the other, he owns the blood of one of those sturdy families from Holland, who settled in this country as artisans, during the reign of James IV., or early in the 16th century. The story of Wallace's escape from assassination, as it is told in the Old Statistical Account of Longforgan, is pretty well known. As he was flying from Dundee, hotly pursued, he sat upon a stone standing by the side of the door of a house belonging to a family named Smith, to rest himself, and was supplied with bread and milk by the good wife of the house, which was the means of restoring his exhausted strength. This stone was such a peculiar one, that there has been no difficulty in identifying till this day the family to whose hospitality we are indebted for aiding in the preservation of the life of the great Scottish Patriot. In Sir John Sinclair's work, mention is made of "a very respectable man in Longforgan," of the name of Smith, a weaver, and the farmer of a few acres of lands, who had in his possession a stone called "Wallace's Stone" with the traditionary story attached. The stone was hollow in the centre, like a large mortar, and had been used in ancient times for making pot barley, its usual place, when out of use, being outside the door. This stone continued in the possession of the same family up till twenty years ago, when it was presented to Mr Paterson, Castle Huntly, by Miss Smith, when she vacated the house which is believed to have been occupied by her ancestors, with some alterations, since the days of Wallace, or for more than 500 years. Miss Smith, who is well advanced in years, is still living in Dundee, and is a second cousin of Mr Cox of Cardean and Clement Park, Mr Cox being related to this notable family through his paternal grandmother. The connection was brought about by the marriage, in 1739, of John Smith to Helen Cock, their daughter, Helen, being Mr Cox's grandmother, while their son was the father of Miss Smith, to whom we have referred as living in Dundee. The pedigree of the family of Cox, as we have indicated, goes back to the beginning of the 16th century, when they settled in the Carse of Gowrie on arriving, as is believed, from Holland. It is thought that from the first they were weavers and bleachers, the trade being carried on from father to son. At all events, they can be traced in this connection for eight generations, the works gradually increasing in their hands until they have attained their present enormous proportions, being the largest works of the kind in the world. Upon the introduction of jute into this country, Mr Cox at once realised its importance, and adopted its manufacture with such success that employment is now given to 5000 workpeople, who are engaged in working up the raw material from bales brought direct from their house in India till it is placed in the market in such widely diversified forms as the coarsest packing and the finest imitation of human hair. The family registers show that the orthography of the name has, like so many other family names, passed through many changes. The first spelling of the name we find in the records is "Cock," the same as that of the valiant Burgomaster of Amsterdam, "Frans Banning Cock," represented in the "Night Watch" by Rembrandt. At other times we find it spelt "Coke," "Coock," "Cuik," "Cok," "Cocks," &c., the present mode of spelling the name having been adopted by Mr Cox of Cardean in 1842. The estate of Cardean was purchased by Mr Cox in 1878, from the heirs of the late Admiral Popham. For about one mile, it is bounded on the north by the Isla and by the Dean, one of the finest trouting streams in Scotland. The Dean takes its rise in the Loch of Forfar, flows through the policies, and presents many charming little "bits" of scenery. Openings have been made at the most romantic points, the beauties of some of which could scarcely be excelled, particularly at those parts which include a view of a very ancient bridge, believed to have been originally constructed in connection with the Roman camp in the neighbourhood. The old name of the property, "Caer-Dean," signifies the camp, or fortress upon the Dean. The camp from which Cardean derives its name was in a good state of preservation between twenty and thirty years ago, but it has since been almost entirely ploughed up. When Mr Cox entered into possession he found the fine old bridge leading to the camp in a state of ruin, the northmost of the three arches having been blown up with gunpowder about 50 years previously, to prevent

A "BIT" ON THE DEAN.

it being used by the public in preference to a new one which had been some years previously built farther down the river. The ruined arch has been restored by Mr Cox in its original style, the stones having been partly taken from the dry arch on the south side, which is now built up. The entire bridge was at the same time thoroughly renovated, and now forms a very picturesque feature in the landscape, as well as a serviceable road. The Dean bounds the property on the south-east for about 2½ miles, and its passage through the policies is the only

OLD ROMAN BRIDGE.

part of its course where the water has a good fall, producing that gentle murmur which is so delightful to the ear. The northern bank stands high above the river, and is richly wooded with good specimens of oak, mountain ash, beech, sycamore, and a few soft-wood trees. When Mr Cox acquired the property, the river flowed through the policies at its own sweet will, a large portion of the southern bank having been washed away, while the walls of a lade which runs along the river side were broken down, so that the water formed little islands where

these were most objectionable. The whole of this portion of
the policies, indeed, was a veritable wilderness, and those who
know it only by its former state will scarcely recognise it now.
The southern bank of the Dean and the channel of the lade have
been substantially built with stone, and the ground has been
tastefully laid out with ornamental trees and shrubs. At the
top of the lade a very fine croy has also been built by
Mr Cox, to enable the salmon, which frequent the river
in great numbers, to ascend. An old quarry, which interfered with the amenity of the place has been filled up, and
converted into a bowling green. Numerous walks have also
been formed, an elegant new bridge has been thrown across
the river, and a gas-work erected. The mansion-house is
comparatively new, having been built in 1852, and is a
beautiful structure, finely situated, and surrounded by a number
of grand old trees. The first of the trees to attract attention is
as magnificent a specimen of the lime as can be seen anywhere.
With a height of 40 feet, it is compactly clothed with rich
foliage to the ground, the branches spreading along the lawn in
a most beautiful manner. The circumference of the branches
is 120 feet, and the tree throughout is in perfect form, adding
not a little picturesque beauty to the house. There are a number
of gigantic beeches on the lawn and neighbourhood. These are
the remains of an avenue which at one time led to the mansion.
There are eight of these old avenue trees still standing, and the
one nearest the house, which we measured, may be taken as a
fair sample of the whole. Clear of the swell of the roots, the
girth is 16 feet 2 inches, and at the narrowest part of the bole
the girth is 15 feet. The bole is only about 6 feet, but the chief
beauty of the tree is its magnificent arms, ten of which break off
from the trunk with a considerable girth, and bring the height
of the tree up to fully 100 feet, making it altogether a truly
noble specimen. An oak, which had formed part of the same
avenue, girths 15 feet 10 inches at 1 foot up, and 11 feet at 5
feet. It has a bole of 9 feet, and carries a very fine head.
There are a few specimens of the newer coniferous trees. The
most notable of these is a nice *Wellingtonia gigantea*, 30 feet
high, and girthing 7 feet 11 inches at the ground. As the soil

is mostly rich loam, the plantations are not very extensive, but in an estate of 2000 acres there are fully 150 acres under wood, which is well mixed with the common varieties of hard and soft timber trees.

THE GREAT BEECH AT CARDEAN.

CLATHICK, in the neighbourhood of Crieff, and the property of Colonel W. Campbell Coloquhoun, also affords a very striking example of what may be done within a comparatively short period to improve the amenities of an estate. The property is a small one, nicely situated between Ochtertyre and Lawers, and enjoys much of the natural advantages of these famous places. The house is handsome and commodious, and possesses every modern convenience. An ample supply of splendid spring water was recently introduced at a cost of £200, the pressure being sufficient to throw a column of water 30 feet above the top of the house. There is very little old timber, although there are a few trees of considerable size. In the North Park, for instance, there is an oak girthing 17 feet 8 inches at 1 foot from the ground, and 12 feet 10 inches at 5 feet, with a good round bole of 11 feet. It is altogether a fine spreading and highly-ornamental tree. A grand ash has a bole of 19 feet, with a girth of 12 feet at 1 foot, and 10 feet at 5 feet. An oak measures 19 feet of bole, with a girth of 13 feet 6 in. at 1 foot, and 11 feet 8 in. at 5 feet. There are several other oaks which reach a similar girth. A beautiful silver fir on the east approach girths 13 feet at 1 foot, and 9 feet 9 inches at 5 feet. There are close upon 150 acres of plantation, the principal of these being the High Wood, 110 acres in extent. It consists almost entirely of larch from 50 to 60 years old. It was originally pretty well mixed with Scots fir, but these did not thrive, probably on account of the variety being inferior to the ones at Lawers, Dunira, and Ochtertyre, where there are so many grand specimens. A good many of the larches in the west portion of the High Wood are suffering from disease. The Home Wood, adjoining the North Park, was planted in 1876, and extends to 20 acres, oak coppice having been cleared out to make way for larch. The plants were bought in as saplings, each one being properly pitted, and they are doing so well that many of them have reached a height of 15 feet. Colonel Colquhoun is strongly of opinion that the best larch seed is not preserved for propagation. It is chiefly, he says, old men who make a livelihood by gathering larch seed, and they are disposed to take the easiest means to secure the cones. Old and effete plants bear the largest number of cones, and these being nearest the ground are the most easily collected. He also believes that an exchange of seed between different counties would be advantageous. By these means he considers that they would be able to grow larches of the Dunkeld and Monzie stamp, instead of seeing them pine away when they become sleeper size. A considerable extent of fencing has been erected within the past few years, notably in 1876. The Colonel has a very high opinion of Morton's galvanised tubular sheet upright wire fencing, of which he had one mile erected in 1876 at a cost of 2s. 4d. per yard. The back stays have proved extremely effective, and almost invisible above the ground. The fence has given him entire satisfaction, being strikingly handsome and strong, although rather expensive. Many other improvements of a useful and ornamental character have been effected since Colonel Colquhoun came into possession of the property, in the management of which he takes a thoroughly practical interest.

CULTOQUHEY, also in the neighbourhood of Crieff, is deservedly famed for its fine collection of newer coniferous trees, in the rearing of which the proprietor, James Maxtone-Graham, takes a very great interest, most of these having been planted by himself. It is only in recent years that the latter appellation has been added to the family name. The name and arms of Graham were assumed when the present laird succeeded to the property of his uncle, Robert Graham of Redgorton, cousin and heir of Thomas Graham, M.P. for Perthshire, the celebrated commander in the Peninsular War, who achieved the memorable victory of Barossa, and was created Baron Lynedoch. The mansion-house at Cultoquhey was built from a design by the late Sir Robert Smirke. It is in the early Tudor style, and is a pure and good example of that great architect's work. Perhaps the grandest of all the trees is a magnificently-furnished *Pinus Cembra*, 40 feet high, and one of the finest trees of the kind in Scotland. The girth of the stem is 9 feet 2 inches at 1 foot, and 6 feet 3 inches at 5 feet. This tree was brought in a pot from Cumbernauld in 1826, and planted in what was then the garden, where it now stands. So little were the coniferæ valued in those days, that the old gardener of the place was in the habit of pruning it annually because it encroached on one of his black currant bushes. The tree bears the marks of this early and severe castigation till the present day. There are also many beautiful specimens of *Wellingtonia gigantea*, the best specimen having a girth of 10 feet 6 inches at the ground, and a height of 35 feet. A fine compact specimen of *Thujopsis dolobrata*, planted in 1871, has a circumference round the branches of 27 feet. The best *Cedrus Deodara* reaches a height of 60 feet. One of the specimens of *Abies Douglasii* has a girth of 8 feet at 1 foot, and 6 feet at 5 feet up. This plant was a seedling of 1844, from the parent tree at Lynedoch, which was given, with others, to Lord Lynedoch, by Mr Douglas, the introducer of the tree. There is a grand plant of *Picea nobilis*, from which large numbers of seedlings have been raised. The girth of the stem is 7 feet 10 inches at 1 foot, and 6 feet 6 inches at 5 feet. There are two fine specimens of *Picea Nordmanniana*, about 50 feet high, and 6 feet 2 inches in girth at 1 foot from the ground. There are also fine specimens of *Picea Orientalis*, *Picea grandis*, *Picea Morinda*, *Picea Webbianna*, *Picea magnifica*, *Cupressus Lawsoniana*, *Cryptomeria Japonica*, &c. Two of the larches are above the average size, the one having a girth of 12 feet 6 inches at 1 foot, and 9 feet 10 inches at 5 feet, and the other having a girth of 10 feet 2 inches at 1 foot, and 8 feet at 5 feet up. The best of the hardwood trees is a beautiful oak near the house. The girth at 1 foot up is 15 feet, and at 5 feet the girth is 12 feet 6 inches, there being a bole of 10 feet. An oak and Spanish chestnut over-hanging the terrace in front of the house, are interesting from the fact that the acorn and chestnut from which they grew were brought from the Hague by the late Mrs Maxtone in 1827. A younger oak on the same terrace is from an acorn off the great oak in Knolle Park, Kent, one of the largest oaks in England, and brought by Mrs Maxtone-Graham in 1851. One of the features of the policies is a splendid collection of rhododendrons. The principal objects of interest, however, as we have already said, are the coniferous plants, and their chief characteristics are their compactness and vigour. There are a variety of soils, but the bulk of the land is strong, resting upon red sandstone. The best of the plants grow upon a slope in front of the house, having a southern exposure, and formerly the kitchen garden, the soil of which is a fine black loam.

MONZIE CASTLE, in the neighbourhood of Crieff, is celebrated for the possession of four of the original larches, which have attained a great size. This estate was long the property of a branch of the noble house of Breadalbane, but is now in the possession of Mr Johnston of Lathrish and Kilwhis, who acquired it some years ago by purchase. The ancestor of the old Monzie family was Archibald, fifth son of Sir Duncan Campbell—created first Baronet of Glenurchy in 1625—by his wife, Lady Jane Stewart, second daughter of John, Earl of Athole. Several of the Campbells of Monzie have occupied prominent positions in the civil and military annals of the country; the last male member of the family who possessed Monzie being Member of Parliament for Argyllshire from 1841 to 1843. The Castle is one of the most magnificent and pleasantly-situated mansions in Scotland, and is of great age, one of the older portions bearing the date 1634. On every side it is surrounded with magnificent timber. The beauty of its sylvan adornments attracted the attention of the writer of the New Statistical Account so much that he was led to remark that "were it not a well-authenticated fact that Dr Johnson regarded Scotland as a region destitute of trees, one might be tempted to conclude that he had visited this spot, and received from it his first impression of the 'Happy Valley' in *Rasselas*." The principal arboreal attractions, however, are the great larches in the garden. Although there are now only four of them, there were originally six, and are said to have been brought or sent by Mr Menzies of Culdares, who brought the ones at Dunkeld. The largest of the four girths 26½ feet at the surface of the ground, at 3 feet the girth is 18 feet, and at 5 feet the girth is 16 feet 3 inches. The height is fully 100 feet, and the tree contains about 380 cubic feet of timber. The tree is splendidly clothed with branches almost to the ground. Some of these branches are about 45 feet in length, and are so fantastically twisted and curved that they have to be supported on forked poles to preserve them from injury. On the south side, the branches are particularly beautiful, hanging over the carriage drive, and dipping into the waters of an ornamental lake. The girth of the second largest one is 22 feet 3 inches at the base, 15 feet 6 inches above the swell of the roots, and 12 feet at 5 feet, measuring from the highest part of the ground —the cubic contents being about 300 feet. The third in size girths 16 feet at the base, 10 feet 1 inch at 3 feet up, and 9 feet 7 inches at 5 feet—the cubic contents being about 270 feet. The smallest of the four has a girth of 14 feet 8 inches at the base, and 9 feet 11 inches at 3 feet up, and 8 feet 8 inches at 5 feet—the cubic contents being about 200 feet. The height of these three trees is very little less than the largest one, all of them approaching 100 feet. Although now about 145 years old, they are exceedingly healthy, and continue to make wood rapidly. The soil is dry, and slopes gently on what was once the side of a shrubbery fence. The second largest tree has a peculiar protuberance, the result of having stood in the way of the fence. The exposure is to the south and east, and the trees are sheltered from the north and west by rising ground and umbrageous trees. The "Planting Duke" of Athole took a particular interest in these rivals of the Dunkeld larches, and annually sent his gardener to report upon their progress. When this functionary reported that the Monzie larches were outstripping the ones at Dunkeld, his Grace would jocularly allege that his servant had permitted General Campbell's good cheer to impair his powers of observation. Besides the larches, there are many other fine trees in the garden. An elm has a girth of 15 feet at 1 foot up, and 13 feet at 5 feet, with a bole of 18 feet. A beech tree, with 23 feet of a bole, has a girth of 17 feet at 1 foot, and 15 feet at 5 feet. A lime in one of the parks girths 17 feet 5 inches at the narrowest part of a bole of 8 feet, and has a spread of branches of 80 feet. The circumference of the tree is altogether 256 feet. At one time it was much greater, but on the removal of a protecting fence it was pruned. Another lime near the Castle girths 14 feet at the narrowest part of a bole of 10 feet, and also carries a magnificent head. There are several fine beeches. One girths 14 feet at the narrowest part of the bole. Another girths 14 feet 8 inches above the swell of the roots, and 13 feet at 5 feet up. There is a very peculiar oak, with a short bole of 5 feet, after which it breaks into two great branches. The bole girths 14 feet 8 inches, and the branches girth 10 feet 6 inches and 9 feet 10 inches respectively. "Eppie Callum's oak" is so named because of its having been planted by that genial old lady of a former generation in a tea-pot, from which it was transplanted to her garden, where it has become a great tree. There is also a grand silver fir in the policies. The girth at the base is 20 feet 8 inches, 18 feet at 1 foot up, 14 feet 6 inches at 3 feet, and 13 feet 10 inches at 5 feet. Her Majesty, when residing at Drummond Castle, paid a visit to Monzie, in commemoration of which she planted three trees. The Castle has been let for a number of years to Mr John Bald, London, who has effected many improvements both in the policies and within the Castle.

DELVINE, the property of Sir Alexander Muir Mackenzie, Bart., is prettily situated in the Stormont, and includes as fine a tract of agricultural land as is to be found in Perthshire. The estate is remarkably well farmed—a circumstance which is largely due to the spirit of enterprise infused into the tenants by the present and recent proprietors. The original name of the family was Muir, but on Alexander Muir succeeding, at the beginning of the century, to the estate of his great uncle, John Mackenzie of Delvine, he assumed the name of Mackenzie, and was created a baronet in 1805. On succeeding to the property he found it in a condition that may be expressively described as a wilderness. He was, fortunately, a man of indomitable energy and enlightened enterprise, and at once recognised in his estate a congenial field for the exercise of these qualities. He first of all set to work in laying out the farms, making roads, planting hedgerows and woods, and otherwise putting the property into something like the condition it is to-day. He had the pleasure of seeing the improvements he had inaugurated fully carried out, and when he died, in 1835, he was succeeded by his son, Sir John William Pitt Muir Mackenzie. He was trained for the Bar, to which he was admitted in 1830, but he devoted a large share of his attention to the management of the estate. The laying out of some 12 miles of walks, and otherwise improving the amenity of the place, was the work of the late Sir John M. Mackenzie, who also, along with Lady

Mackenzie, laid out the tastefully formed gardens and shrubberies. He died in 1855, when he was succeeded by his son, the present laird, who possesses, in a large degree, the characteristics of his grandfather. Time, as a matter of course, wrought its ravages upon the farm buildings, and modern experience has worked quite a revolution in agriculture since Sir Alexander's grandfather carried out his improvements. Sir Alexander Muir Mackenzie, however, has more than progressed with the times, and at the present moment his steadings and farm buildings are about the most complete in the country. He has, at a very considerable expense, had all his farm buildings overhauled and enlarged, and the whole of the cattle courts covered with corrugated iron roofs. Special attention has also been paid to the conveniences of the farm. Water has been introduced in abundance wherever necessary, and the various additions have been so arranged as to reduce labour to a minimum. Altogether, the estate, from an agricultural standpoint, may be described as a model one ; and that the efforts of the laird to maintain it in this honourable position are not unappreciated by the tenants, may be inferred from the remark of one of them, who merely gave expression, in a few words, to the sentiments of all with whom we have come in contact, "We have a grand farm, and a grand landlord." While Sir Alexander Muir Mackenzie has, very properly, devoted the largest share of his attention to the maintenance of his agricultural subjects, he has not been unmindful of his woods. He prides himself upon being his own forester, and a pretty minute examination showed that the woods are not suffering at his hands, either in the policies or the plantations. The policies, or rather the North Bank adjoining the policies, are adorned with three splendid larches, believed to have been planted at the same time as the parent ones at Dunkeld. They are all nearly 100 feet in height, and the girth of the largest one is 15 feet 4 inches at 1 foot, and 11 feet 2 inches at 5 feet. The others girth 14 feet 10 inches at 1 foot, and 11 feet 2 inches at 5 feet ; and 14 feet 6 inches at 1 foot, and 10 feet 2 inches at 5 feet. The whole of this North Bank is filled with splendid timber trees, the beeches being especially good—one which was measured girthing 12 feet 5 inches at 1 foot, and 10 feet 10 inches at 5 feet. On the South Bank there are also a large number of splendid trees, including beech, oak, ash, larch, and Scots fir. These banks form the sides of Inchtuthil, a flat of about 200 acres, on the north point of which Delvine House is beautifully situated. This place derives its name from its having at one period been surrounded by the waters of the Tay. Tradition even says that Delvine was, in comparatively recent times, on the south side of the river, instead of the north as at present; and the names of some of the places, such as "Laird Young's Cast," indicate that salmon have been caught at spots which are now far from the river. Mr Crerar, who has been landsteward at Delvine for close upon sixty years, pointed out to us a spot where he has frequently killed salmon, with the present laird's grandfather, but which is now beyond the reach of the river except when there is a very high flood. Inchtuthil was an important Roman station, and is supposed by some to be the station *In medio*, to which Tacitus says that his father-in-law, Argicola, led his troops after the battle of Mons Grampius. Prior to the Roman invasion, Tulina, one of the principal towns of the Picts, stood upon this quondam island, and vestiges of Pictish defences are still to be seen. The Romans had a large camp on the north-east portion of the island, the remains of which are also still tolerably complete, and which have yielded several valuable relics, including a Roman bath in good preservation. The approach to Delvine House is through a splendid avenue, about one mile long, leading from Dunkeld Road. At the entrance the trees are of different varieties—beech, oak, and sycamore chiefly—ranging about 10 or 12 feet in girth, but towards the house both sides are planted with finely-shaped lime trees in the full vigour of their growth. The policies are adorned with several trees of considerable size. The beeches include a gnarled but very picturesque and branchy tree, with a short bole about 13 feet in girth; and another beautiful timber tree girthing 12 feet 10 inches at 1 foot, and 11 feet 2 inches at 5 feet. There is also a magnificent purple beech, with a great spread of branches. There are several fine silver firs, the largest one girthing 13 feet at 1 foot, and 11 feet at 5 feet up. The best of the spruce trees has a girth of 11 feet 8 inches at 1 foot, and 9 feet 4 inches at 5 feet. The newer coniferous trees are chiefly represented by young specimens, although there is a well developed *Wellingtonia gigantea* girthing 9 feet at the ground, and some handsome araucarias. Sir Alexander Muir Mackenzie has done a good deal of planting, chiefly in strips, since he came into possession of the property. The plantations extend to about 350 acres. The most interesting of these is the Hill of Caputh, where there are over 120 acres of wood. This hill rises to a considerable elevation, and commands a wide stretch of the fairest country in Perthshire. About six miles of the Tay, from Dalpowie to near Meikleour, are seen at a glance; and on the southern bank of the river the whole of the Murtly property, with its magnificent castle, rare avenues and park trees, extensive plantations, trackless moors, and highly-cultivated fields, are spread out before us. To the north and west the view is bounded by the Birnam and Logiealmond Hills, and to the south the eye wanders over a variety of interesting properties until it rests upon the towers and chimneys of the "Fair City." The wood here, as in the other plantations is chiefly larch, spruce, Scots fir, and *Abies Douglasii*, with a sprinkling of hardwood, the bulk of the timber being of mature age. Sir Alexander has the additional honour of having added considerably to the family possessions, the adjoining property of Glendelvine having been purchased by him about ten years ago. It is a very pretty little estate, with a nice house, and lies right in the heart of Delvine itself, so that its possession was very desirable to square the property.

SNAIGOW, which marches with Delvine, was purchased in 1874, from the widow of Mr Keay, a well-known advocate, by William Cox, brother of Mr Cox of Cardean, and a member of the firm of Cox Brothers, Dundee. Mrs Keay occupied the property for a considerable number of years before it came into the hands of Mr Cox, who found it very much out of repair. Like his brother at Cardean, Mr William Cox at once commenced to effect a transformation on the estate, and has spent large sums of money upon improvements of all kinds. He has reset about thirty miles of dykes, drained all the wet parts of the property, and made numerous broken roads passable. For the mending of the roads he utilised many loads of stones that had been accumulating for years at the corners of fields, and by the sides of dykes, where they were an eyesore, and had them broken up by a stone-breaker which he procured for the purpose. At many parts of the estate there were no dykes, but these places have now been enclosed with wire-fencing. All the farm steadings were out of repair, and these Mr Cox has put in first-rate order, and erected covered reeds, without any charge being made on farmers under existing leases. There were about 30 parks laid out in grass when Mr Cox purchased the property. These fields he caused to be ploughed up and put through a course of cropping, and afterwards sown out with grass, without a grain crop, the result being pasture lands of unusual richness, rendered all the more valuable for grazing by the nicely undulated character of the ground, which causes burns to flow alongside almost every field and supply the stock with abundance of water. About one-fourth of the property was covered with wood, principally matured larch, when it was purchased by Mr Cox, who has since planted about half-a-million of trees, chiefly Scots fir, spruce, and larch, on the rougher ground and the worst of the fields. In the policies there are a number of very good beech, lime, elm, and oak trees about 150 years of age. The largest of the trees, however, are at Dungarthill, a nice-lying place near Stenton, and where there was at one time a mansion. A number of the larches here are about the same age as the parent ones at Dunkeld, and have attained considerable size. The three best larches measure as follows:—No. 1, 113 feet high; girth at 1 foot up, 14 feet 2 inches; at 5 feet up, 9 feet 9 inches; at 20 feet, 8 feet 6 inches; and at 100 feet, 2 feet:—No. 2, 105 feet high; girth at 1 foot, 12 feet 8 inches; at 5 feet, 9 feet 3 inches; and at 20 feet, 7 feet 6 inches:—and No. 3, over 100 feet high; girth at 1 foot, 13 feet; at 5 feet, 10 feet; and at 20 feet, 7 feet 8 inches. There are also a few fine old larches varying in height from 80 to 100 feet, and having an average girth of from

SNAIGOW—STENTON.

8 to 9 feet at 5 feet up. An elm has a girth of 16 feet at 5 feet up. There are also a number of ash, beech, and lime trees, girthing from 12 feet 6 inches to 16 feet at 5 feet up, and having a height of from 30 to 40 feet. On the farm of Cults on the higher lands of Snaigow, there are three Druidical stones highly prized by antiquarians, the best of them being about 10 feet high. This part of the estate is exceedingly attractive from a scenic point of view, as it is sufficently high to command a view of the beautiful chain of lochs between Dunkeld and Blairgowrie.

STENTON.

STENTON, the property of Thomas Graham Murray, W.S., brother of the late John Murray Graham of Murrayshall, and uncle of the present proprietor, is one of the most sweetly-situated residences on the banks of the Tay. Wedged in on either side by Delvine and Dungarthill, it directly overlooks, with a southern exposure, the lordly domain of Murtly, and is backed by a tree-clad hill and cliff of singular beauty. Looking from the river, it would be difficult to conceive a more charming picture than is presented by this quiet retreat. The name is supposed to be derived from "stane toun," or the town (*Anglicé*, buildings) beside the rock, the place having been long the residence of a family of Stewarts of Stenton. The Rock of Stenton rises to a height of 300 feet above the house, and the summit, reached by a gently winding walk through a wood, has the remains of several ditches belonging to a former stronghold, known in the country as "The Robber's Hold," from which, tradition reports, pilgrims to Dunkeld Cathedral were descried and plundered. The views from two points of the summit are most extensive and varied. To the east, the view looks over the Stormont, and the windings of the Tay, with part of Strathmore, to the Sidlaw and Ochil ranges, and the Lomonds of Fife beyond. The view to the west commands further windings of the river, with Birnam Hill, the entrance to Dunkeld, &c. The oldest existing part of the house bears the date of 1745, and it is said that the slaters were roofing it when Prince Charles Edward passed along the road. The larger portion of the present house was built by Mr Murray in 1860, and is in the Italian style. The garden is beautifully laid out at the foot of the Rock, which forms an unusual and picturesque background. There are in the grounds some good specimens of the newer coniferous trees, and some good timber trees, although not sufficiently large to be specially mentioned.

At BIRNAM, the property of John Guthrie Lornie, there are some very fine specimens of beech, sycamore, and other trees, as may be understood from the fact that the estate is situated in the centre of the original forest of Birnam, "the last of Great Birnam Wood," as represented in the frontispiece, being in front of the mansion-house, although on the Murtly property. In one part of the estate, there are twelve beeches having a girth, at 5 feet from the ground of from 8 feet 6 inches to 16 feet 2 inches; and two sycamores girthing 9 feet and 14 feet 6 inches respectively. They have a splendid spread of branches, and are two of the most beautifully-proportioned trees in the district. There are also numerous young oaks, limes beeches, chestnuts, cedars, larches, and firs in a most healthy state. At Pitcastle in Strathtay, also the property of Mr Guthrie Lornie, there are a number of very good trees, although few of them are above the ordinary size. The situation of this estate is exceedingly picturesque, and in the "Broad Shot," a graceful bend of the Tay, it possesses one of the best angling stretches of the river. There are about 30 acres of woodlands, including many fine oak standards, as well as good beech, elm, sycamore, and other trees. Both at Birnam and Pitcastle, Mr Guthrie Lornie contemplates adding considerably to the plantations.

At LOGIERAIT, there is an enormous ash in the hotel-keeper's garden. It attracted the attention of the writer of the *New Statistical Account*, who gives the girth at the ground as $53\frac{1}{2}$ feet; at 3 feet the girth was 40 feet; and at 11 feet from the ground, 22 feet, the height of the tree being 60 feet. The height is said to have been 90 feet at one time, but the top was blown away. There is an opening of 5 feet 9 inches on the south side, and the trunk is perfectly hollow. Maimed as it is, it has still the enormous girth of 47 feet 7 inches at 1 foot from the ground, and 32 feet 5 inches at 5 feet, so that in its perfect state, it must have been a very great tree indeed. In spite of the wide gap on the south side of the trunk, the north side is clothed with luxuriant foliage, springing from numerous branches. The main trunk is thickly covered with ivy. Advantage has been taken of the opening in the trunk to form a summer-house, which is large enough to hold a large party.

BONSKEID, the seat of George Freeland Barbour, is deservedly classed amongst the most beautiful properties in Perthshire, and is of considerable interest from an arboricultural point of view. It occupies a delightful situation between Blair-Athole and Pitlochry, and is surrounded by an amphitheatre of fine hills. On the south it is bounded by the classic Tummel, and on the east by the Garry and the famous Pass of Killiecrankie. It was to the woods around Bonskeid that King Robert the Bruce retired after the Battle of Methven in 1306, and it was as a reward for the hospitality then shown to the defeated Scots that the lands were granted to the ancestors of the family still in possession. The building of the mansion has a curious history attaching to it. One Sabbath, about a hundred years ago, the house was burned to the ground while the family were at church. Thirty years afterwards, Lady Bath and her husband, Sir James Pultney, were so much charmed with the place that they took a lease of it, and set about building a new house, taking up their abode in a tent, that they might enjoy the scenery and superintend the work. Lady Bath died before the house was finished, but Sir James completed the work, and furnished the house, but he never had the good fortune to inhabit it, as, while testing a firelock which he had purchased, the piece went off and inflicted a fatal wound. The present proprietor took down the greater part of this house, and rebuilt it in 1868, from plans prepared by Mr A. Heiton, Perth. The site of the house commands an extensive view of the grand valley of the Tummel, the Howe of Moulin, &c. Perhaps the most notable of the trees is that known as the "Stewart Larch," planted by Dr Stewart upwards of 100 years ago, and is the only survivor of three planted at the same time. At 1 foot from the ground the girth is 12 feet, and at 5 feet the girth is 9 feet ; and at 12 feet from the ground the girth is 8 feet 9 inches. The height of the tree is 97 feet, and it contains about 190 cubic feet of timber. We are indebted to Mr Michie, author of *The Larch*, for our illustration of this fine tree. Several other larches on the estate are considerably over 100 feet in height, but, although they are splendid timber trees, they cannot be compared with this one for symmetry. There are also several fine specimens of the Scots fir. A number of the newer coniferous trees have done very well. One of the specimens of *Cupressus Lawsoniana* belongs to the first batch of seedlings raised by Mr Lawson, and a more symmetrical tree could not be found. The height is 38 feet, and the tree has never been pruned. *Abies magnifica* does very well, reaching a height of 33 feet, and beautifully clothed to the ground. A *Wellingtonia gigantea* reaches a height of 37 feet ; and *Abies Albertiana* and *Araucaria imbricata* to a height of 30 feet.

Shrubs which do not stand the winter at places further south are here found to thrive very well, including such plants as

THE STEWART LARCH.

Escallonia macrantha, Catalpa syringæfolia, &c. The plantations extend to 760 acres, and consist of natural birch, larch,

Scots fir, and oak. The birches present one of the grandest features of the estate. The trees, which are chiefly of the weeping variety, are of fine form, and in the autumn especially, when they are taking on their golden hues, their light tassellated branches waving in the air present a most lovely sight. A good many *Abies Douglasii* have been planted out, and are doing well.

The estate of LUDE, in the Athole district, the property of Mr M'Inroy, contains several trees of more than ordinary size. The most prominent of these are the larches. On the Beech Walk there is a larch, planted in 1737, having a girth of 9 feet 9 inches at 1 foot, and 8 feet 5 inches at 5 feet, with a height of 75 feet. At the west end of the Long Terrace, there is another aged larch, probably about the same age, having a girth of over 12 feet, and a height of 50 feet. On the Garden Walk there are the following fine specimens of the larch :—No. 1, 11 feet 6 inches at 1 foot, and 9 feet 6 inches at 5 feet, and 85 feet high ; No. 2, 11 feet at 1 foot, and 8 feet at 5 feet, with a height of 85 feet; and No. 3, 9 feet at 1 foot, and 8 feet at 5 feet, with a height of 85 feet. There are also a number of very good Scots firs, about 130 years of age, one of them, which may be taken as an average, having a girth of 11 feet at 1 foot, and 9 feet 6 inches at 5 feet, with a height of 80 feet. There are several silver firs over 100 feet in height and of corresponding girth,—one of those 100 feet high girthing 12 feet at 1 foot, and 10 feet at 5 feet. Some fine standard oaks are scattered over the parks and policies ; and on the Lower Terrace, which extends to about 680 yards in length, there is a row of fine lime trees, while immediately to the right of the present mansion, are four very old sycamores of respectable size. About 80 acres of moorland have been enclosed with the intention of planting larch, and a few Scots firs, in the spring of 1884.

KILBRYDE CASTLE, about three miles from Dunblane, is the seat of Sir James Campbell of Aberuchill and Kilbryde, the representative of that branch of the Campbells whom we have already referred to in connection with Lawers. The barony of Kilbryde, with its fine historic castle, a seat of the ancient Earls of Monteith, was acquired by Sir Colin Campbell of Aberuchill in 1669. Sir Colin took an active part in public business. After the Revolution he was appointed an ordinary Lord of Session, taking his seat on the Bench in November 1689. In January following, he was made a Lord of Justiciary and Privy Councillor, in which offices he was continued upon the succession of Queen Anne. Sir Colin frequently appears in the Acts of Parliament as Commissioner of Supply for the county of Perth. From 1690 to 1702 he represented Perthshire in Parliament. In 1689, he is mentioned as being appointed a Commissioner for ordering the Militia, and as overseer of the election of the Magistrates and Town Council of Perth. In 1670, a Charter under the Great Seal erected Aberuchill, and all his other landed property, in a free barony to hold of the Crown, and erected Inneruchill into a burgh of barony. In 1680, the Bishops of Dunblane granted to Sir Colin the chaplainry of Trinity Altar in the cathedral there, for life. He suffered very severely during the Revolution. In 1690 we read of Parliament passing an Act in favour of Sir Colin on his petition, showing that the rebels then in arms in the Highlands had wasted his property, and remitting to the Sheriff-Depute of Perthshire, the Steward Depute of Monteith, and the Sheriff Depute of Stirlingshire to value and liquidate the damage. Three years later, he presented another petition, showing that this remit had not been attended to, and another Act was passed empowering the Commissioners of Supply to assess the damage. They estimated it at £17,201, 12s. 4d. Scots, and the case was sent by Parliament to the Privy Council for recommendation to the king. This also appears to have been ineffectual, as in 1695 a petition by Sir Colin craving that his losses may be stated was read in Parliament, and remitted to the Committee for Private Affairs. Their report, which was equally favourable, having been read, Parliament recommended his case to the favourable consideration of the king, but with what success is not known. There is a tradition in the family that Sir Colin, who ranked among the most considerable personages of his time, was offered, but declined, the honour of a peerage, probably as a recompense for the losses he suffered at the Revolution, and for which he had so long and so vainly tried to obtain compensation. Sir James Campbell, the fifth and present baronet, is also heir-male of the Earls of Loudoun as well as of the Campbells of Lawers. He is Deputy Surveyor of the Forest of Dean in Gloucestershire, and resides at Whitemead Park int hat county, of which he is a Magistrate. He was born, on 5th May 1818, and married, 28th July 1840, Caroline, daughter of Admiral Sir Robert Howe Bromley of Stoke Hall, Nottinghamshire, representative of the elder branch of Lord Carrington's family, and has a son, Major Alexander Campbell, R.A., born 10th August 1841. Kilbryde was at one time a separate parish, but was united to Dunblane in 1618. The castle was built between 1400 and 1450, and was a place of great strength. Five years ago, Sir James had begun to re-roof and repair the castle, when one of the main walls gave way, and he had to rebuild a great part of it, the plans being prepared by Mr Heiton, Perth, who has retained the main characteristics of the old castle, and has most successfully worked in the best parts of the original structure. The castle is delightfully situated on an eminence commanding a view of a wide tract of country, and is surrounded by richly-wooded policies and a large extent of plantation. Many of the trees around the castle are very old. The most noticeable feature in the policies is a number of splendid sycamores, the best of which girths 14 feet 9 inches at 1 foot, and 15 feet 6 inches at 5 feet up. There are several oaks of great size, the best of them girthing 16 feet at 1 foot, and 12 feet 9 inches at 5 feet. The best specimen of the other varieties measured, are as follows :—Ash, 10 feet 7 inches at 1 foot, and 10 feet at 5 feet ; beech, 14 feet at 1 foot, and 13 feet at 5 feet ; lime, 11 feet 4 inches at 1 foot, and 10 feet 6 inches at 5 feet ; silver fir, 14 feet 4 inches at 1 foot, and 11 feet 6 inches at 5 feet ; Scots fir, 7 feet at 1 foot, and 6 feet 6 inches at 5 feet ; and larch, 11 feet at 1 foot, and 9 feet at 5 feet from the ground.

ARGATY, near Doune, the property of George-Home-Munro-Binning Home, is exceedingly well wooded. Up till about 100 years ago, the whole of this district was destitute of timber, the present laird's grandfather, Mr George Home, being the first to take away the barren aspect of the landscape by planting strips of Scots fir, larch, oak, &c., on the Drum of Argaty, a round hill of considerable height. Mr George Home was also a noted agriculturist in his day, and was the first in the district to plant turnips and potatoes in drills. Mr David Munro-Binning, father of the present laird, continued the work of planting on a much larger scale, and Mr Binning-Home has also done a little, principally for shelter. The mansion-house, indeed, is completely surrounded by wood, with the result that the climate here has been very considerably ameliorated within the memory of those still living. The present mansion is the third which has been built. The original house was a noted stronghold, the tradition being that it was the largest castle in the district after that of Doune. Cromwell burned the first house, some of the remains of which were found in the second, which the present laird took down in 1857, and replaced it by a splendid baronial mansion, of which he was himself the architect. The largest of the trees are those immediately surrounding the castle. The finest of all is a great ash girthing 18 feet at 1 foot up, and 14 feet 8 inches at 5 feet, with a bole of 30 feet. Another very old ash girths 15 feet at 1 foot, and 14 feet at 5 feet. Amongst the other trees mention may be made of an elm girthing 11 feet 3 inches at 5 feet, a sycamore girthing 13 feet at 5 feet, with a bole of about 20 feet; and a beech girthing 12 feet 6 inches at 5 feet. There are many trees about the same size scattered throughout the policies and the woods, the extent of the latter being about 100 acres. The families represented by the present laird are of great antiquity and historic importance. Regarding the family of Binning, tradition relates that, in 1308, William Bynnie, took the Castle of Linlithgow from the English by concealing himself and seven sons in a waggon of hay, which was stopped in the threshold, when the porter and those with him were instantly slain, while the garrison was put to the sword by a large party lying in ambush. As a reward for his gallant adventure, he received a gift of lands. There is a want of information as to who the first Sir Patrick Home was, but the belief is that he married an Edmonstone of Duntreath in Stirlingshire, and got the estate of Argaty from the Crown when the Earldom of Monteith was forfeited, the first charter being dated 1446. The Homes of Polwarth, were at that time a very powerful family, and seem to have been put down at Argaty and Rednoch for the express purpose of watching the keepers of Doune Castle, and the residence on the island in the Lake of Monteith. The second Sir Patrick of whom mention is made was married, first, to Margaret Sinclair, from whom are descended the present Lord Polwarth and Sir Hugh Hume Campbell, through the Lords Polwarths and the Earls of Marchmont. Sir Patrick's second wife was Helen Shaw, widow of Archibald Haliburton, Lord Dirleton, whose only son, George, married the Lady Margaret Erskine, second daughter of Robert, third Lord Erskine of Mar, and accompanied his father-in-law to Flodden, where Lord Erskine was slain, and George Home escaped. Harry Home, who succeeded in 1629, was a strong supporter of Charles I., and, in consequence of great losses on behalf of the king wadsetted the lands of Rednoch and Inchinnoch, to Gilbert Graham of Auchlye, of which wadset the present laird has a copy. At that time the Argaty family was much reduced and nearly ruined by Cromwell. It is somewhat singular that had the present laird been Home of Argaty by male descent, he would now have been Earl of Marchmont, and had the Earldom of Morton gone to heirs general, he should now have been Earl of Morton. As it is, he is the oldest heir of line of the Macdonalds, Lords of the Isles, Earls of Ross, and Lord Skye.

LANRICK CASTLE, in the parish of Kilmadock, is one of the most beautiful and interesting estates in that district. It was originally a seat of a branch of the Haldanes of Gleneagles, but as the laird of Lanrick of the day took part in the Rebellion of 1745, his estate was confiscated by the Crown and acquired by the York Building Company. It was subsequently purchased by Sir John Macgregor, whose son, Major-General Sir Evan John Macgregor, K.C.B., and G.C.H., Governor of the Windward Isles, sold it, in 1840, to William Jardine, M.P., a founder and partner in the well-known mercantile firm of Jardine, Mathewson, & Co. The present proprietor is Robert Jardine,

M.P. for Dumfriesshire. The property is kept in exceptionally good order, and extensive improvements have been carried out since it came into the possession of the Jardines. There are three fine approaches to the castle from the north, the south, and the west. The northern approach leads over the Teith by a highly ornamental and substantial iron bridge, 44 yards long, with one graceful span, and erected about four years ago. The castle is situated on the north bank of the Teith, at a spot where the river flows very rapidly through over-hanging woods, and is a massive and picturesque structure. It has evidently been built at various dates, the centre part being of considerable antiquity, with walls of massive thickness. The policies are adorned with many valuable hard and soft-wood trees. In a secluded part of the policies there is a monument of a very unique design, and of special interest to arborists. It is built in imitation of the trunk of an oak tree, 35 feet high, and has a girth of 46 feet at the base. It is altogether a splendid piece of masonry, the curves about the swell of the roots and the knots, being most artistically carved, a good tree having evidently been taken as a model. At the top of the trunk there is an ornamental frill springing from a band, which supports a number of columns, awnings, &c. There is no inscription upon the monument, nor does there appear to be any record to show its origin. It is, however, comparatively modern, the state of the masonry indicating a probable age of seventy or eighty years. There are a variety of large trees growing in the neighbourhood of the castle. Quite close to the castle are four ash trees of the following dimensions;—No. 1, on the west side of the castle, 19 feet 6 inches at 1 foot up, and 13 feet 6 inches at 5 feet, with a bole of about 30 feet; No. 2, on the north side, near the river, 18 feet 6 inches at 1 foot, and 15 feet 6 inches at 5 feet, with a bole of about 30 feet ; No. 3, on the east side, 14 feet both at 1 and 5 feet up, with a short bole running out into two large branches ; and No. 4, on the north-east side, with a short bole running out into four large branches, the trunk girthing 15 feet 6 inches at 2 feet from the ground. Another fine ash grows close to the burying ground, the girth being 18 feet at 1 foot, and 13 feet at 5 feet, with a bole of 12 feet, supporting two great limbs. At the burying-ground, which lies a little to the west of the Castle, there is a grove of ancient yews forming a complete circle, the largest of the trees girthing 7 feet at 3 feet up. The sycamores include one in front of the castle having a girth of 13 feet 6 inches at 1 foot, and 12 feet at 5 feet, with a bole of 10 feet, and carrying a splendid head; and another, near the burying-ground, having a girth of 11 feet 3 inches at the narrowest part of a bole of 6 feet. A very picturesque Spanish chestnut near the castle girths 15 feet 6 inches at 1 foot, and 14 feet 4 inches at 5 feet, with a bole of about 20 feet. There are over 700 acres of plantation, principally larch, spruce, and Scots fir, with a mixture of oak. The greater part of this wood is a first crop, the country having been previously almost destitute of trees. The oldest of these plantations is between eighty and ninety years, about one-third of the whole being about that age. The remainder of the plantations range from forty to sixty years of age. The bulk of the soil, although a little stiff, is of good quality for growing timber. The plantations have been all thoroughly drained. There are some very good hedges, chiefly on the roadsides, but wire and bar fencing have been principally used. On curved and angular ground, a novel and effective fence is used, the invention of Mr Dinwoodie, the forester, and the patent for which has expired. The peculiarity of this fence is that it adapts itself to curves and angles without the aid of side or end stays. It is a wire fence with a three-quarter inch top rod, the latter being the backbone of the fence. The principle upon which it is constructed causes it to bulge out, and to keep its perpendicular, instead of yielding to the strain. The standards are 3 feet apart, and are held in the ground by two forks 14 inches in depth, with flat rests 6 inches by 3 inches, with a catch in the inside, turning up 3 inches to prevent them from sliding. The effect of this mode of construction is that the greater the curves and irregularities of the ground, the greater the strength of the fence.

CAMBUSMORE, two miles east of Callander, is an exceedingly pretty and interesting property, its historic associations being as rich as its natural beauties. John Buchanan of Arnprior, and

formerly of Auchleshie, to whom Strathyre (yet to be referred
to) was restored, married Murray Kynnymond Edmondstone of
Old Newton, near Doune, whose family were hereditary Standard-
bearers to the Royal Stuarts. At Cambusmore there is preserved part of the staff on which the Royal Standard floated at the
battles of Killiecrankie and Sheriffmuir, as well as a wineglass
out of which Prince Charles Edward drank as he passed the
house of Old Newton on his way from Dunblane. This John
Buchanan was a great narrator of the events of those stirring
times, and from him Sir Walter Scott, who was a constant
visitor at Cambusmore gathered many of the anecdotes to be
found in his works. Along with the daughters of the house,
Sir Walter roamed over the then almost unknown district of
the Trossachs, and from Cambusmore, where a great part of
The Lady of the Lake was written, he took the ride,
immortalised as Fitz-James' in that poem. A few of Sir
Walter's letters are treasured at Cambusmore, but most of
them were, unfortunately, destroyed some years ago. John
Buchanan of Arnprior, to whom we have referred, had three sons
and four daughters. His youngest son, James Edmondstone,
Scots Guards, fell at the battle of Talavera in 1808; his
second son, Thomas, married a daughter of General Sir Ralph
Abercromby, and was father of the present John Buchanan of
Powis. John Buchanan of Arnprior died in 1817, and was
succeeded by his eldest son, Alexander, late 39th Regiment, a
Deputy Lieutenant of Perthshire, well known in the district for
half a century as "The Major," and in whose time, as in his
father's, the hospitality of Cambusmore was a household word.
His only son, Alexander, survived his father but two years,
leaving, in 1848, an only child, Catherine E. Grace, the
present possessor. She married, in 1869, John Baillie
Hamilton of the family of Haddington, J.P., Commissioner
of Supply, and a Deputy-Lieutenant of the County, who
has assumed the name of Buchanan in addition to his own.
The mansion-house, which is partly old and partly new, is
exceedingly picturesque and commodious. The original portion
appears to be about 200 years old, and seems to have been
added to about 100 years ago. In 1880 a very extensive

addition to the mansion was commenced, the work not being completed till 1882. The practical part of the work was carried out under the superintendence of Mr Baldie, Glasgow. The design, which was sketched by Mrs B. Baillie Hamilton, not only shows considerable taste and skill, but not a little ingenuity. At the west front there is a natural dip in the ground, and advantage is taken of this to construct a terrace of six arches, in the Gothic style. Underneath the terrace two ancient stones belonging to the original house or castle have been inserted for preservation. Although the weather has made itself felt to a large extent, it can still be seen that the stones have been elaborately carved, and bear the date 1580. A prominent feature in the newer portion of the house is a massive Gothic tower with an arched porch. Various styles of architecture have been introduced, but care has been taken to make them harmonise. The whole of the stone—a nice hard, bluish sandstone—was obtained on the property, within one mile of the house. Cambusmore consists of about 5000 acres, of which 1000 are moor and 350 woodland, chiefly larch, Scots fir, and spruce, with a mixture of oak coppice. Formerly, the woods were much more extensive, as may be seen from the remains still visible. Indeed, during the last 30 years upwards of £20,000 worth of timber are known to have been cut down and sold, and the probability is that the destruction was much greater. These blanks are being gradually filled up by the present proprietor, who has added 40 acres to the plantations during the last five years alone, the additions consisting of a mixture of larch, Scots fir, silver fir and spruce, the whole having an outside fringe of sycamore and beech as a protection from the wind. The policies are enriched by many fine specimens of the best known varieties of trees. Immediately behind the house there is a magnificent specimen of the fern-leaf beech, there being an exceptionally grand spread of branches. A sycamore at the garden gate girths 10 feet at the narrowest part of a bole of 12 feet, or at 5 feet from the ground. A larch within the policies girths 10 feet at 1 foot from the ground, and 7 feet 10 inches at 5 feet, the height being about 100 feet without a single strong branch. A silver fir on the edge of the approach girths 13 feet 5 inches at 1 foot, and 11 feet 10 inches at 5 feet, with a height of 100 feet. About 8 feet up there is a strong off-shoot growing straight up to the full height of the tree and serves to balance it on the weakest side. About 12 years ago a slight upheaval of the ground was observed round this fine tree, and the proprietor had a cairn of heavy blocks of stone placed on the weak side as a counterpoise, and there is not now the slightest appearance of the tree giving way. An oak in the park girths 15 feet 7 inches at 1 foot, and 11 feet 5 inches at 5 feet. This is an exceedingly well-balanced and healthy tree, with particularly fine branches. In the cow park, close to the public road there is a splendid Scots fir, with a height of 80 feet, and a bole of 25 feet before a branch is reached. The girth at 1 foot up is 9 feet 4 inches, and at 5 feet the girth is 8 feet 1 inch. There are also a number of very good ash trees, one of these at Drumvaich, close to the roadside, girths 21 feet at 1 foot, and 15 feet 6 inches at 5 feet, with a magnificent bole of 24 feet. At Easter Coilechat, near the march with the Lanrick Castle property, there is another splendid ash with a girth of 21 feet at 1 foot, and 14 feet 9 inches at 5 feet, after which the bole swells very considerably, and breaks out into a number of huge branches, at a height of 15 feet from the ground. Near these ash trees there are four sycamores of considerable size, and although they are not so large as the one in the garden, they present such a gorgeous mass of foliage to the roadside that they at once attract attention. Besides the trees mentioned, the policy parks contain many trees of noble proportions and fine foliage, the chief varieties being beech, oak, birch, and lime. The place also owes not a little of its attractions to the Keltie, a fine mountain stream which flows from the well-known Falls of Bracklyn through the woods and grounds of Cambusmore till it joins the River Teith. The Barony of Strathyre, to which reference has already been made, has been in the possession of the Buchanans since the 15th century, but there is little worthy of note connected with the property till the exciting time of '45, when its possessor, Francis Buchanan of Arnprior, at that time residing at Leny, which belonged to his wife, was seized and carried off to Carlisle, where he was hanged for complicity in the Rebellion. There is mention of him in most of the histories of that time. Though not himself actively engaged in the Rebellion (he having been prevented by his friend and neighbour, the Laird of Blair-Drummond, from personally joining the standard of Prince Charles), he had sent all the Buchanans he could gather, under his cousin and heir, a youth of seventeen, to join the rebels; and it was well known that he had kept up a secret correspondence with the exiled Royal Family, and in order to assist them sold the third part of the lands of Buchanan to the Duke of Montrose in 1744. On the restoration of his forfeited estates, the remaining portion of his land at Arnprior, Kippen, were sold by John Buchanan of Auchleshie to Mr Stirling of Garden, to defray the expenses due to the Government. The family, however, still retain the superiority, and the lands of Strathyre and Cambusmore are now held together as the Arnprior estates. The Disannexing Act restoring Strathyre to John Buchanan of Auchleshie as cousin and male heir of the attainted Francis Buchanan of Arnprior includes the right of the salmon fishings of Loch Lubnaig, and the proprietors have also the right, over the adjacent lands of Stank, of lowering the level of the loch. Strathyre is 7000 acres in extent, of which 150 acres are in wood. The trees in the plantations are of exceptionally good quality. A splended larch wood at Creegan of 26 acres, 60 years old, was cut in 1872, and replanted with Scots firs in 1875, the young wood being in a most thriving and vigorous condition. Immervoulin Wood, 64 acres in extent, contains, besides some thriving young larch and other trees, a number of very fine specimens of old Scots fir the character of which may be gathered from the following table, showing the size of twelve of those measured at random as fair specimens of the lot:—

EXTREME HEIGHT IN FEET.	HEIGHT OF BOLE.	GIRTH AT 1 FOOT.	GIRTH AT 5 FEET.
56	35	8' 3'	7' 7"
66	20	8' 3"	7' 3"
58	35	6' 8"	6' 0"
71	38	9' 3"	8' 3"
76	35	10' 11"	10' 5"

Extreme Height in Feet.	Height of Boll.	Girth at 1 Foot.	Girth at 5 Feet.
66	40	7' 4"	6' 8"
70	40	6' 6"	5' 11"
72	20	10' 9"	9' 8"
69	27	11' 10"	9' 8"
70	35	9' 9"	8' 9"
68	25	12' 10"	11' 5"
60	18	12' 0"	11' 0"

Invertrossachs, the property of George Addison Cox, brother of Mr Cox of Cardean and Mr Cox of Snaigow, is not only an exceedingly pretty place, but it illustrates in an eminent degree, what has been done to improve the country by planting within recent years. It extends along the greater part of the southern shore of Loch Vennachar, or for about 4½ miles out of the 5 miles to which the loch extends. The road from Callander leads directly to Invertrossachs House, where it ends, the road to the Trossachs being on the northern side of the loch. In approaching the loch from Callander, we pass Coilantogle Ford,

"Clan Alpine's outmost guard,"

where the combat took place, according to Sir Walter Scott, between Roderick Dhu and Fitz-James. A bridge now spans the river near the place where it was formerly forded. Within the past few years the road from Callander has been re-constructed. Formerly it ran quite close to the water, but when the Glasgow Water Commissioners were taken bound to raise the level of the loch 8 feet, so as to maintain a full supply for the Teith, the road had to be shifted to a higher level, and the old road now forms part of the bed of the loch, but can still be traced by the stumps of the trees with which it was lined. Invertrossachs House stands on an eminence about a mile and a half from the top of the loch, of which it commands a complete view, with Ben Ledi in the foreground, and Ben Venue in the background. Its beautiful and retired situation commended it to the attention of Queen Victoria, who resided here for a fortnight in September 1869, and made numerous tours to the places of interest in the neighbourhood. The house was originally built in 1843, but it has been twice enlarged.

Although there was at this time an old oak wood along a portion of the loch side, it was not of great extent, and about 24 years ago it was cut down, a few of the best trees, of about 150 years old, being retained as standards amidst the coppice. There had, evidently been at one time a good deal of old timber in the neighbourhood of the house, principally sycamore and ash, as the stools of these trees to be met with in the young plantations are of considerable size. With the building of the house, systematic planting was commenced, and the work is still being continued at regular intervals. Mr Cox purchased the property, in 1875, from Mr Stuart Macnaughton, now of Southampton, and at that time there were about 300 acres of well matured timber. Since then 200 additional acres have been planted on the hillside. The main crop is larch, but Scots fir and spruce are liberally mixed, and a few hardwood trees have been planted on the dryer portions. These young plantations are growing splendidly, and in the course of a few years will effect a great improvement upon the appearance of the estate, and afford excellent shelter for cattle and sheep. Throughout the whole property, spruce trees thrive amazingly, and grow in unusually fine form. The grazing is entirely in Mr Cox's own hands. As many as 2500 black-faced sheep and 50 Highland cattle are kept, and as he has drawn upon the best strains of blood for sires, a considerable improvement upon both sheep and cattle has been effected within the past few years. The whole of the land has also been much improved by surface drainage, and substantial dykes and wire-fencing have been extensively erected.

LENY, which lies a little to the west of Callander, is interesting both on account of the specimens of the rarer trees which it contains, and also on account of the extent of planting which has been done in recent years. The mansion-house, which occupies a situation excelled in picturesque beauty by few residences in Scotland, is surrounded by parks that are notable alike for their extent, and for their tasteful diversity of timber. The original home of the lairds of Leny was about a quarter of a mile to the south of their present residence, and is said to have been burned down shortly after the battle of Pinkie in 1547. A portion of the original building has been preserved and incorporated in the existing house. The proprietor, John Buchanan-Hamilton, is the Chief of the Buchanans, and the representative of the Hamiltons of Bardowie, his property also including that of Spittal in the county of Dumbarton and Bardowie in Stirlingshire. Both families took a prominent part in the making of Scottish history, but perhaps the most eminent member was Dr Francis Buchanan-Hamilton, author of a number of works on India, and father of the present laird. Dr Hunter, in a postscript to his *Imperial Gazetteer of India*, expresses his obligations to Dr Francis Buchanan-Hamilton, remarking that he was by far the greatest man who during the first century of our Indian rule devoted himself to the study of the country and the people. He was born in 1762, and took his degree as Doctor of Medicine in Edinburgh in 1783, and went out to India in 1794, in the service of the East India Company. He returned in 1816, and succeeded, on the death of his brother, to the Buchanan estates of Leny and Spittal, together with the property of his mother, who represented the Hamiltons of Bardowie. He was a rural investigator, and a man of science. During his Indian career, he served for some time on the personal staff of the Marquis of Wellesley, Governor-General. It was under the promptings of Dr Buchanan-Hamilton that an establishment was formed in the Governor-General's park at Barrackpore for investigating the natural history of India. His chief work was the Statistical Survey of Bengal. His labours were so highly appreciated by the Marquis of Wellesley, that, on the retirement of the doctor in 1817, his lordship wrote that no part of his governorship of India afforded him more matter of satisfactory reflection than the opportunities he had of rendering the abilities and knowledge of Dr Buchanan-Hamilton useful to the world. " In discharging this public duty, the intimate acquaintance and friendship which was established between us enabled me to appreciate the integrity, independence, and frankness of your character, and the manly spirit of truth and honour which animated your intercourse with all persons in power." Dr Hunter further states that Dr

Buchanan-Hamilton knew more about India than any European of his time, and always stated his views freely, whether they were pleasing or displeasing to those who sought his opinion. He died at Leny in 1829, at the age of 67, and was succeeded by his son, the present proprietor. During his stay at Leny, Dr Buchanan-Hamilton planted very extensively about the glens, and in clumps, and 200 or 300 acres of land on the verge of the moors were reclaimed. The present proprietor has planted between 400 and 500 acres on the hills between Callander and the famous Pass of Leny, a work which has vastly improved the appearance of this part of the country. Perhaps the most notable plant in the policies is an Irish yew, said to be the second largest in Scotland, and a splendid bushy plant. The tree is about 25 feet high, and about 18 feet in diameter. There are fine specimens of the American walnut, plain-leaf ash, and fern-leaf beech, the latter girthing 4 feet 10 inches at 5 feet up. The best of the sycamores girths 12 feet 6 inches at 5 feet, the oaks about 10 feet, and limes and beeches about 11 feet, at 5 feet from the ground. There is a very good collection of newer coniferous trees, although none of the specimens are particularly large.

EDINCHIP, the property of Sir Malcolm Macgregor of Macgregor, chief of the clan Macgregor, at present a minor, ten years of age, is beautifully situated in Balquhidder, the name signifying "the hill in the shape of a shoemaker's last." It is noteworthy as possessing a number of grand old Spanish chestnuts. These, numbering 23 in all, grow on the west side of the house at the very edge of a high bank overhanging the river Ceann Drom, and they are said to have been planted at the same time as the famous ones at Inchmahome. The four largest girth as follows:—No. 1, 15 feet 10 inches at 1 foot from the ground, and 12 feet at 5 feet; No. 2, 15 feet 5 inches at 1 foot, and 11 feet 7 inches at 5 feet; No. 3, 13 feet 6 inches at 1 foot, and 10 feet 10 inches at 5 feet; and No. 4, 13 feet 4 inches at 1 foot, and 10 feet at 5 feet. Of the remaining 19, all but 3 girth from $8\frac{1}{2}$ to 10 feet at 5 feet from the ground. A little farther down the river, there are seven characteristic and self-

EDINCHIP—CONCLUSION.

sown Scots firs, the two largest of which girth 13 feet 2 inches at 1 foot, and 9 feet 4 inches at 5 feet ; and 11 feet 4 inches at 1 foot, and 9 feet 5 inches at 5 feet respectively. These trees we regret to say, suffered severely in the Tay Bridge gale. On the east side of the house is a row of (for the Highlands), fine horse chestnuts, planted early last century, and to the north of the house there are some very good sycamores of about the same age.

We have now concluded the work which we set before ourselves, but we have by no means exhausted the subject. With so much material at our disposal, this we could never hope to do, but we trust we have succeeded in accomplishing what we chiefly had in view. Our main object has been to point out what has been done to improve the face of the country by planting within the past century, and we have endeavoured to show that Perthshire now presents a very different aspect from what it did in comparatively modern times ; while we have sought to collect and preserve whatever information we could obtain regarding those old, remarkable, and historic trees which have a special charm for all who have a love for that sylvan scenery which, Lord Beaconsfield has said, never palls. Our labours, we need hardly say, have afforded us genuine pleasure. They have led us into all parts of the county, and brought us into contact with the woods in their various moods and aspects ;—in the delightful freshness of spring, the season which " unlocks the flowers to paint the laughing soil ; " in summer, when the trees are clothed in all their leafy beauty ; in the rich and " yellow autumn, wreathed with nodding corn ; " and in winter, the stern " ruler of the inverted year." In sunshine and in shower, in calm and in storm, our task has taken us into the heart of the woods, which have never failed to yield a pleasure that amply rewarded us for whatever trouble may have been endured.

INDEX TO OLD AND NOTABLE TREES.

Abies Albertiana—Pages 123, 136, 211, 280, 359, 399, 546.

Abies Douglasii—77, 103, 112, 123, 186, 215, 216, 288, 321, 322, 343, 361, 465, 487, 500, 535.

Araucaria imbricata—67, 78, 175, 205, 217, 281, 343, 487, 492, 493, 518, 546.

Ash—68, 123, 125, 126, 133, 160, 164, 175, 187, 188, 204, 231, 232, 238, 259, 265, 287, 298, 314, 321, 340, 344, 378, 388, 398, 441, 454, 455, 463, 471, 488, 494, 502, 506, 517, 525, 533, 545, 550, 551, 553, 557.

Beech—52, 69, 75, 120, 122, 125, 143, 151, 160, 162, 175, 203, 214, 237, 252, 287, 298, 308, 314, 320, 340, 379, 388, 397, 442, 455, 463, 471, 476, 486, 489, 494, 498, 506, 507, 513, 517, 525, 531, 537, 539, 550, 551.

Black Italian Poplar—127, 187, 287, 344, 346, 490, 500, 513.

Cedar of Lebanon—124, 135, 142, 204, 234, 287, 314, 320, 454, 488.

Elm—122, 124, 134, 143, 194, 234, 259, 287, 298, 308, 314, 320, 342, 343, 442, 463, 471, 478, 486, 488, 490, 493, 494, 495, 498, 506, 517, 519, 524, 537, 551.

Horse Chestnut—125, 134, 142, 233, 287, 342, 345, 489, 490, 506.

Larch—51, 53, 54, 60, 61, 68, 122, 164, 233, 255, 280, 287, 300, 314, 345, 346, 352, 362, 389, 390, 398, 423, 455, 476, 479, 486, 488, 494, 498, 535, 536, 539, 542, 546, 548, 550.

Lime—78, 141, 157, 175, 234, 254, 341, 353, 379, 388, 442, 463, 490, 498, 531, 537, 550.

Oak—51, 53, 68, 73, 74, 102, 103, 110, 112, 121, 122, 125, 133, 142, 145, 151, 162, 165, 175, 187, 194, 195, 204, 218, 230, 232, 251, 282, 287, 298, 308, 314, 321, 340, 344, 371, 398, 457, 461, 471, 489, 490, 493, 495, 498, 513, 517, 531, 533, 535, 550, 557.

Scots Fir—106, 122, 150, 163, 165, 218, 233, 254, 259, 282, 298, 344, 345, 353, 415, 443, 456, 463, 502, 550, 557, 558, 563.

Sycamore—73, 102, 103, 115, 121, 124, 151, 171, 175, 187, 194, 204, 230, 233, 240, 254, 259, 264, 265, 266, 285, 287, 297, 298, 314, 321, 341, 343, 371, 379, 389, 396, 463, 486, 489, 493, 494, 495, 502, 506, 525, 550, 551, 554, 556, 562.

Silver Fir—54, 69, 112, 142, 151, 160, 164, 170, 171, 175, 186, 215, 234, 236, 257, 266, 281, 287, 314, 342, 345, 352, 378, 464, 488, 494, 526, 533, 538, 540, 556.

Spanish Chestnut—50, 54, 69, 122, 124, 141, 160, 165, 218, 233, 254, 282, 311, 319, 342, 343, 344, 378, 388, 395, 455, 463, 485, 490, 498, 519, 554. 562.

Spruce—60, 124, 135, 151, 162, 234, 321, 352, 360, 479, 486, 541.

Tulip Tree—145, 287, 371, 400, 519.

Walnut—54, 69, 151, 194, 203, 232, 259, 350, 490, 498.

Wellingtonia gigantea—79, 143, 219, 237, 288, 299, 343, 398, 454, 477, 487, 493, 498, 500, 524, 531, 546.

Yew—53, 187, 219, 238, 287, 308, 343, 432, 461, 498, 502, 508, 513, 554.